Footprints of Chaos
in the Markets

Footprints of Chaos
in the Markets

Analyzing non-linear time series
in financial markets and
other real systems

RICHARD M. A. URBACH

FINANCIAL TIMES
Prentice Hall

London   New York   San Francisco   Toronto   Sydney
Tokyo   Singapore   Hong Kong   Cape Town   Madrid
Paris   Milan   Munich   Amsterdam

PEARSON EDUCATION LIMITED

Head office:
Edinburgh Gate,
Harlow CM20 2JE
Tel: +44 (0)1279 623623
Fax: +44 (0)1279 431059
Website: www.pearsoned-ema.com

London office:
128 Long Acre, London WC2E 9AN
Tel: +44 (0)207 447 2000
Fax: +44 (0)207 240 5771
Website: www.business-minds.com

First published in Great Britain 2000

© Richard Urbach 2000

The right of Richard Urbach to be identified as
Author of this Work has been asserted by him in accordance
with the Copyright, Designs and Patents Act 1998

ISBN 0 273 63573 5

British Library Cataloguing in Publication Data
A CIP catalogue record for this book can be obtained
from the british library.

1 3 5 7 9 10 8 6 4 2

Typeset by Focal Image Ltd, London
Printed and bound in Great Britain by
Biddles Ltd, Guildford & King's Lynn

*The Publishers' policy is to use paper manufactured
from sustainable forests.*

About the Author

Richard Urbach is a leading US expert in the application of quantitative methods to investment banking and trading. Now based in London, he has co-managed a $110 million investment fund at Panther Capital Management, and held senior positions at UBS Securities, Morgan Stanley and Citibank. He has a PhD, in Mathematical Statistics and Operations Research, from Columbia University, New York.

Acknowledgments

I owe a deep debt of gratitude to Michael Cohn of Panther Capital Management Ltd. for his moral and material support, without which this book may not have been written. I am also indebted to Bob Chamberlain, and again Michael Cohn, for reading the early manuscript and giving me many helpful comments. Any errors or omissions are of course my own. I would also like to thank Joseph Metcalf for helping me transcribe many of the figures in the book. Finally, I am grateful to Imperial College, London, for allowing me generous use of their libraries to do the research reflected in this book.

Contents

Preface

A dynamical system is a system whose forward evolution is determined by a mathematical function of its present state. Thus, if the system is presently in state s, then its state at some time t in the future is given by $f[s, t]$. As far as concepts go, that is pretty simple. However, these systems can exhibit behavior that can be called anything but simple, and paradoxically, even though we can predict in theory exactly where the system will be at any time in the future given its present state, in practice prediction can be very limited.

Classical mathematical analysis of dynamical systems goes back at least to Isaac Newton's development of celestial mechanics and continued into the early 20th century with the work of H. Poincaré. But Poincaré was perhaps the first to realize the extent and nature of the behavioral complexity possible in these systems. He apparently also despaired at the difficulties in giving a complete analytic description of that complex behavior. The alternative of the numerical approach was also out of reach at the time due to the massive computation resources (man-hours in those days) needed to simulate and analyze the dynamics. Poincaré was not mistaken. Analytical results were hard to come by, and progress was slow until the introduction of cheap computing devices circa 1980. From then on research in the subject grew rapidly. The past twenty or so years has brought results of both theoretical and practical importance, especially for systems that exhibit low dimensional chaotic behavior. However, much of what is practical in those results has been slow to emerge in a form that is readily understandable and usable by the general practitioner of time series analysis. The purpose of this book is to make available to the general practitioner the concepts and tools of chaotic dynamical systems analysis developed since the early 1980s. The credit for the work reported here belongs to the many referenced researchers, and of course, the responsibility for any errors in reporting their work rests solely with myself. If I have made a contribution it is in organizing and presenting the material in such a way that it is relatively easy to understand and apply by the non-cognoscenti.

Specifically, this book is about the numerical analysis of chaotic time series originating with nonlinear dynamical systems. The objective of that analysis is to develop a model that explains or predicts the behavior of the 'real' system that generated the time series. A 'real' system is one that has a material presence in the real world. Real systems can hurt you if they run into you, or fall on your toe. But they can also hurt you in less physical ways. For example, a political system can restrict your freedom, and an economic system or market can seriously impact your bank account.

Each chapter of the book treats a different concept in dynamical systems analysis and is organized into three sections. The first section explains the subject with simple ideas and figures that should not be difficult to understand by any quantitatively oriented reader. A reader only wishing to obtain a feeling for the subject could read the first section of each chapter without making contact with the theory. The second section covers enough theoretical material to underpin the intuition of the first section and support the algorithms developed in the third section. With respect to the sections on theory and algorithms, some assumption has to be made concerning the knowledge of the intended audience, and that will be a basic grounding in differential calculus, linear algebra, probability and statistics.

A more detailed description of the topics covered and the way in which they are organized would make little sense here without having first developed at least some concepts that are basic to dynamical systems, but perhaps not familiar to many readers. That will come in the first chapter along with statements of the key operational and technical assumptions that will apply throughout the book. The operational assumptions delimit what we know about the system under study. On the one hand, those assumptions make life difficult in that we will not be in a position to establish with mathematical rigor whether or not many of the theoretical results hold for the systems we are interested in. On the other hand, as the operational assumptions are realistic and the objective is numerical analysis, we have little choice but to just adopt those technical assumptions which are reasonable and lead to tractable computations of quantities that have theoretical meaning. If the quantities so obtained enable us to build a model that explains or predicts the behavior of the system we are studying, then we have arrived at a useful result, and it really does not matter what assumptions we made to get there, if a useful result was the objective. In other words, if a model explains and predicts what we observe in a system, then it fulfills its purpose, regardless of the means used to get it, or whether it duplicates 'nature's' computations.

The examples included in the algorithms section illustrate the effectiveness of each method in meeting the objectives of the algorithm on data generated from chaotic dynamical equations or obtained from measurements on real chaotic systems. Most of those systems have been heavily analyzed and their characteristics are well documented. The reason for including such intensely studied systems as examples is not to prove that they are chaotic, since they are known to be so, but rather to demonstrate the ability of a particular algorithm to reproduce known characteristics. If an algorithm can do that, it is a valuable tool. But, as with any tool used on unfamiliar material, when these methods are applied to less well understood systems, a good deal of the quality of the result will depend on the ingenuity and skill of the user.

There have been a number of papers and books published on the application of dynamical systems analysis to financial time series with the intention of demonstrating the existence or non-existence of determinism in capital markets. As far as I can tell, all the published results are inconclusive or unconvincing. There are a number of reasons why the reader should not be disheartened by reports of failed attempts to discover a deterministic price setting mechanism, in other words, predictability in financial markets. One reason is that research in this area will almost surely suffer from selective publishing. While I am not surprised at seeing inconclusive reports in print, I would be more than mildly surprised to see a report describing a deterministic model of a capital or commodity market. It just beggars the imagination to think that any individual intent on discovering determinism in a market's price setting mechanism would, having found the model, turn around and publish the details. Perhaps I am being too cynical, but I think that if someone has discovered something in this area that has profit potential, then they are quietly exploiting the discovery for their own gain, resisting the temptation for acclaim at least until their memoirs come out.

If the reader does not buy that argument, then there is another reason to remain optimistic. Some problems remain unsolved for long periods of time despite the efforts of the best minds. Then one day someone solves the problem, not necessarily because they had better tools to work with, but often because they saw the problem from a novel perspective or in a different context, or just applied the tools in an unconventional manner.

In memory of
Edward Ignall
inspirational teacher and colleague

1

Dynamical systems

INTUITION

In this chapter we undertake to describe what a dynamical system is, what chaos is, and what makes a dynamical system chaotic. We will also make explicit the operational assumptions that apply throughout the book. In the theoretical section we present some of the key technical concepts and theorems that underpin the analysis of dynamical systems and delineate the technical assumptions that will be adopted in the rest of the book.

A physical system is a piece of the real world circumscribed by an imaginary boundary. Thus, a system is a set of connected or interacting parts that form a unitary whole. The *state* of a system at a particular point in time is defined to be the condition of all of its parts at that time. If the system has m parts, then its state is given by an ordered list of m specifications describing the condition of each part[1]. We will assume that, in principle, the condition of each part of the system is quantifiable in the sense that we can assign a number to each admissible condition.

An m-dimensional space is a collection of 'points' each of which is defined by an ordered list of m attributes, say m real numbers. Thus, we can identify a state of a system, consisting of m parts, with a point in an imaginary m-dimensional space where every possible state of the system is represented by a unique point. The imaginary space just described is called the *state space* of the system. A system with an m-dimensional state space is called an m-dimensional system. If we let each number in the list of numbers defining a point in the state space represent the distance along a coordinate axis, then the states of the system are endowed with a notion of location in the state space. Clearly, if anything causes some part or parts of the system to change condition, then a change of state occurs and the new state is located at a position that is different from the original state. A sequence of such state changes then defines a corresponding sequence of points in the state space. If we think of these points as being connected by a line, then we obtain a 'curve' in the state space which is called an *orbit* or *trajectory* of the system (see Figure 1.1).

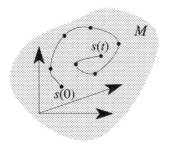

Figure 1.1 A trajectory $s(t)$ with initial state $s(0)$ in a 3-dimensional state space M.

---

[1]Some of the parts of the system may not be able to change independently of the others so that a system's state can be specified by giving the condition of only some of its parts. The number of such elemental conditions that are necessary to specify a state is called the number of *degrees of freedom* in the system.

Now imagine that part of the system is a mechanism that causes the system to change state over time. Such a system would be dynamic, in fact self-propelling, and could be represented for all time by an orbit in its state space. A ***dynamical system*** is a self-propelling system whose mechanism for state space change is given by a mathematical function of its state. Hence, we can write the (forward) evolution of a dynamical system, that is, its trajectory, as an equation of the following type[2].

$$s_t = f_\mu(s_0, t) \qquad\qquad \textbf{Eq. 1.1}$$

where μ is a parameter of the system, and s_t[3] represents a trajectory giving the state of the system at time t. In particular, s_0 represents the ***initial state*** or ***initial condition*** of the system. By the ***trajectory of a point*** s we mean the trajectory of the dynamical system beginning with the initial state s. Actually, we require a bit more than Eq. 1.1, namely that

$$s_0 = f_\mu(s_0, 0)$$

and if u is an intermediate point in time, then

$$s_{t+u} = f_\mu(s_u, t) = f_\mu(f_\mu(s_0, u), t)$$

so that the trajectory from an initial state s_0 is unique[4]. If the orbit ever intersects itself, that is, if ever $s_t = s_u$, $t \neq u$, then the orbit is necessarily ***closed*** or ***periodic*** in the sense that it will keep revisiting the same states in the same order over and over again. If the periodic orbit consists of a single state, then the system becomes static, that is, it ceases to change state, in which case it is said to be in ***equilibrium*** and its state is called an ***equilibrium*** or ***fixed point*** of the system.

Now, suppose we are in possession of a device that is able to measure the condition of some part of an m-dimensional system, and that we perform an experiment in which we use the device to observe the system at times $t_1, t_2, \ldots, t_{N_d}$. The experiment yields N_d measurements $x(t_1), x(t_2), \ldots, x(t_{N_d})$ which we will call a ***measurement*** or ***observational time series***. Geometrically (see Figure 1.2), the measuring device is, at discrete points in time, projecting the m-dimensional trajectory of the system onto a one-dimensional axis of the state space. In the figure, the projection is denoted by the function h, and the axis of projection is the x-axis. If we now adjoin to that axis another one-dimensional axis to encompass the element of time, and pull the measurements off the first axis in the direction of the time axis, then like the unfolding of an accordion we get the usual picture of a time series.

---

[2]In the sequel we will refer to both Eq. 1.1 and f_μ variously as the ***state transition function, the dynamical equation*** or simply ***the dynamics*** of the system. We will continue to use the term ***dynamical system*** even when the dynamics have a modest stochastic (random) component. The terminology is not universal.

[3]We will often alternate between two equivalent notational forms for indexed variables, for example s_t and $s(t)$, to avoid proliferation of parentheses or embeddings of subscripts.

[4]In practice the situation is quite different. As real measurements on a real system are subject to finite accuracy, the measurement error causes whole collections of nearby trajectories to become identified.

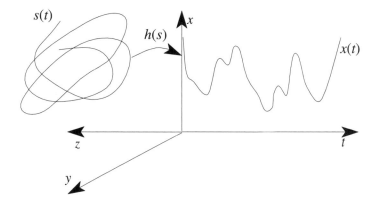

Figure 1.2 A measuring device whose output is given by a function h is used to take one-dimensional snapshots of the trajectory $s(t)$. The process yields a measurement time series $x(t) = h[s(t)]$.

The following operational assumptions apply throughout this book, and are collectively referred to as *the experimental situation*.

Experimental situation

1. The system that we are studying came into existence at some time in the distant past under some unknowable conditions and is now on an immutable trajectory determined by those initial conditions.
2. We do not know the dynamical equation that governs the motion of the system.
3. The actual state space and trajectory of the system are unobservable.
4. The only information we have is a measurement time series.

These assumptions just make explicit the usual or likely state of affairs in practice. The first assumption exposes the easily overlooked implications of the fact that many real systems are difficult, if not impossible, to stop and restart. Namely that, if we cannot restart the system, then its current trajectory is the one-and-only trajectory that we will ever observe, and any observations (measurements) we make on the system are observations taken from that same trajectory.

Time series plots often share similar qualitative features even when they derive from very different systems, but it would not be unreasonable to expect that a specific system would leave a distinctive footprint even on a one-dimensional projection. That is not to say that such a footprint is obvious, say to the naked eye, but it might be. For example, Figures 1.3 to 1.6 show time series of the S + P stock market average on different time scales. It does not take a lot of imagination to see a sort of similarity between the figures, in the sense that each seems to vaguely replicate the gross features of the others, but on different time scales. Coincidence? Perhaps, but if it is persistent, then it may be the footprint of the action of a deterministic mechanism.

Chaotic dynamical systems often have trajectories that collect in a structure that has self-similarity on many scales, and that feature can be transmitted to measurement time series of the systems. In any case, the eye is a poor tool to use to uncover the cause of any apparent self-similarity. The human brain loves to find patterns, and is

5

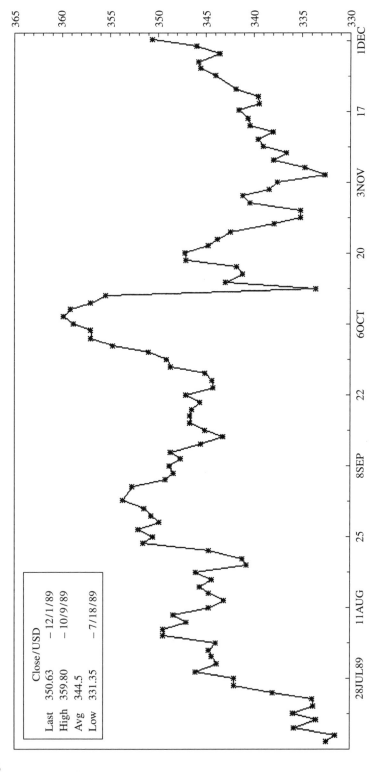

Figure 1.3 Standard and Poor's stock index. Daily data from 7 July 1989 to 1 December 1989.

Source: Adapted from Bloomberg L.P. ©1998.

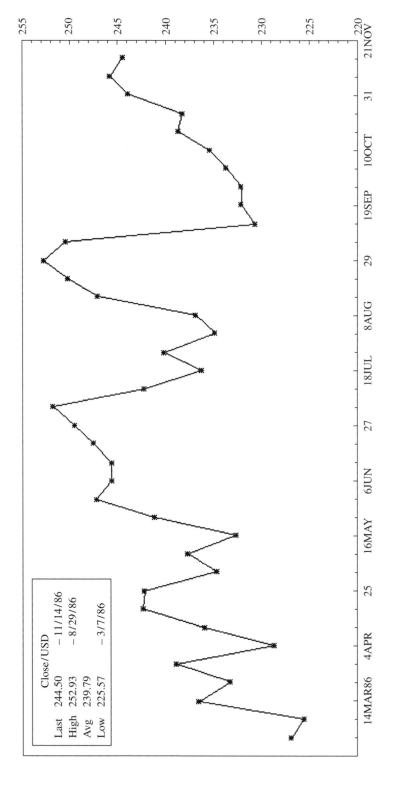

Figure 1.4 Standard and Poor's stock Index. Weekly data from 28 February 1986 to 14 November 1986.

Source: Adapted from Bloomberg L.P. ©1998.

7

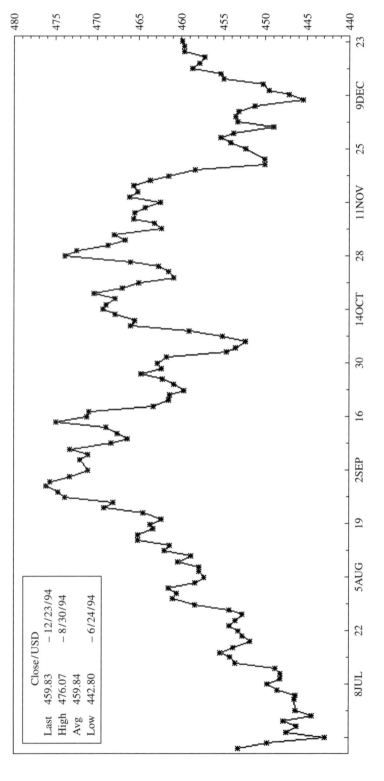

Figure 1.5 Standard and Poor's stock Index. Daily data from 22 June 1994 to 23 December 1994.

Source: Adapted from Bloomberg L.P. ©1998.

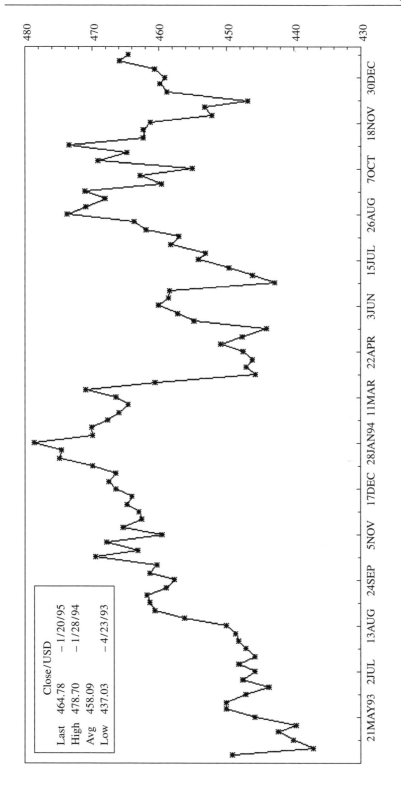

Figure 1.6 Standard and Poor's stock Index. Weekly data from 16 April 1993 to 20 January 1995.

Source: Adapted from Bloomberg L.P. ©1998.

The chart legend reads:

Close/USD

Last	464.78	– 1/20/95
High	478.70	– 1/28/94
Avg	458.09	
Low	437.03	– 4/23/93

far too easily led to attributing unrelated causes to those patterns. What is needed, and what we will undertake to present in the following chapters, are concepts and methods of numerical analysis of observational time series that can objectively establish the existence of specific characteristics of the underlying process that generated the time series.

When a dynamical system is allowed to operate for a long period of time, its orbit may settle down to wandering over some relatively small subset of the state space. If that happens, the subset of the state space onto which the orbit settles is called an *attractor* of the system. Implicit in this description of an attractor, is the dependence on the initial state that gave rise to the orbit that evolves to the attractor. Hence, different sets of initial states can give rise to different attractors. The set of initial states that give rise to a particular attractor is called the **basin of attraction** of that attractor, and while a system can have different attractors, their basins of attraction are distinct.

Attractors can be very simple or very complex objects. Simple attractors occur when the system's trajectory evolves onto an equilibrium point or a periodic orbit. Alternatively, the system's trajectory may evolve in an aperiodic fashion so that we wind up with an orbit that is not only confined to a small region of the state space, but is also destined never to run into itself. When trying to imagine this situation the image of a bowl of tangled spaghetti comes to mind. And, as if that situation were not complicated enough, we note that we have been talking about the evolution of a single trajectory of the system. Typically, the trajectories of a large portion of the points of the state space suffer the same fate of being drawn into a single attractor. Thus, it is not just a single orbit that must occupy the confined space of the attractor without running into itself, but many orbits have to occupy that small region of the state space never running into themselves *and* never running into one another. Given that such an attractor can materialize, we should expect the resulting tangle of trajectories to possess rather special properties. For one, even though the orbits are precluded from intercepting themselves, being confined to a bounded region of the state space implies that they must come (arbitrarily) close to intercepting themselves over and over again, and these close encounters will take place everywhere on the attractor. That means that, not only will every trajectory revisit the neighborhood of every state on its own path infinitely often, but it will also visit the neighborhood of every state along the path of every other trajectory infinitely often. This kind of behavior is suggestively called *recurrence*. If the statistical properties of the spatial distribution of a trajectory are generally independent of the particular trajectory, then a system with the recurrence property is, not so suggestively, called *ergodic*[5]. Ergodicity implies a persistent invariant recurrence of a trajectory throughout the attractor and it permits some important computational simplifications as we shall see later.

The control parameter μ of a dynamical system is like the rich/lean fuel mixture setting on the carburetor of an automobile engine. At one setting the engine idles smoothly, while at another setting the idle is rough and may even cause the engine to stall. Similarly, the control parameter of a dynamical system can have a profound effect on the nature of the attractor that the system aspires to. For a given set of initial states for the system, some settings of μ might yield an equilibrium point or periodic

---

[5] See the section of this chapter entitled *Theory* for more on ergodicity.

attractor, while for other values of μ we get aperiodic attractors which are more or less 'disorganized'. In the latter case, for example, it is possible for an orbit to wrap itself around a torus[6] in such a way that it is aperiodic and recurrent. Such orbits are called quasi-periodic. Quasi-periodic orbits are not simple, but they are in some sense neatly organized. If we pick a small region on the torus as a set of initial states, then the trajectories emanating from those states will remain close over time because each winds its way around the torus in the same way. For that to happen, the trajectories of the system have to be 'sociable' things willing to coexist in small confined regions of the state space. They like, or at least do not mind, being squeezed up like sardines in a tin (see Figure 1.7).

Figure 1.7 On the left, the beginning of a quasi-periodic orbit wrapping itself around a torus. On the right, a 'patch' B_0 of initial states is carried over the surface aperiodically and without being deformed. After one trip around the torus, B_0 returns to the vicinity of its initial position as the patch B_t.

At a different setting of the control parameter μ, however, the orbits may become antisocial relative to the position of nearby orbits and begin 'repelling' each other along some directions. As a result, the orbits emanating from two nearby states diverge, so the evolution of the system becomes sensitive to the location of the initial state (see Figure 1.8).

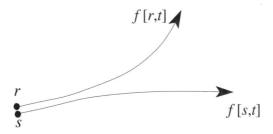

Figure 1.8 Divergent trajectories of two nearby points, r and s.

If the trajectories on an attractor are repelling, that is, sensitive to the initial state, then the attractor and the orbits giving rise to it are called ***chaotic***[7]. Based on this

---

[6] Surface of a donut.

[7] Chaotic attractors are also called ***strange attractors***, although this terminology is not universal.

definition of chaos, 'chaotic attractor' may sound like a contradiction in terms. After all, how can all these orbits be repelling one another and still be recurrently circulating in a confined region of the state space? In simple terms the answer is that implicit in the state transition function f_μ is a mechanism which folds the diverging trajectories back on themselves in such a way that they never collide. Characteristically, chaotic attractors arise through systems that operate to pull apart, fold back, and then squeeze trajectories together again. The phenomenon is known as ***phase space deformation*** and is illustrated in Figure 1.9.

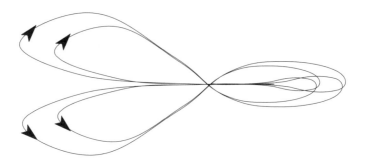

Figure 1.9 In order to keep the trajectories in a bounded region, the dynamics of the system operate to fold divergent trajectories back on themselves without causing them to intersect.

We will see more on phase space deformation later in Chapter 2, Entropy. For the moment we want to look at what chaotic dynamics implies about predictability. From what has already been said, chaotic attractors have a complexity which just beggars the imagination. If we obtain a measurement time series from a chaotic orbit, which is in effect a projection of the orbit into a one-dimensional space, then we are adding obscurity to complexity, and we should not be surprised to find that the observed time series has statistical properties in common with a time series generated by a random (indeterminate) system[8]. But this is apparent or pseudo randomness and is not in principle a problem in the prediction game, since, as we will show later, with a lot of highly accurate data, we can work our way back from a pseudo random observational time series to a reasonable model of the dynamical system. The real prediction problem does not have a solution because it stems from limits imposed by nature.

The notion that we will ever be in possession of a lot of highly accurate data in the experimental situation is, to begin with, unrealistic, but even if we were to be fortunate enough to have it or, better yet, lucky enough to correctly guess the exact dynamical equation of the system, we would still have a serious prediction problem in the presence of chaos. The root of the trouble lies with the repelling characteristic of chaotic orbits. Namely that even if two states of the system are located very close together in the state space, the two trajectories that emanate from those points, that is,

---

[8] By indeterminate we mean that we get a different trajectory each time we start a random system from the *same* initial state. In contrast, deterministic systems yield the same trajectory when started from the same initial state.

the subsequent movements of the system, become very different very quickly. Now, in the real world, perfect precision is unobtainable and every measurement we make has some error. If we attempt to make a prediction from an incorrect initial state we get a trajectory that goes off in one direction while the actual system, using the correct initial state, goes off in another. The accuracy of a prediction for a chaotic system will depend on both the error in specifying the current (initial) state of the system, and the speed with which the trajectories of the system move apart. Note that the uncertainty inherent in a chaotic system is of a very different nature than the uncertainty inherent in a random system. While, by definition, random systems are indeterminate full stop, the reliability of predictions made in the presence of chaotic dynamics only degenerates, albeit fairly quickly, with the passage of time. Having said that, some predictability is a lot better than no predictability, and so we soldier on.

To carry the military metaphor a bit further, we now outline the campaign for the remaining chapters of the book. The obvious first step in modeling the dynamics of the system would be to obtain a sample of one of its trajectories. With a trajectory in hand, we could draw on a massive body of numerical function estimation techniques for estimating the dynamical equation of the system. Unfortunately, in the experimental situation all we have to work with is an observational time series which is at best a highly encrypted image of the original trajectory. A bit of thought should suggest that retrieving the original trajectory from its ciphered time series image is unlikely to be a trivial undertaking. We take up that quest in Chapter 3, Phase space reconstruction. If the application of the methodology developed in that chapter is successful, we will be in possession of a *representation* of the original trajectory which we call the **experimental trajectory**. The shape of the experimental trajectory will almost certainly not resemble the original, but a properly reconstructed experimental trajectory will be just a smooth distortion of the original, and it will faithfully duplicate the important geometrical and dynamical characteristics of the original. Geometrical and dynamical characteristics which are left unaltered by the process of phase space reconstruction are called **system invariants**. We will be interested primarily in system invariants in this book precisely because they are the system characteristics that are not changed by the unavoidable process of reconstructing a trajectory. Dimensionality and sensitivity to initial states are two characteristics that are crucial to understanding the complexity of a model for the system's dynamics, and the limitations of predictions made from that model. Fortunately, those characteristics are system invariants.

To get an estimate of the number of degrees of freedom in the system we will look at the dimension of its attractor in Chapter 4, Fractal dimension. To obtain limits on the predictability of the system we will look at the rate at which trajectories of nearby states diverge in Chapter 5, Lyapunov exponents. The two subjects are not unrelated and Chapter 2, on entropy, another system invariant, provides the unifying concept. Given a chaotic system's sensitivity to initial states, contaminated or 'noisy' data is a big problem, and we address that issue in Chapter 6, Noise reduction. Finally, in Chapter 7, Concluding remarks, we develop a procedure for weighing the evidence for, and against, the proposition that chaotic dynamics are responsible for the observed data.

THEORY

Dynamical systems

Dynamical systems are defined formally in terms of a function relating the system's past to its future or vice versa. The function effectively expresses the rule that the system uses to make transitions from one **state** to the next. Since that **state transition rule** is a mathematical function, a dynamical system is a **deterministic** system in the sense that given its present state precisely, the system's future is determined for all time and, if the state transition rule is invertible then, the system's past is also determined for all time. Implicit in the definition of a dynamical system is that the system is propelled through some m-dimensional **state space** M by applying the transition rule repeatedly after each change of state. As the system moves through the state space it traces out an **orbit or trajectory** that temporally links the states that it visits. When the state transitions are continuous in time the simplest model for the system is an **autonomous** differential equation of the form,

$$\frac{ds(t)}{dt} = F_\mu[s(t)] \qquad t \geq 0 \qquad\qquad \text{Eq. 1.2}$$

where $s(t) \in M$ is the state vector at time t, F_μ is the **'velocity' vector** or **vector field** of the system, and μ is a vector of parameters for the system. Since the explicit reference to μ implies that it is a variable, Eq. 1.2 really defines a family of **continuous time** dynamical systems. Under mild assumptions (basically that F is Lipschitz), there exists a unique time-continuous solution, $s(t) = f_\mu[s(0), t]$, to Eq. 1.2 for every **initial condition or initial state** $s(0)$. Clearly, $s(t) = f_\mu[s(0), t]$ is a trajectory of the system, though in the case of a continuous time system it is also known as a **flow**. If the vector field F_μ is an explicit function of time, then the system is defined by the **nonautonomous** differential equation

$$\frac{ds(t)}{dt} = F_\mu[s(t), t] \qquad t \geq 0 \qquad\qquad \text{Eq. 1.3}$$

A system defined by a nonautonomous equation is **nonstationary**, that is, the equation governing its movement in the state space is changing over time. For the time being we will assume that the system is stationary in the above sense. The restriction will be eased after we discuss 'bifurcations' below.

We can also define a vector field for a continuous time dynamical system that depends on more of the system's past than just its current state $s(t)$. For example,

$$\frac{ds(t)}{dt} = F_\mu[s(t), s(t - T)] \qquad t \geq T \qquad\qquad \text{Eq. 1.4}$$

where the lag time T is another system parameter, is known as a **delay differential equation**. In effect, the 'state' of a system defined in this way is the whole continuous trajectory segment from $s(t - T)$ to $s(t)$. Hence, delay differential equations define infinite dimensional systems. We will see an example of such a system a little later on.

If the state transitions of the system are not time-continuous, then the system is called a **discrete** time system and is defined by a difference equation of the form

$$s(n + 1) = f_\mu[s(n)] \qquad n = 0, 1, 2, \ldots \qquad\qquad \text{Eq. 1.5}$$

Clearly, $f_\mu[s(n)]$ is the n-fold iteration of f_μ with itself, that is

$$\overleftrightarrow{n\text{-}times}$$
$$f_\mu[s(n)] = f_\mu \circ f_\mu \circ \cdots \circ f_\mu[s(0)] = f_\mu^n[s(0)] \qquad \textbf{Eq. 1.6}$$

Note that in the case of the flow $f_\mu[s(0), t]$, we can define the **time-T map**

$$s(1) = f_\mu[s(0), T] = f_{\mu,T}[s(0)]$$

$$s(n) = f_\mu[s(0), nT] = f_{\mu,T}^n[s(0)] \qquad \textbf{Eq. 1.7}$$

which, in effect, **discretizes** the flow, that is, extracts a discrete time map from a continuous time system. The time-T map is also known as a **stroboscopic map**. The device of discretizing the flow in this way is sometimes useful, and we will use it, for establishing a result for a continuous time system that is only proven for a discrete time system. There are other ways, as we shall see shortly, to discretize the flow of a continuous time system which also reduce the dimension of the dynamics.

A dynamical system is called **differentiable** if $F_\mu[s]$ (continuous time) or $f_\mu[s]$ (discrete time) is differentiable in the state variable s. In the sequel we will assume that the system we are looking at is a differentiable dynamical system. Since continuity in the state variable is then assumed, we will only use the term continuous system to denote a system that is continuous in time. To keep the notation simple, we will use f (and occasionally f_μ) to denote either a flow or a discrete map, and f^t to denote the time t flow or the convolution of the discrete map. Also if f is invertible, then f^{-t} is the inverse of f^t, and the time index in Eq. 1.2 through Eq. 1.7 can be negative.

Attractors and limit sets

Definitions
- A point c is a **limit point** of a state s if for every neighborhood U of c, $f^t[s] \in U$ infinitely often as $t \to \infty$. The set $L(s)$ of all limit points of s is called the **limit set** of s.
- A limit set Q is an **attracting limit set** if there exists an open neighborhood U of Q such that $L(s) \subset Q$ for all $s \in U$. The union of all such open neighborhoods of Q is called the **basin of attraction** of Q.
- There does not seem to be a definitive technical definition of an attractor, though in spirit they all seem to be close to the concept of an attracting limit set. We will adopt the following operational definition. An **attractor** is a subset of the state space upon which trajectories of a system accumulate in the long term. The set of all initial states whose trajectories evolve onto the attractor is the **basin of attraction** of the attractor. Attracting limit sets are attractors. Attractors are invariant in the sense that, for A an attractor, $f(A) = A$. Thus, any trajectory that starts out on the attractor will remain on the attractor.
- An orbit that is a limit set, or that is in the basin of attraction of, and close to an attractor, will be called a **stationary orbit**. A stationary orbit is one which is recurrent, but not necessarily on an attractor. However, if a stationary orbit is non-attracting, then it has no basin of attraction, and the only way to get onto the orbit is to start out at some state on the orbit.

We assume that the system under observation has an attractor and that it is on a stationary orbit. One implication of that assumption is that the system will behave in the future as it has in the past. The basin of attraction of an attractor is likely to occupy a large portion, if not almost all of the state space, but the attractor will typically occupy only a small volume of the state space. In fact, we hope that the attractor's dimension is very small, certainly less than 10, as that implies that the asymptotic motion of the system only involves a few degrees of freedom, regardless of the dimension of the state space. If that turns out to be the case, then we will only need to consider a handful of variables in modeling the system. We make no *a priori* assumptions about the dimension of the system's attractor as that amounts to just wishful thinking in the experimental situation. The methods developed in Chapter 4, on fractal dimension, will let the data do its own talking. However, we note that if both the true dimension of the attractor, and the maximum average rate at which orbits diverge, are too large, then we may not be able to obtain, even in principle, enough relevant data to reconstruct a trajectory for the system[9].

Reality check
The state space of a dynamical system is littered with a zoo of latent stationary orbits that are sometimes thought of as being 'observable' only when they are attracting. Used in that way 'observable' is defined by a process. Thus, suppose that we could subject the system to laboratory conditions so that we could stop and restart the system from different initial states. Then we are in a position to carry out an experiment where we pick a state 'at random' from the state space, start the system off from that state, and observe what sort of, if any, stationary orbit is pursued by the system. For concreteness, assume that the state space is $\Re^3$ and that the probability of a random choice of initial state falling into some specific region of the state space is proportional to the volume of that region[10].

We can carry on the above experiment as many times as we like and we will have zero probability of ever observing a particular non-attracting stationary orbit, and a high probability of observing one with a large basin of attraction. The point of this digression is that even if the system we are observing has an attractor whose basin of attraction has probability 1 (in the sense of the laboratory trials), the world may have conspired to give the system one of those nil-probable initial states that put the system on a non-attracting stationary orbit. The reason for assuming that the system that we are observing has an attractor is to give ourselves the intellectual comfort of knowing that, in that case, almost any initial state would have resulted in the trajectory that we actually observe being one that is on or near the putative attractor. The fact is that the system we observe is on whatever trajectory it is on, and if that is not a trajectory which is asymptotic to the assumed attractor, then we cannot improve our chances of it being so by restarting the system at some other initial state.

There follow some examples of simple and chaotic attractors drawn in ***phase portrait*** form. Phase portraits are diagrams depicting the global behavior of a system's orbits. The ambient space of a phase portrait is called the ***phase space***. A phase space is a state space without an explicit time coordinate. Because we assume that the

---

[9]The point will be taken up again in Chapter 3, Phase space reconstruction.

[10]We use $\Re^3$ to denote the continuous 3-dimensional physical space that we live in (or at least perceive) and $\Re^n$ to denote the generalization of that space.

dynamics of the system of interest are time independent (autonomous), phase space and state space have the same meaning.

Some simple attractors

An *equilibrium or fixed point* of a dynamical system is a state p such that $f^t[p] = p$ for all t. A system that is in an equilibrium state is motionless, that is, it has zero velocity. Hence, if the system is continuous, then its vector field F_μ vanishes at p (see Figure 1.10).

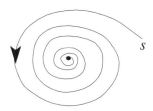

Figure 1.10 Equilibrium point attractor p (dot at center of the figure). Any point s in p's basin of attraction has an orbit which is asymptotic to p.

A *periodic orbit* (of period Γ) of a dynamical system is a limit set C such that $f^t[c] = f^{t+\Gamma}[c]$ for all t and $c \in C$. A *limit cycle* is an attracting periodic orbit (illustrated in Figure 1.11).

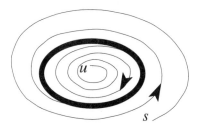

Figure 1.11 A limit cycle C (heavy ellipse). The basin of attraction for C includes points like u and s that can be interior to C or exterior to C.

The asymptotic behavior of continuous dynamical systems in two or less dimensions is very limited, as the following theorem demonstrates.

Theorem (Poincaré–Bendixon)
A differentiable two-dimensional continuous dynamical system has only fixed point and limit cycle attractors.

Hence, chaotic dynamics cannot arise in continuous time systems in less than three dimensions. Discrete time systems, however, can be chaotic in two or less dimensions.

17

Some chaotic systems with their attractors

Hénon map[11]

$$x_{n+1} = 1 - ax_n^2 + y_n$$
$$y_{n+1} = bx_n$$

$\mu:$ $a = 1.4, b = 0.3$

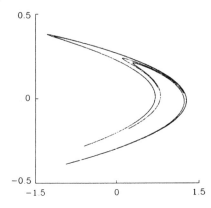

Figure 1.12 Hénon attractor.

Source: Adapted from Hammel (1990) p. 424.

Ikeda map[12]

$$x_{n+1} = \kappa + \beta(x_n \cos \omega_n - y_n \sin \omega_n)$$
$$y_{n+1} = \beta(x_n \sin \omega_n + y_n \cos \omega_n)$$
$$\omega_n = 0.4 - \alpha/(1 + x_n^2 + y_n^2)$$

$\mu:$ $\alpha = 5.4, \beta = 0.9, \kappa = 0.92$

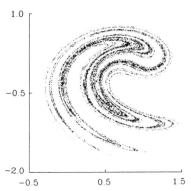

Figure 1.13 Ikeda attractor.

Source: Adapted from Hammel (1990) p. 423.

[11]Hénon (1976).

[12]Hammel, Jones and Moloney (1985).

Rössler flow[13]

$$\frac{dx}{dt} = -(y + z)$$

$$\frac{dy}{dt} = x + ay$$

$$\frac{dz}{dt} = b + z(x - c)$$

μ: $a = 0.15, b = 0.20, c = 10.0$

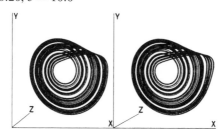

Figure 1.14 Stereoscopic view (parallel projections) of the Rössler attractor. The left-hand picture is meant for the right eye, and the right-hand picture for the left eye. If you stand back about 25–30 centimeters and you let your eyes cross a bit, the two figures will merge into a three-dimensional like image of the attractor.

Source: Adapted from Rössler (1976) p. 398.

Lorenz I flow[14]

$$\frac{dx}{dt} = \sigma(y - x)$$

$$\frac{dy}{dt} = x(\rho - z) - y$$

$$\frac{dz}{dt} = xy - \beta z$$

μ: $\sigma = 16.0, \rho = 45.92, \beta = 4.0$

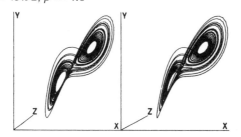

Figure 1.15 Stereoscopic view (parallel projections) of the Lorenz I attractor. The left-hand picture is meant for the right eye, and the right-hand picture for the left eye. If you stand back about 25–30 centimeters and you let your eyes cross a bit, the two figures will merge into a three-dimensional like image of the attractor.

Source: Adapted from Rössler (1976) p. 397.

[13]Rössler (1976).
[14]Lorenz (1963).

Mackay-Glass flow:[15] *Delay differential equation*

$$\frac{dx}{dt} = \frac{ax(t+T)}{1 + [x(t+T)]^c} - bx(t)$$

μ: $a = 0.2, b = 0.1, c = 10.0, T = 31.8$

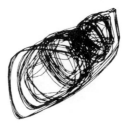

Figure 1.16 Two-dimensional projection of Mackay-Glass attractor.
Source: Adapted from Pawelzik and Schuster (1991) p. 1810.

Bifurcations

We have already remarked that a system can have more than one attractor. In fact it can have different types of attractors (limit cycles, fixed points), and non-attracting stationary orbits depending on which initial state the system starts out from. But we have also hinted at the fact that, as μ is varied, even the asymptotic (long term) behavior of the trajectory of a given initial state can change radically. Changes in a system's attractor type are called ***bifurcations*** and the values of μ for which they occur are called ***bifurcation points***. We made the assumption earlier that the equation governing the system's motion is autonomous. In particular, that means that the system parameter μ is constant. Realistically, that will rarely be the case. So what we will assume is that while μ may wobble a bit over a small region of the parameter space, that region does not contain any bifurcation points.

Reducing flows to discrete maps

Any numerical analysis of a system's dynamics that can be done in even one less dimension will have a significant computational impact. It can also have a big impact on our ability to visualize the geometry of the system's trajectory. Hence, the following device can be very useful.

Suppose that the attractor Q of the system lies in some m-dimensional space M. Now imagine taking an $(m - 1)$-dimensional surface and inserting it through the attractor in such a way that the system's trajectories always intersect the surface transversely. Such a surface is called a ***Poincaré section or surface of section***. By construction, as the system evolves along its trajectory it eventually intercepts some point on the Poincaré section, after which it evolves for a period of time before it again intercepts a point on the surface, and so on. An $(m - 1)$-dimensional surface

---

[15]Mackay and Glass (1977).

in an m-dimensional space can be thought of as having a 'front' and a 'back' side, so that any transverse interception of the surface by the trajectory occurs either as the trajectory passes from the front to the back of the surface, or as the trajectory passes from the back to the front of the surface. Hence, we can associate with each point of interception one of two directions the trajectory had as it hit the surface. Call these the plus and minus directions, so we can label each point a plus intercept or a minus intercept. Pick a direction, it does not matter which: say, minus. Now collect all the minus intercepts in the time order of their occurrence and denote their location in the Poincaré section by r_i, $i = 1, 2, \ldots$ (as shown in Figure 1.17).

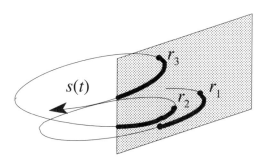

Figure 1.17 Poincaré surface of section (shaded area) showing the first three 'minus' intercepts (r_1, r_2, r_3) with the trajectory $s(t)$. These intercepts occur as the trajectory passes from the back to front of the surface as drawn. Note that we ignore the 'plus' intercept that occurs between r_1 and r_2.

The sequence of interception points r_i defines a discrete $(m - 1)$-dimensional map

$$r_{i+1} = P(r_i)$$

P is called the Poincaré *first return map* of the continuous system. Many of the important characteristics of the original continuous system are preserved in its first return map.

Manifolds and tangent spaces

While we do not require that the state space M of the dynamical system we are analyzing be $\mathfrak{R}^m$, we will assume that M is a manifold. An m-dimensional manifold is a set of points that locally looks like $\mathfrak{R}^m$, that is, it is an object which we can think of as being constructed out of small (perhaps overlapping) pieces of $\mathfrak{R}^m$. Obviously, $\mathfrak{R}^m$ is itself an m-dimensional manifold. The surface of a sphere or a torus are examples of two-dimensional manifolds. Pretzels and bagels are examples of three-dimensional manifolds[16].

Suppose that the state space manifold is the surface of a sphere. Then the trajectories of a continuous system will form curves on the spherical surface and the vector

---

[16]For a more formal definition see Chapter 3, Phase space reconstruction.

field of the system will define velocity vectors that are tangent to each trajectory at every point on the trajectory. Vectors that are tangent to a curve on the sphere are also tangent to the sphere. Hence, the vector field of the dynamical system defines a set of vectors, called **tangent vectors**, that are tangent to the state space manifold at every point of the phase space. Another dynamical system defined on the same state space manifold will define a different set of tangent vectors. At any point common to a trajectory for both systems, the velocity vectors will be generally different, but still tangent to the manifold at the given point. Those two tangent vectors span a two-dimensional plane that is itself tangent to the manifold at the point (see Figure 1.18).

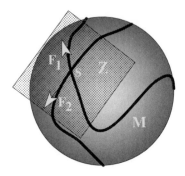

Figure 1.18 Tangent vectors F_1 and F_2 define the tangent space Z, to M, at s.

In the general situation, the set of all possible tangent vectors at a given point x on an m-dimensional manifold defines (spans) an m-dimensional linear space $\mathcal{T}_x$ called the **tangent space** of the manifold at x. The definition of a manifold's tangent spaces is independent of the concept of a dynamical system, but in the context of a dynamical system with dynamics f, the space derivative (Jacobian) $D_x f$ defines the tangent spaces of the state space manifold. The tangent spaces $\mathcal{T}_x$ possess a kind of invariance with respect to $D_x f$, the Jacobian at x. Specifically, $D_x f$ maps the tangent space at x to the tangent space at $f[x]$ (see Figure 1.19).

Over small regions of the state space manifold M, M and $\mathcal{T}_x$ are nearly coincident at x, and a small vector δ on the manifold near x is approximately a tangent vector in $\mathcal{T}_x$. It follows that, over small regions of the manifold, differentiable dynamics are approximately linear[17], that is, for a small perturbation vector δ at x

$$f[x + \delta] - f[x] \approx D_x f \cdot \delta$$

or

$$\delta_1 \approx D_x f \cdot \delta_0$$

---

[17]We use this fact to get a local linear model of the system's dynamics in Appendix 5, Linearized dynamics.

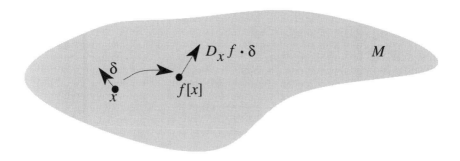

Figure 1.19 Small tangent vectors at x get mapped by $D_x f$ to small tangent vectors at $f[x]$.

From Linear Algebra theory, the eigenspaces of the linearized dynamics, $D_x f$, form a decomposition of $\mathcal{T}_x$. The action of the dynamics on a small vector in one of these eigenspaces will result in the vector being either expanded, contracted, or left unaffected. In what follows it will be convenient to define a space $E^u[x]$ formed from the direct sum ($\oplus$) of all the eigenspaces corresponding to expansion of tangent vectors, and another space $E^s[x]$ defined similarly for the eigenspaces corresponding to contraction of tangent vectors. $E^s[x]$ and $E^u[x]$ are called, respectively, the **stable** and **unstable eigenspaces** at x, and are invariant with respect to $D_x f$ in the same sense as the $\mathcal{T}_x$ are invariant with respect to $D_x f$.

From here we do not have far to go to develop some insight into the amount (lack) of predictability in the system. Consider the orbits emanating from two nearly identical initial states $s(0) \approx u(0)$, and define the drift or separation between the two orbits after a time t by,

$$\delta(t) = f^t[u(0)] - f^t[s(0)] \approx D_{s(0)} f^t \cdot \delta(0)$$

Since the two initial states are assumed to be nearly identical, $\delta(0) = u(0) - s(0)$ is very small and, to good approximation, is a tangent vector at $s(0)$. As such, it can be decomposed into components which lie in the eigenspaces of $D_{s(0)} f$. If any component of $\delta(0)$ lies in an expanding eigenspace, then the orbits from $u(0)$ and $s(0)$ are separating at an exponential rate in the direction of that eigenspace. If the orbits are separating at an exponential rate in *any* direction at all, then prediction will be limited, and the severity of that limitation will depend on the magnitude of the rate of separation. If two nearby states give rise to exponentially separating trajectories, anywhere on the state space manifold, then the dynamical system is said to be **sensitive to initial states**. A system which is sensitive to initial states is called **chaotic**. In general, the exponential rates at which orbits 'attract' or 'repel' one another are related to the system's Lyapunov spectrum, a topic we take up in Chapter 5, on Lyapunov exponents.

Stable and unstable manifolds

We now turn to yet another pair of objects, called stable and unstable manifolds, for two reasons. The first is that they will come up in the development of algorithms in Chapter 6, Noise reduction. The second reason for considering stable and unstable manifolds is that they introduce a fascinating picture of the way chaos can arise and the complex trajectories it can produce.

As already mentioned, the eigenspaces that decompose the tangent space, in particular $E^s[x]$ and $E^u[x]$, are linear and tangent to the state space manifold. Corresponding to these, there are two *nonlinear* objects, $W^s[x]$ and $W^u[x]$, defined on the *state space* called, respectively, the **stable and unstable manifolds** of a given point x. Loosely speaking, they are the sets of points in the state space which have, respectively, the same future or past as x. For x on a periodic orbit or x an equilibrium point, and B a neighborhood of x, the stable and unstable manifolds of x are defined by the relations

$$W^s[x] = \left\{ y \in B \,\middle|\, \lim_{t \to \infty} \left| f^t[x] - f^t[y] \right| = 0 \right\}$$

$$W^u[x] = \left\{ y \in B \,\middle|\, \lim_{t \to \infty} \left| f^{-t}[x] - f^{-t}[y] \right| = 0 \right\}$$

Because $W^s[x]$ and $W^u[x]$ are submanifolds in the state space, and contain the point x, they must be tangent to the tangent space at x. In fact, each is tangent to the corresponding stable and unstable eigenspaces of the tangent space at x, $E^s[x]$ and $E^u[x]$, respectively (see Figure 1.20).

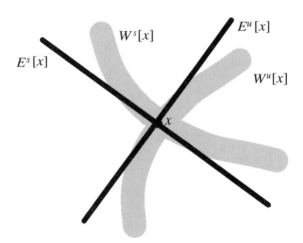

Figure 1.20 The stable manifold at x, $W^s[x]$, is tangent to the stable eigenspace at x, $E^s[x]$, and the unstable manifold at x, $W^u[x]$, is tangent to the unstable eigenspace at x, $E^u[x]$.

$W^s[x]$ and $W^u[x]$ are invariant with respect to the dynamics f in the sense that

$$f[W^s(x)] = W^s[f(x)]$$

and

$$f^{-1}[W^u(x)] = W^u[f^{-1}(x)]$$

Similar ideas are used to obtain stable and unstable manifolds for an arbitrary point in the state space, and the stable and unstable manifolds of trajectories are defined by the unions of the stable and unstable manifolds of the points on those trajectories. Points or trajectories that have only a stable (unstable) manifold or eigenspace are called *stable (unstable)*. If a point or trajectory has both stable and unstable manifolds or eigenspaces, then it is called a *saddle*. The existence of the manifolds $W^s[x]$ and $W^u[x]$, and their relationship with the eigenspaces $E^s[x]$ and $E^u[x]$, can only be established rigorously for systems with certain special properties[18] which most real systems do not seem to possess. We will assume that, for the system under study, the above objects exist locally, and the stated relationships hold approximately. As Chapter 6, Noise reduction, also discusses approaches that do not involve $W^s[x]$ and $W^u[x]$, the failure of the assumptions we make here concerning them is not fatal to the objectives of that chapter.

Stable manifolds *cannot* intersect either themselves or other stable manifolds. The same is true of unstable manifolds. However, a stable manifold *can* intersect an unstable manifold. When a stable manifold and an unstable manifold of a saddle point intersect transversely, the intersection is called a *homoclinic intersection*[19]. Since a point of homoclinic intersection necessarily belongs to both invariant manifolds, it will be mapped by the dynamics to another point which belongs to both manifolds (see Figure 1.21).

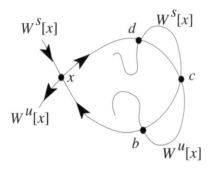

Figure 1.21 Homoclinic point c is mapped by the dynamics f to the homoclinic point b, and mapped by the inverse f^{-1} to the homoclinic point d.

Hence, one homoclinic point of intersection implies infinitely many others in the past and future of the point's trajectory. If a trajectory has a homoclinic point then the future and past of the trajectory will approach the saddle equilibrium point or saddle periodic orbit which gave rise to the intersecting manifolds. The path of such an orbit

---

[18] Among those is the property of **hyperbolicity** discussed further in Chapter 6, Noise reduction.

[19] Keep in mind that we are talking about intersecting manifolds here and not intersecting trajectories which are impossible. Manifolds can intersect without the trajectories on them intersecting.

is most easily visualized by looking at a Poincaré section of a saddle periodic orbit with homoclinic intersections as depicted in Figure 1.22.

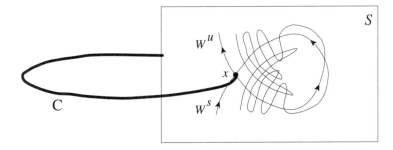

Figure 1.22 Poincaré section S showing the homoclinic tangle of the saddle periodic orbit C. Note that only the first few intersecting loops of the tangle on the Poincaré section are shown in the diagram. The complete tangle is far more complex.

The complicated interweaving of the stable and unstable manifolds on the Poincaré section in Figure 1.22 is known as a ***homoclinic tangle***. The full picture of the saddle periodic orbit's homoclinic tangle is obtained by tracing out the Poincaré section of the tangle, as the Poincaré section is rotated around and perpendicular to the orbit[20]. Any trajectory that passes through (but not necessarily on) that homoclinic tangle will be influenced by it, and will be both very complicated and vaguely periodic as it circulates for a time near the saddle orbit. It is also possible for the stable and unstable manifolds of one point to intersect their opposites for another point. The resulting tangle, called ***heteroclinic***, looks like back-to-back homoclinic tangles.

The phase space could be littered with latent saddle periodic orbits and their homoclinic and heteroclinic tangles. In fact it turns out that chaotic attractors can have a densely embedded skeleton of saddle type periodic orbits. If the trajectory of the system is not on the stable manifold of any of those saddle orbits, then it can endlessly circulate from the influence of one tangle to that of another. That is indeed a complicated trajectory.

Natural measure

One of the most fundamental properties of a dynamical system is the spatial distribution of points (states) on its attractor (trajectory.) By spatial distribution we mean the geometric relationship among the points on the attractor as reflected in the density of points in small neighborhoods of the attractor. We can express the idea of point density mathematically as a probability measure. To that end, define the indicator function χ_B of a set B by,

$$\chi_B(s) = \begin{cases} 1 & \text{if } s \in B \\ 0 & \text{otherwise} \end{cases}$$

[20]For some diagrams see Abraham and Shaw (1992).

If a dynamical system has an attractor $Q \subset M$ and $s(0)$ is any point in its basin of attraction with trajectory $s(t) = f^t[s(0)]$, then for any open set $B \subset Q$ we define the **natural measure** of B by[21]

$$\rho(B) = \underset{T \to \infty}{\text{Lim}} \frac{1}{T} \int_0^T dt \chi_B[s(t)] \qquad \text{Eq. 1.8}$$

If ρ exists, then for any continuous function $\phi : M \to \mathfrak{R}$, the quantity

$$E(\phi) = \int_M \rho(dx) \, \phi(x) \qquad \text{Eq. 1.9}$$

is called the **expected value or space average** of ϕ with respect to the measure ρ, and

$$\overline{\phi} = \underset{T \to \infty}{\text{Lim}} \frac{1}{T} \int_0^T dt \phi[s(t)] \qquad \text{Eq. 1.10}$$

is called the **time average** of ϕ with respect to the dynamics f, on the basin of attraction of Q. The conditions necessary for the existence of the limits in Eq. 1.8 through Eq. 1.10, and the equality of the space and time averages is the subject of Ergodic Theory[22]. Basically, the theory says that if the natural measure exists and is ergodic, then the space average is equal to the time average, and vice versa. Chaotic systems are expected to be ergodic, but proving that the conditions for the ergodic theorems hold in the experimental situation can be difficult. Fortunately, it appears that most physical systems have stable time averages which suggests the existence of a natural ergodic measure. In any case, in this book we use the time average

$$P(x; \varepsilon) = \frac{1}{T} \int_0^T dt \chi_{B_\varepsilon[x]}[s(t)] \qquad \text{Eq. 1.11}$$

as a probability measure for a neighborhood $B_\varepsilon[x]$ of x, assume it is ergodic, and proceed to equate time and space averages. The probability measure $P(x; \varepsilon)$ is basic to the invariants of a dynamical system, concepts developed in the chapters on entropy, fractal dimension and Lyapunov exponents. In those chapters $P(x; \varepsilon)$ appears in the equivalent (discrete) form

$$P(x; \varepsilon) = \frac{1}{N} \sum_{k=1}^{N} \Theta(\varepsilon - \|s(k) - x\|) \qquad \text{Eq. 1.12}$$

where Θ is the Heaviside function defined by

$$\Theta(z) = \begin{cases} 1 & \text{if } z > 0 \\ 0 & \text{otherwise} \end{cases}$$

Some closing words on ergodicity. To say that a set A is **invariant** with respect to the dynamics f, is to say that $f(A) = A$. Thus, a set A is invariant if trajectories starting in A never leave A. To say that the natural probability measure ρ is **ergodic**

---

[21]The formulation is for a continuous system; for discrete systems replace the integral with a summation. For an alternative form using the Dirac delta see the Discrimination algorithm in Chapter 6, Noise reduction.

[22]See for example, Katok and Hasselblatt (1995); or see general references.

is to say that, if A is invariant, then either $\rho(A) = 0$ or $\rho(A) = 1$. In other words, if an invariant set has positive natural measure, then it has all the measure: it is an all or nothing affair. Now, a chaotic attractor is an invariant set, and ergodicity implies that the trajectories of the system are almost always recurrent throughout the attractor and the statistical properties of the recurrence, that is, the spatial distribution of a trajectory's points, is (in the long term) independent of the particular trajectory. It follows that ρ almost every trajectory will cut a dense path through the attractor, and the amount of time that those trajectories spend in any given region of the attractor is proportional to the size of the region. The starting states which do not have a dense orbit on the attractor add up to a set of ρ measure zero, and include the unstable and saddle periodic orbits and fixed points of the system. Since the natural measure is defined by the system's dynamics f, f is called ergodic if its natural measure is ergodic.

We remark that chaos means something more than ergodicity since the quasi-periodic motion on a torus discussed earlier is ergodic but not chaotic. The deformation of phase space, characteristic of chaotic systems, implies that the end-points of the trajectories emanating from any set of positive ρ measure will be scattered randomly (in accordance with the measure ρ) over the whole attractor within a short period of time depending on the system's sensitivity to initial states.

REFERENCES AND FURTHER READING

Abraham, R. H. and Shaw, C. D. (1992) *Dynamics: The Geometry of Behavior*, Addison-Wesley, Redwood City.

Auerbach, D., Cvitanovic, P., Eckmann, J., Gunaratne, G. and Procaccia, I. (1987) 'Exploring chaotic motion through periodic orbits', *Physical Review Letters* 58.23, pp. 2387–2389.

Cvitanovic, P. (1988) 'Invariant measurement of strange sets in terms of cycles', *Physical Review Letters* 61.24, pp. 2729–2732.

Eckmann, J. and Ruelle, D. (1985) 'Ergodic theory of chaos and strange attractors', *Reviews of Modern Physics* 57.3, pp. 617–656.

Hammel, S. M., Jones, C. K. R. T. and Moloney, J. V. (1985) 'Global Dynamical Behavior of the Optical Field in a Ring Cavity', *J. Opt. Soc. Am.* B 2, pp. 552–564.

Hammel, S. M., (1990) 'A noise reduction method for chaotic systems', *Physical Letters A* 148.08–9, pp. 421–428.

Hénon, M. (1976) 'A Two-Dimensional Mapping with a Strange Attractor', *Comm. Math. Phys.* 50, pp. 69–77.

Katok, A. and Hasselblatt, B. (1995) *Introduction to the Modern Theory of Dynamical Systems*, Cambridge University Press, Cambridge.

Lorenz, E. N. (1963) 'Deterministic Non periodic Flow', *J. Atmos. Sci.* 20, pp. 130–141.

Mackay, M. C. and Glass, L. (1977) 'Oscillation and Chaos in Physiological Control Systems', *Science* 197, pp. 287–289.

Parker, T. and Chua, L. (1987a) 'Chaos: a tutorial for engineers', *Proceedings of the IEEE* 75.08, pp. 982–1008.

Parker, T. and Chua, L. (1987b) 'INSITE –a software toolkit for the analysis of nonlinear dynamical systems', *Proceedings of the IEEE* 75.08, pp. 1081–1089.

Pawelzik, K. and Schuster, H. (1991) 'Unstable periodic orbits and prediction', *Physical Review A* 43.04, pp. 1808–1812.

Rössler, O. E. (1976) 'An Equation for Continuous Chaos', *Physics Letters A* 57, pp. 397–398.

Tufillaro, N., Solari, H. and Gilmor, R. (1990) 'Relative rotation rates: Fingerprint for strange attractors', *Physical Review A* 41.1, pp. 5717–5720.

2

Entropy

Entropy: *A measure of the amount of thermal energy in a system that is not available for conversion to mechanical work; a measure of the frequency with which an event occurs within a system; a measure of the rate of transfer of information in a message; a measure of disorganization or randomness; a hypothesized tendency toward uniform inertness.*

INTUITION

The above definitions for entropy seem to describe a variety of things, so it is no wonder that there is often some confusion over its meaning. Since entropy is intimately connected with many of the defining features of chaotic systems, it is worth spending a bit of time on developing a feeling for it. We will concentrate on the relationship between dynamical systems and the concepts of event frequency, randomness, and information transfer.

Let us start with the notions of event frequency and disorganization, or randomness. Consider a see-through container filled with an equal number of small black or white colored particles, and suppose there is some device which causes the particles to change position randomly. Further, assume that the container is illuminated with a stroboscopic lamp so that the instantaneous state of the system is observed only at discrete points in time. Using our eyes as an observational device, we watch the different configurations of the particles as they evolve. This exercise is likely to degenerate rapidly into extreme boredom as the container will appear to remain in a stationary state; a uniform spatial distribution of black and white. Clearly the configuration is changing; it just looks stationary. Like all physical measuring devices our eyes have limitations on the degree to which they can discern the location of the particles or the spatial relationship among black and white particles; any uniform distribution of black and white looks pretty much the same as any other.

Now every configuration is unique and has the same probability of occurrence *a priori*, that is, if this system had been in operation for say a million years before we came along, then the initial state that we observe is, with equal probability, any admissible configuration. But an overwhelming majority of the possible configurations are close to something that appears to be a uniform distribution of black and white particles, so the system will almost always be in one of these states. Hence, even if we start the system out with all the black particles at one end of the container and all the white particles at the other end, a highly 'organized' and discernible arrangement, it will rapidly degenerate into a 'stationary state' where it cycles through 'disorganized' uniform patterns of black and white. It is possible to return to the original arrangement, or some other highly organized pattern, but that would require an extraordinary sequence of events; a highly unlikely occurrence by definition.

A system found in one of the organized low probability states is thought of as having low entropy, and one found in the high probability collection of disorganized states is thought of as having high entropy. It is not surprising then that, in the above sense, systems with low entropy always seem to eventually evolve to systems with high entropy, or go from organized to disorganized states, and then remain there.

To see the relationship between entropy and information transfer, suppose we discover a discrete dynamical system where we know the dynamical rule (mapping) $f : x_n \rightarrow x_{n+1}$, but not the initial condition from which the system started. *A priori*, the system could be in any admissible state by the time we discover it. If we take a measurement to obtain its state x_d at the time of discovery, we gain a great deal of information. Indeed, the maximum amount of information, as an initial state is the only thing we do not know about the system. Having obtained x_d, we now take another measurement one unit of time later to obtain x_{d+1}. The amount of information we gain from the second measurement is zero, because knowing x_d and the dynamical rule f allows us to predict the outcome beforehand. That is, the second measurement was pointless since we knew it would yield $x_{d+1} = f[x_d]$. In this context, entropy is related to our *a priori* knowledge of the outcome of taking a measurement, or equivalently, to the amount of new information gained from taking the measurement. If prior knowledge (the prior probability distribution) of the result of a measurement is vague, then the outcome is highly uncertain, the information gained is high, and the associated entropy is high. Conversely, a prior knowledge of the probability distribution for the result, that implies low uncertainty in a measurement, has low entropy.

Returning to the original example for the moment, suppose now that we are viewing the particles in the container through a telescopic lens with a zoom feature. If we zoom in, then we effectively inflate the volume in the container, so that some formerly indistinguishable aspects of the particle distribution become distinguishable. If we reverse the zoom feature some formerly distinguishable aspects now become indistinguishable as we have effectively compressed the volume in the container. When we expand the volume, information that was on length scales below the eye's resolution capacity (the **micro scale** or **heat bath**), can become resolvable. Conversely, when the volume is contracted some previously resolvable (**macro scale**) features are reduced to micro length scales where they are no longer distinguishable.

A volume expanding system is one which acts as if it had a zoom-in feature focused on its own state. Information that is initially hidden (unobservable) is revealed by the system's state transition function as we move forward in time. In particular, different, but indistinguishable, initial states eventually become distinguishable under the action of the state transition function. Thus, information is being created[1] as details of the heat bath are brought into macroscopic expression (see Figure 2.1).

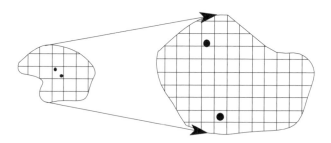

Figure 2.1 In a volume expanding system, information that is initially hidden (unobservable) is revealed as we move forward in time; information is being created. The grid represents the maximum resolution of the measuring device. Points that are closer than the dimension of a grid rectangle are indistinguishable.

In order for indistinguishable points to become distinguishable, the mapping must be operating on information beyond the resolution of the measuring device so as to

---

[1]While the words create and destroy are often used in this way to describe how a system is operating on information, the words reveal and conceal, respectively, are surely more accurate.

make it measurable; it is creating information in the sense that the distinguishing information was not in the initial data; the new information becomes available, that is, measurable, only through the iteration of the mapping. Consequently, later observations on the state of the system reveal more than earlier observations as any micro-scale uncertainty in an initial measurement is amplified and propagated into the future. The amplification of measurement uncertainty in these systems is referred to as *sensitive dependence on initial conditions* or *initial states*.

A *dissipative* system contracts volumes so that details of an earlier state are eventually contracted to a volume smaller than the resolution of the measuring instrument. Hence, in a dissipative system information is being destroyed as we move forward in time in the sense that the information in the initial data is no longer accessible after iteration of the map. Consequently, later observations on the state of the system give you less information than earlier observations as distinguishing macro-scale features are lost to the heat bath (see Figure 2.2).

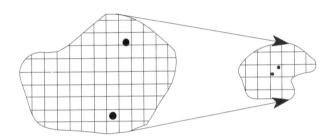

Figure 2.2 In a dissipative system information is being destroyed as we move forward in time.

Some dynamical systems have the capacity to expand or contract volumes in phase space. In fact they are capable of simultaneously expanding and contracting volumes in different directions, as Figure 2.3 illustrates.

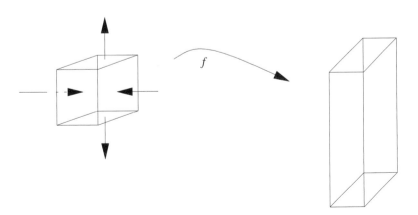

Figure 2.3 Simultaneous expansion and contraction of volumes in different directions.

What we mean when we say a system is expanding or contracting phase space volumes is that the trajectories of the system are, respectively, diverging, being stretched apart, or converging, being squeezed together (see Figure 2.4).

Figure 2.4 A 'bundle' of trajectories in phase space is stretched in some directions while squeezed in others.

Source: Adapted from Shaw (1981) p. 94.

Systems that expand and/or contract phase space are called non-conservative as they do not conserve volume. The mechanism can be understood quite simply by visualizing what happens to a two-dimensional rectangle, with sides Δx and Δy, under the action of the mapping $f : (x, y) \rightarrow (10x, 0.01y)$. With a single application of the rule f, the 'volume', in this case an area Δx by Δy, is both expanded in the x direction and contracted in the y direction to an area $10\Delta x$ by $0.01\Delta y$. Suppose we can only measure the location of a point in this two-dimensional area to an accuracy of 10^{-3}, and imagine two points a and b (see Figure 2.5) close together inside the rectangle; so close that their separation in the x direction is about 10^{-4} (though we cannot measure it) and 10^{-2} in the y direction. To our measuring device they appear to have the same x coordinate but different y coordinates. Now visualize what happens to these two points when the above rule is applied a couple of times to the area containing the two points. They now appear to have the same y coordinate but different x coordinates; the reverse of the initial situation.

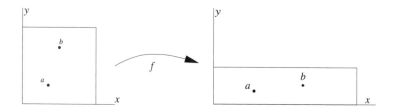

Figure 2.5 Initially, the points a and b are distinguishable in the y-direction, but not in the x-direction. In time, the dynamics f takes the initial points to two new points which are now distinguishable in the x-direction, but no longer distinguishable in the y-direction.

Even if the two points are initially so close that we cannot resolve them in either coordinate, that is, they are identical for all we can tell with our measuring device, the repeated application of the map f would soon reveal any distinction by drawing up the micro scale differences in the x coordinate to a measurable scale.

Of course this simple rule will expand any area indefinitely in the x direction so that, if the system is to remain bounded, the mapping must also possess a mechanism to 'fold' the x coordinate back onto itself (see Figure 2.6).

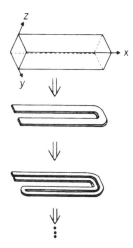

Figure 2.6 The stretching process requires a folding mechanism if the system's long-term behavior is to remain bounded.
Source: Adapted from Grassberger and Procaccia (1983) p. 190.

In the case of the Rössler attractor, (Figure 2.8) we can easily see from Figure 2.7 how these stretching, squeezing, and folding mechanisms operate on three-dimensional trajectories to produce a quasi two-dimensional object.

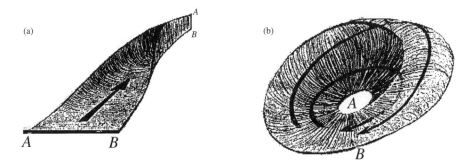

Figure 2.7 (a) Trajectories are stretched, folded, squeezed... (b) and then sutured back together to form a seamless quasi 2-dimensional object in 3-dimensional phase space. If the ribbon of trajectories in panel (a) is given a half-twist before it is sutured, the result resembles the Rössler attractor shown in Figure 2.8.
Source: Adapted from Shaw (1981) p. 94.

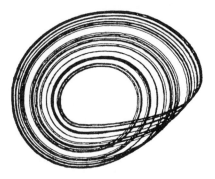

Figure 2.8 2-dimensional projection of the actual Rössler attractor.
Source: Adapted from Shaw (1981) p. 95.

This stretching, squeezing and folding of phase space volumes by a system is a hallmark of *chaotic dynamics* and often causes trajectories to collect along a folded *ribbon* that has a *Cantor-set like*[2] structure in the direction parallel to the direction of contraction. Structures that are Cantor-set like in some directions are called *fractal*. Fractal sets seem to have lacunarity or white space on all scales which gives them a non-integer dimension called *fractal dimension* discussed in a later chapter. Such sets often have structure, and sometimes self-similarity, on all scales (see Figure 2.9).

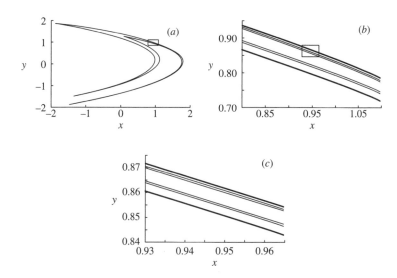

Figure 2.9 The Hénon attractor, with successive magnification of the areas enclosed by the rectangles, showing the self similar nature of the attractor.
Source: Adapted from Ott (1993) p. 14.

---

[2]The Cantor set is obtained by removing the middle third of the unit interval, then the middle third of each of the remaining intervals, and so on ad infinitum.

It should be clear by now that, in non-conservative systems, there is a flow of information between the macro and micro scales. The size and spatial direction of the flow at any location will depend on the orientation and rate of expansion and contraction at that location. The rate of the flow of information is measured in bits (binary digits) per unit time and is closely related to the *Lyapunov exponents of the map* to be discussed in a later chapter. The rate is defined to be positive if the flow is from the micro to macro scales when information is being created, and negative if the flow is in the reverse direction when information is being destroyed.

Finite measurement accuracy implies that there are only finitely many bits of information available to describe the trajectory of a bounded system. As new information from the low order bits (the micro scale) manifests itself through positive information flow, it must do so at the expense of the high order bits which contain the initial information. Eventually all knowledge of the initial condition is lost. Hence, the rate of information creation in a system determines a time frame over which we have any knowledge of its future behavior. After that time, the state of the system is determined by information which is initially hidden in the heat bath. Since that information is unknowable, so is any state that depends on it.

We might hope that, at least in principle, the consequences of uncertainty in our knowledge of system dynamics or the state of the system, as discussed in the above paragraphs, would be mitigated by increased accuracy from technological advances in the devices we use to observe and measure real systems. Unfortunately, the Uncertainty Principle implies that, even in theory, observational error is bounded away from zero, so we are condemned to live in an imprecise world where knowledge is finite and systems that create information have limited predictability[3].

As an interesting aside, there appear to be deterministic systems which are unpredictable even when we have infinitely accurate knowledge of the initial state of the system and are able to compute its evolution infinitely accurately. These systems, certain cellular automata for example, have the computational capacity of a Turing machine, and it follows from the undecidability of the Halting Problem that there is very little we can say about what they will do when started from an arbitrary initial condition. The only way to find out is to start the system up and just watch to see what happens.

THEORY

Shannon entropy

In an operational sense, information occurs as a message or measurement with an *a priori* probability distribution $\{p_i\}$ on its form or structure. We define the *information content* or *Shannon entropy of a distribution* $\{p_i\}$ as

$$H \equiv - \sum_i p_i \, \mathrm{Log}_2 \, p_i \quad \text{bits} \qquad \textbf{Eq. 2.1}$$

[3]This refers to the human condition. Nature (natural sytems) may in fact see and compute with infinite precision.

In the context of a dynamical system, we can take p_i as the probability of observing the system in state i. If the measurement is *a priori* certain, then $H = 0$; we learn nothing from observing the system. If, on the other hand, an observed measurement is *a priori* equally likely to fall anywhere in, say, the unit interval $[0, 1]$ and the interval is divided up into n equal measurable parts, then $p_i = 1/n$, and the information content of the measurement is

$$H = \sum_i \frac{1}{n} \text{Log}_2 n = \text{Log}_2 n \qquad \qquad \textbf{Eq. 2.2}$$

Since we assumed a uniform prior (probability distribution), Eq. 2.2 is an upper bound for the entropy of any prior. As $n \rightarrow \infty$, the resolution increases, and the information content of a measurement increases, but only up to the level dictated by measurement accuracy. Hence, the knowledge we obtain about the state of a system from any measurement is bounded by a computable number.

Let the state space be divided into disjoint regions each the size of the maximum resolution of the measurement tool used to observe the system. If there are $N(0)$ such regions, then we have $N(0)$ observable states. With a uniform prior (probability distribution), $p_i = 1/N(0)$, so the information content of an observation is[4]

$$H(0) = - \sum_1^{N(0)} p_i \text{Log}_2 p_i = \text{Log}_2 N(0) \qquad \qquad \textbf{Eq. 2.3}$$

If the system is contracting then the number of observable states decreases because of the finite resolution of the measurement tool (as Figure 2.10 illustrates).

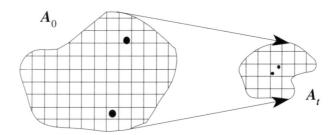

Figure 2.10 The number of observable states declines as volumes are contracted to scales below the resolution of the measuring device.

If we know that the system state is in A_0 with a uniform prior $p_0(i)$ initially, then determinism dictates that we know it is in A_t with a uniform prior $p_t(i)$ at some later time t, and that uniform prior is the conditional probability distribution

$$\{p_t(i|A_0) = 1/N(t)\}$$

[4]In the sequel we will drop the subscript 2 to indicate the base of the logarithm, and just assume that Log means Log_2. If any other base is used, entropy, and any quantity that depends on it, will be measured in units other than bits. For many of the examples in the book the base is taken to be the natural number e and the units of measurement are called nats.

where $N(t)$ is the number of observable states in A_t. Hence, the information content of an observation taken at time t is

$$H(t) = - \sum_{1}^{N(t)} p_t(i|A_0) \, \text{Log}_2 \, p_t(i|A_0) = \text{Log}_2 \, N(t)$$

The difference in the information content of two observations made t units of time apart is then

$$\begin{aligned} \Delta H(t) &= \text{Log } N(t) - \text{Log } N(0) \\ &= \text{Log}\,[N(t)/N(0)] \end{aligned}$$

Since volume $A_t <$ volume A_0, $N(t) < N(0)$, and ΔH is negative. A similar argument and result hold for expanding systems, but in that case ΔH is positive.

To gain some insight into how these ideas fit into the bigger picture, we digress for a while and consider the one-dimensional discrete map f. Suppose the system is contaminated with uncorrelated noise so that there is some uncertainty (error) Δx in our knowledge of the initial state of the system. Note that this error is not caused by measurement accuracy, which we denote by Δh. Under the dynamics, an interval of size Δx gets mapped to an interval of size $|\Delta f(x) = f(\Delta x)|$, so if the error is initially Δx, then the error after one iteration will be $|\Delta f(x)|$. In the interval $|\Delta f(x)|$ the number of observable states is approximately $|\Delta f(x)|/\Delta h$, and similarly for the interval Δx it is $\Delta x/\Delta h$. Hence, the information change per iteration is

$$\Delta H_x = \text{Log}\left|\frac{\Delta f(x)/\Delta h}{\Delta x/\Delta h}\right| = \text{Log}\left|\frac{\Delta f(x)}{\Delta x}\right|$$

The ratio $|\Delta f(x)|/\Delta x$ is the amplification of an initial error after one iteration. For small Δx, we can approximate that ratio with $|df(x)/dx|$, the magnitude of the slope of the map f, and write

$$\Delta H_x = \text{Log}\left|\frac{df(x)}{dx}\right|$$

Using the ergodic natural measure[5] induced by f, ρ, we can compute the **average information change** $\langle \lambda \rangle$ over the domain of f with

$$\langle \lambda \rangle = \int dx \, \rho(x) \, \text{Log}\left|\frac{df(x)}{dx}\right|$$

Ergodicity of the map then allows us to replace the integral in the above expression with the time average

$$\begin{aligned} \langle \lambda \rangle &= \lim_{n \to \infty} \frac{1}{n} \sum_{t=1}^{n} \text{Log}\left|\frac{df(x)}{dx}\right|_{x_{t-1}} \quad\quad\quad \textbf{Eq. 2.4} \\ &= \lim_{n \to \infty} \frac{1}{n} \text{Log}\left|\frac{df^n(x)}{dx}\right|_{x_0} \end{aligned}$$

[5]See Chapter 1.

Hence, for ergodic maps, the average information change per iteration over the attractor is the log geometric average of the error growth along a particular trajectory. By way of a preview to a later chapter we remark that, in the one-dimensional case, the average information change, as defined above, is the same as the Lyapunov exponent of the map f[6]. Actually, the above results do not depend on the introduction of noise into the system. In a noiseless system the initial error $\Delta x = \Delta h$, and we get the same results, so we will drop the noise assumption for the time being.

If we can measure to an accuracy of Δh bits, then an initial condition contains an amount of information given by

$$H_I = -\mathrm{Log}\ \Delta h$$

Since $|\langle\lambda\rangle|$ is the rate at which information is being created or destroyed per iteration, all the initial information, H_I, will be lost after $\lfloor z$ iterations[7], where z satisfies

$$H_I - z|\langle\lambda\rangle| = 0$$

which implies that

$$z = \frac{-\mathrm{Log}\ \Delta h}{|\langle\lambda\rangle|}$$

If the system is expanding phase space volume, then the information content of the initial observation is replaced by information drawn up from the heat bath. If the system is contracting phase space volume, then the initial data is lost to the heat bath. In either case, the state of the system after z iterations is independent of the initial state. As $\langle\lambda\rangle$ approaches zero it takes longer and longer for this independence to set in, but it will do so in finite time. The initial data affects the system indefinitely only when $\langle\lambda\rangle = 0$.

In the general situation, a Poincaré surface of section[8] will yield the average change in information $\langle\lambda\rangle$ per iteration, for the induced Poincaré return map f[9].

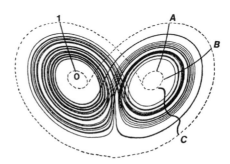

Figure 2.11 Various Poincaré surfaces of section through the Lorenz I attractor. For the return map induced by the Poincaré surface of section at 0–1, $\langle\lambda\rangle = 0.98$ bits per iteration.

Source: Adapted from Shaw (1981) p. 96.

[6] At least for ρ continuous. See Chapter 5, Lyapunov exponents.

[7] $\lfloor z$ denotes the integer part of z.

[8] See Chapter 1.

[9] We are again abusing notation here. We have been using f to denote either a discrete map or the flow of a continuous system with vector field F. Here we are also using f to denote either the map of a discrete system or the induced discrete Poincaré return map of a continuous system.

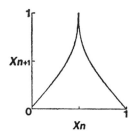 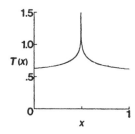

Figure 2.12 Poincaré return map $x_n \to x_{n+1}$ and return time $T(x)$ in seconds, for the Poincaré surface of section at 0–1 in Figure 2.11.

Source: Adapted from Shaw (1981) p. 96.

To get the average rate of information change per unit time λ_T, the rate of information change per iteration of the Poincaré return map must be normalized by $T(u)$, the time it takes for the system to return to the surface of section. Thus

$$\langle \lambda_T \rangle = \int du \frac{\rho(u)}{T(u)} \text{Log} |D_u f|$$

where ρ is the induced measure on the Poincaré surface of section, and $D_u f$ is the Jacobian of f at u. Ergodicity now implies

$$\langle \lambda_T \rangle = \lim_{n \to \infty} \frac{1}{n} \sum_1^n \frac{\text{Log} |D_{x_i} f|}{T(x_i)} \qquad \text{Eq. 2.5}$$

For the Lorenz I system of Figure 2.11, $\langle \lambda \rangle = 0.98$ bits per iteration, and $\langle \lambda_T \rangle = 1.19$ bits per second. Note that if the system is discrete, then $T(x) = 1$, and if it is one-dimensional, then $D_x f = df/dx$, so we recover Eq. 2.4 from Eq. 2.5.

After making some further generalizations of entropy below, we will have more to say about $\langle \lambda_T \rangle$ when we turn to asymptotic results. But we can make two important observations now, without any new concepts. First, from the range of the Poincaré return map, and the measurement resolution, we can compute the initial information H_I for a uniform prior. Namely, from Eq. 2.2

$$H_I = \text{Log} \frac{\text{Range}}{\text{Resolution}}$$

Now,

$$t^* = \frac{H_I}{|\langle \lambda_T \rangle|}$$

is the time after which the information in the initial data is lost. A map of the process after t^* units of time should look (relative to the natural measure) randomly spread out over the extent of the attractor if the system is creating information.

Next, note that for a uniform prior

$$\frac{\Delta H}{\Delta t} = \frac{\text{Log} N(t + \Delta t) - \text{Log} N(t)}{\Delta t} \xrightarrow[\Delta t \to 0]{} \frac{d \text{Log} N(t)}{dt}$$

43

If $N(t)$ follows a power law, so that $N(t) \sim t^{\alpha}$, then

$$\frac{dH(t)}{dt} = \frac{d \operatorname{Log} N(t)}{dt} = \frac{\alpha}{t} \xrightarrow[t \to \infty]{} 0$$

Since this implies that the change in the information content of an observation goes to zero, eventually new observations give no new information. That is, the rate of information creation/loss goes to zero, so the uncertainty in the state stays constant, and the system's behavior is predictable for an indefinite period of time. So, whether or not a system is predictable seems to depend on whether $N(t)$ is a polynomial or an exponential function of time. If $N(t)$ is a polynomial, then the uncertainty in an initial condition remains constant along a trajectory. If $N(t)$ is an exponential with a positive rate, then $dH(t)/dt$ remains positive, the uncertainty in the initial condition grows along the trajectory, and the system is eventually unpredictable.

Note that for a fixed resolution grid

$$\frac{N(t)}{N(0)} = \frac{\text{Volume } A_t}{\text{Volume } A_0} \equiv \frac{V(t)}{V(0)}$$

implies that

$$N(t) = N(0)\frac{V(t)}{V(0)}$$

It follows that, for a continuous time system, the rate of change of the information content of an observation is given by

$$\frac{\Delta H(t)}{\Delta t} = \frac{\operatorname{Log} N(t) - \operatorname{Log} N(0)}{\Delta t} = \frac{\operatorname{Log} V(t) - \operatorname{Log} V(0)}{\Delta t}$$

which implies that, after taking limits,

$$\frac{dH(t)}{dt} = \frac{d \operatorname{Log} N(t)}{dt} = \frac{1}{V(t)}\frac{dV(t)}{dt}$$

The relative change in the volume element on the right hand side is the **divergence** or **Lie derivative** of the vector field F that defines the dynamics of the system. That is,

$$\frac{1}{V(t)}\frac{dV(t)}{dt} = \nabla \cdot F = \frac{\partial dx_1/dt}{\partial x_1} + \frac{\partial dx_2/dt}{\partial x_2} + \cdots$$

Hence, where the number of distinguishable states is proportional to the volume

$$\frac{dH(t)}{dt} = \nabla \cdot F$$

It is not hard to imagine situations where $N(t)$ is not proportional to $V(t)$. In Figure 2.13 we could have $V(t) = V(0)$, i.e. a conservative system, where $N(t) > N(0)$, i.e. where the *effective* volume grows.

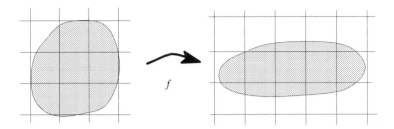

Figure 2.13 The volume remains constant, but spreads out over the state space such that the number of observable states grows.

Conditional entropy

One way to generalize the entropy concept is to condition an observation's prior probability distribution on previous observations on the system. To wit, let $S = S(t) = \{s_i\}$ denote the possible (measurable) states of the system at some time t, and let $\{P_S(s_i)\}$ be the probability distribution for $S(t)$. Further, let $Q = Q(t) = \{q_j\} = S(t + T)$, so that Q is the set of measurable states at time $t + T$. Now consider the coupled system

$$(Q, S) = (S(t + T), S(t)) = \{[q_j, s_i]\}$$

with probability distribution P_{QS}. Let $P_{Q|S}(q_j|s_i)$ be the distribution of $q_j \in Q$ conditioned on $s_i \in S$ and define the pointwise **conditional entropy**

$$H(Q|s_i) = -\sum_j P_{Q|S}(q_j|s_i) \operatorname{Log} P_{Q|S}(q_j|s_i) \qquad \text{Eq. 2.6}$$

The pointwise conditional entropy measures the information gained from making a measurement at time $t + T$, given the information obtained from an observation at the earlier time t, that is, given an earlier observation *which yielded the value s_j*. The expected value of the information gained from a measurement at time $t + T$, given the information from a measurement at time t, is then

$$
\begin{aligned}
H(Q|S) &= \sum_i P_S(s_i) H(Q|s_i) \\
&= -\sum_i P_S(s_i) \sum_j P_{Q|S}(q_j|s_i) \operatorname{Log} P_{Q|S}(q_j|s_i) \\
&= -\sum_{i,j} P_{Q,S}(q_j, s_i) \operatorname{Log} \frac{P_{Q,S}(q_j, s_i)}{P_S(s_i)} \qquad \text{Eq. 2.7} \\
&= -\sum_{i,j} P_{Q,S}(q_j, s_i) \operatorname{Log} P_{Q,S}(q_j, s_i) \\
&\qquad + \sum_{i,j} P_{Q,S}(q_j, s_i) \operatorname{Log} P_S(s_i) \\
&= H(Q, S) - H(S)
\end{aligned}
$$

45

Note

$$H(Q|S) = 0 \quad \Leftrightarrow \quad H(Q, S) = H(S)$$
$$\Leftrightarrow \quad P_{Q|S} \in \{0, 1\}$$
$$\Leftrightarrow \quad S \Rightarrow Q$$

and

$$H(Q|S) = H(Q) \quad \Leftrightarrow \quad H(Q, S) = H(S) + H(Q)$$
$$\Leftrightarrow \quad P_{QS} = P_Q P_S$$
$$\Leftrightarrow \quad Q \text{ is independent of } S$$

$H(Q|S)$ gives us the informational value of taking a new measurement at time $t + T$ having already taken one T time units earlier at time t. If $S(t) \Rightarrow S(t + T)$, then there is no value in taking the new measurement because we can predict the outcome from the previous measurement. Hence, the earlier measurement contains all the information that would be available in the later measurement. On the other hand, if $S(t + T)$ is independent of $S(t)$, then the earlier measurement contains none of the information that would be available to the later measurement. Since $H(Q)$ is the number of bits of information contained in an isolated measurement at time $t + T$, and $H(Q|S)$ is the number of bits in that same measurement after having seen $S(t)$, we can see that the quantity

$$R(Q, S) \equiv H(Q) - H(Q|S) \qquad\qquad \textbf{Eq. 2.8}$$

is just the amount of information that $S(t)$ contains about $S(t + T)$. Thus, $R(Q, S)$ gives the number of bits we can predict about $S(t + T)$, given knowledge of $S(t)$. As such, it is also a measure of the degree to which $S(t + T)$ depends on $S(t)$. $R(Q, S)$ is called the ***mutual information*** in (Q, S). It is symmetrical

$$\begin{aligned} R(Q, S) &= H(Q) - H(Q|S) \\ &= H(Q) + H(S) - H(S, Q) \\ &= R(S, Q) \end{aligned}$$

and can be expressed in terms of the underlying distributions

$$R(Q, S) = \sum_{i,j} P_{QS}(q_j, s_i) \operatorname{Log} \frac{P_{QS}(q_j, s_i)}{P_Q(q_j) P_S(s_i)} \qquad\qquad \textbf{Eq. 2.9}$$

In this presentation we have chosen Q and S to represent observations from a single process seperated in time, that is, they represent the same time series, but one lags the other by a fixed time interval T. However, the meaning and validity of $R(Q, S)$ hold in general when Q and S represent any vector spaces of observables. In particular, Q and S can represent two spatially separated observables as well as two temporally separated observables, though the generalization is only useful if we can postulate a sensible spatio-temporal connection between them. For example, a simple variation is the mutual information between $x_i \in X$ and a lagged $y_i \in Y$, called the ***cross redundancy***

$$I(X, Y; k) \equiv H(X) + H(U) - H(X, U) \qquad \text{Eq. 2.10}$$

where $U = \{y_{i-k}\}$.

Block entropy

The generalization of the Shannon entropy to n variables v_i, $i = 1, \ldots, n$, is given by the ***block entropy***

$$H(V_1, \ldots, V_n) \equiv - \sum_{v_1, \ldots, v_n} P(v_1, \ldots, v_n) \operatorname{Log} P(v_1, \ldots, v_n) \qquad \text{Eq. 2.11}$$

where the V_i are vector spaces and $v_i \in V_i$. We have a choice in generalizing mutual information depending on which interpretation we choose to generalize. The generalization of the mutual information among two variables to the information among n-variables is called the ***redundancy*** and is given by

$$R(V_1, \ldots, V_n) \equiv \sum_i H(V_i) - H(V_1, \ldots, V_n) \qquad \text{Eq. 2.12}$$

The redundancy vanishes if and only if there is no dependence among the v_i. So the interpretation generalized here is that of statistical dependence. We can also generalize the notion of predictability by defining a quantity $\mathfrak{R}$, called the ***marginal redundancy***, which measures the expected value of the amount of information about v_n contained in the measurements $v_1, \ldots, v_{n-1}$

$$
\begin{aligned}
\mathfrak{R}(V_1, \ldots, V_{n-1}; V_n) & \\
\equiv \quad & R(V_1, \ldots, V_n) - R(V_1, \ldots, V_{n-1}) \\
= \quad & H(V_n) + H(V_1, \ldots, V_{n-1}) - H(V_1, \ldots, V_n) \qquad \text{Eq. 2.13} \\
= \quad & H(V_n) - H(V_n | V_1, \ldots, V_{n-1})
\end{aligned}
$$

On the other hand, we get a global measure of the increase in predictability obtained by including an observation from X_1, over and above the level already obtained from $X_2, \ldots, X_{m-1}$ by defining

$$
\begin{aligned}
R^*(m) & = \mathfrak{R}(X_1, \ldots, X_{m-1}; X_m) - \mathfrak{R}(X_2, \ldots, X_{m-1}; X_m) \\
R^*(2) & = R(X_1, X_2) \qquad \text{Eq. 2.14} \\
R^*(1) & = -H(X_1)
\end{aligned}
$$

At this point it should be clear that there exists a close relationship between entropy, dependence, and predictability. We explore the relationship further later in the chapter when dealing with the Green–Savit algorithm.

Finally, we have been discussing entropy in the context of deterministic systems, but entropy is a valid concept for stochastic systems as well. We may want to test the assumption of determinism by comparing some of the entropy statistics derived from the data, with the same statistics when they are produced by a stochastic process[10]. In the algorithm section we give an expression for computing some of the entropy statistics when the distribution of the observables is Gaussian, and discuss a statistical test based on a discriminator called the BDS statistic.

Renyi entropy

The most general formulation of entropy is known as the **_Renyi entropy_** defined by

$$H_q(V) \equiv \frac{1}{1-q} \text{Log} \sum_{v \in V} [P_V(v)]^q \qquad \qquad \textbf{Eq. 2.15}$$

With the help of L'Hôpital's rule and some algebra, it is not difficult to show that the Renyi entropy includes the Shannon entropy as a specific case. That is,

$$H_1(V) \equiv \lim_{q \to 1} H_q(V) = - \sum_{v \in V} P_V(v) \text{Log} P_V(v) \qquad \qquad \textbf{Eq. 2.16}$$

For increasing values of q, the main contributions to the Renyi entropy come from the high density regions of the probability distribution. Hence, the larger the value of q, the more H_q measures the entropy of high probability events. Conversely, the smaller - that is, the more negative the value of q, the more H_q measures the entropy of low probability events. The H_q of a uniform probability distribution is insensitive to q. The entropy derivatives R, $\Re$, and $R^*(m)$ have higher order versions formulated with H_q in place of $H = H_1$, and are denoted by R_q, $\Re_q$, and $R_q^*(m)$ respectively.

It should be kept in mind that entropy and its derivatives are computed relative to a prior probability distribution P_V, and the most appropriate prior is the ergodic natural measure ρ.

Noise and measurement error

At this point we have to reintroduce a reality we have conveniently ignored because it simplified the notation. As stressed earlier, all measurement is finite, so in practice all real computations of entropy and its derivatives are a function of the maximum resolution of whatever device we use to observe the system. To make this dependence explicit we will from now on add a measurement accuracy parameter ε to the notation and point out that it was implicit in the foregoing discussion.

While we are on the topic, we should also make note that a purely deterministic dynamics probably never obtains in real life systems because they are open to exogenous influences (noise) that inevitably alter their behavior[11] and contaminate measurements. At the extreme, purely random (white noise) systems exhibit very distinctive behavior; their trajectories tend to explore fully the state space, that is, they tend to fill up whatever space they live in. This follows from the fact that if the state space coordinates are independent and identically distributed, then given any initial state, the state a short time later will be isotropically distributed in a small sphere

---

[10]See also Appendix 7, Surrogate data and non-parametric statistics.

[11]See Chapter 6, Noise reduction.

centered at the initial state. As time passes the sphere just grows isotropically filling out the state space. Hence, in an m-dimensional space V, the distribution of orbital points around the initial condition will be roughly proportional to the volume of an m-dimensional sphere centered on the initial condition, which implies that, within a small radius r of a point v

$$P_V(v; r) \underset{r \to 0}{\sim} Cr^m$$

for some constant C. If the system is actually deterministic and it is just our measurements of the system that are contaminated with noise, then entropy is a function of measurement accuracy only as long as the resolution is coarser than the noise level. If the accuracy is finer than the noise level, then

$$H_1(V; \varepsilon) = -\text{Log } P_V(v; \varepsilon) \sim -m \text{ Log } \varepsilon$$

so entropy scales like noise until the scale exceeds the noise level.

Asymptotic results

To keep the meaning clear, we will use X_i in place of V_i whenever we wish to emphasize that the elements of the V_i correspond to a sequence of observations on a system sampled at some fixed time interval τ, that is, when they represent a time series, and we define Y_i and Z_i similarly to differentiate among different observables. In this context, the observables $X_1, \ldots, X_{m-1}$ are sometimes referred to as delay variables (with respect to X_m) with delay time τ.

The gain in (Shannon) information obtained from one additional observation on the system is given by

$$H_1(V_1, \ldots, V_m; \varepsilon) - H_1(V_1, \ldots, V_{m-1}; \varepsilon) = H_1(V_m | V_1, \ldots, V_{m-1}; \varepsilon)$$

It follows that the rate of information creation per unit time between the m-1st and the mth observation is

$$\frac{H_1(X_m | X_1, \ldots, X_{m-1}; \varepsilon)}{\tau}$$

The limit of this quantity as $m \to \infty$ and $\varepsilon \to 0$, denoted by K_1, is called the **Kolmogorov–Sinai** or **metric entropy**

$$K_1 = \underset{\substack{m \to \infty \\ \varepsilon \to 0}}{\text{Lim}} \frac{H_1(X_m | X_1, \ldots, X_{m-1}; \varepsilon)}{\tau} \qquad \text{Eq. 2.17}$$

Substituting higher order entropies H_q into Eq. 2.17, we get expressions for higher order entropy rates denoted by K_q. By way of tying entropy in with another important system invariant, we remark that multi-dimensional systems possess a spectrum of Lyapunov exponents which characterize the rates at which the system stretches and squeezes phase space in different directions. In directions where phase space is being stretched, the exponents are positive. If a system is stretching phase space, then it is creating information. Since K_1 is the average rate at which the system is creating information, it must be closely related to the positive Lyapunov exponents. The nature of that relationship is explored in more detail in Chapter 5, on Lyapunov exponents.

From Eq. 2.16 and ergodicity

$$H_1(V; \varepsilon) = -\sum_{v \in V} P_V(v; \varepsilon) \operatorname{Log} P_V(v; \varepsilon) = -\frac{1}{N_r} \sum_{i=1}^{N_r} \operatorname{Log} P_V(v_i; \varepsilon)$$

Now, suppose that, at some sufficiently fine scale, in almost every neighborhood

$$P_V(v_i; \varepsilon) \underset{\varepsilon \to 0}{\sim} C\varepsilon^{D_I}$$

That is equivalent to saying that on small enough scales the attractor has a self similar structure, and if it does then

$$\begin{aligned} -H_1(V; \varepsilon) &\approx C' + \operatorname{Log} \varepsilon^{D_I} \\ &= C' + D_1 \operatorname{Log} \varepsilon \end{aligned}$$

so

$$D_I = \operatorname*{Lim}_{\varepsilon \to 0} -\frac{H_1(V; \varepsilon)}{\operatorname{Log} \varepsilon} \qquad\qquad \textbf{Eq. 2.18}$$

D_I is called the ***information dimension***. Hence, we see that entropy underpins yet another important system invariant. By replacing H_1 with H_q in the right hand side of this expression we get the scaling properties of the order-q Renyi entropy

$$D_q \equiv \operatorname*{Lim}_{\varepsilon \to 0} -\frac{H_q(V; \varepsilon)}{\operatorname{Log} \varepsilon} \qquad\qquad \textbf{Eq. 2.19}$$

The D_q are called the ***fractal dimension spectrum*** of the system, and are discussed later in Chapter 4, Fractal dimension. Obviously, $D_I = D_1$, so the information dimension is a specific case of fractal dimension.

ALGORITHMS

1 Correlation integral

We need a practical way of evaluating entropy and its derivatives in real situations where our knowedge of the system consists only of a collection of observations. The observations could be sequential measurements on the same part of the system; or they could be a sequence of simultaneous measurements at different locations; or they could be some combination of spatio-temporally related measurements.

It should be clear that the evaluation problem boils down to estimating the joint probability distributions of the variables corresponding to the observations. An intuitive approach makes use of the ***kernal density estimator*** given by

$$W_V[v; \varepsilon] = \frac{1}{N_V} \sum_{u \in V} \Theta(\varepsilon - \|v - u\|) \qquad\qquad \textbf{Eq. 2.20}$$

where $W_V[v; \varepsilon]$ estimates the probability distribution of $v \in V$, V is a vector (product) space, $\| \cdot \|$ is the norm (max or Euclidean), N_V is the number of elements in V, and Θ is the Heaviside function defined by

$$\Theta(z) = \begin{cases} 1 & \text{if } z > 0 \\ 0 & \text{otherwise} \end{cases}$$

$\Theta(\varepsilon - \|v - u\|)$ is an indicator function for the ε-neighborhood[12] of the point v, and $W_V[v; \varepsilon]$ is an estimate of its expected value. Hence, one would expect that, typically, $W_V[v; \varepsilon] \to P_V(v)$ as $N_V \to \infty$ and $\varepsilon \to 0$.

Recall the definition of Shannon entropy, Eq. 2.11

$$-H(V; \varepsilon) = \sum_{v \in V} P_V(v; \varepsilon) \text{Log}[P_V(v; \varepsilon)]$$

If the process is ergodic, then we can approximate $H(V; \varepsilon)$ by substituting a time average on the right hand side and write

$$-H(V; \varepsilon) \approx \frac{1}{N_r} \sum_{\text{data points } k} \text{Log}[P_V(v_k; \varepsilon)]$$

Now we approximate $P_V(v; \varepsilon)$ with $W_V[v; \delta = \varepsilon/2]$ and write

$$-H(V; \varepsilon) \approx \frac{1}{N_r} \sum_k \text{Log } W_V[v_k; \delta]$$

$$= \text{Log}\left(\prod_k W_V[v_k; \delta]\right)^{\frac{1}{N_r}}$$

Finally, we define

$$C_1(V; \varepsilon) \equiv \left(\prod_k W_V[v_k; \delta]\right)^{\frac{1}{N_r}}$$

so that

$$H(V; \varepsilon) = -\text{Log } C_1(V; \varepsilon) \qquad \qquad \text{Eq. 2.21}$$

The statistic $C_1(V; \varepsilon)$ is the geometric mean of the frequencies $W_V[v; \delta]$ and it is called the *generalized correlation integral of order 1*. If V is a product space, then the foregoing holds for block entropies, and hence, the entropy derivatives of order 1. Numerical algorithms for computing its value from experimental data are discussed in Appendix 3, Correlation integral.

---

[12] An ε-neighborhood of a point v is the set of all points u which satisfy $\|v - u\| < \varepsilon$.

2 Higher order correlation integrals

Assuming ergodicity, we can expand the summation in the definition of the Renyi entropy, Eq. 2.15, as follows

$$\sum_{v \in V}[P_V(v; \varepsilon)]^q = \sum_{v \in V} P_V(v; \varepsilon)[P_V(v; \varepsilon)]^{q-1}$$

$$\approx \frac{1}{N_r} \sum_{k}[P_V(v_k; \varepsilon)]^{q-1}$$

Using the same arguments as we used in the derivation of the order-1 correlation integral, we substitute $W_V[v; \delta]$ for $P_V(v; \varepsilon)$ and write

$$H_q(V; \varepsilon) \approx \frac{1}{1-q} \text{Log} \left\{ \frac{1}{N_r} \sum_{k} \left[\frac{1}{N_r} \sum_{j} \Theta(\delta - \|v_k - v_j\|) \right]^{q-1} \right\}$$

$$\equiv -\text{Log}\, C_q(V; \varepsilon)$$

The quantity $C_q(V; \varepsilon)$ is called the ***generalized correlation integral of order*** q, $C_1(V; \varepsilon)$ being a special case of $C_q(V; \varepsilon)$. Numerical algorithms for computing its value from experimental data are discussed in Appendix 3, Correlation integral. From a computational point of view, $H_2(V; \varepsilon)$ is very tractable, and that is the reason that actual numerical estimates of entropy and its derivatives are often of order 2.

3 Estimation of entropy and its derivatives by order-q correlation integrals

Each of the entropy statistics can now be estimated from experimental data by writing them in terms of the correlation integral as follows.

The ***Renyi entropy*** of order q, Eq. 2.15, gives

$$H_q(V_1, \ldots, V_m; \varepsilon) \approx -\text{Log}\, C_q(V_1, \ldots, V_m; \varepsilon)$$

The ***redundancy*** of order q, Eq. 2.12, gives

$$R_q(V_1, \ldots, V_m; \varepsilon) \equiv \sum_{l}^{m} H_q(V_i; \varepsilon) - H_q(V_1, \ldots, V_{m-1}; \varepsilon)$$

$$\approx -\text{Log}\, \frac{\prod^m C_q(V_i; \varepsilon)}{C_q(V_1, \ldots, V_m; \varepsilon)}$$

The ***marginal redundancy*** of order q, Eq. 2.13, gives

$$\Re_q(V_1, \ldots, V_{m-1}; V_m; \varepsilon)$$

$$\equiv H_q(V_m; \varepsilon) - H_q(V_1, \ldots, V_m; \varepsilon) + H_q(V_1, \ldots, V_{m-1}; \varepsilon)$$

$$\approx -\text{Log}\, \frac{C_q(V_m; \varepsilon)C_q(V_1, \ldots, V_{m-1}; \varepsilon)}{C_q(V_1, \ldots, V_m; \varepsilon)}$$

The **cross redundancy**, Eq. 2.10, gives

$$I_q(X, X; k, \varepsilon) \equiv H_q(X; \varepsilon) + H_q(U; \varepsilon) - H_q(X, U; \varepsilon)$$

$$\approx -\text{Log} \frac{C_q(X; \varepsilon)C_q(U; \varepsilon)}{C_q(X, U; \varepsilon)}$$

where $U = \{x_{i-k}\}$

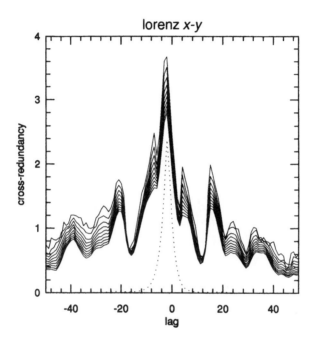

Figure 2.14 Cross-redundancy (solid curves) between the x and y components of the Lorenz equations as a function of time lag k. Top curve is for $\varepsilon = 0.01\sigma$, the next is for $\varepsilon = 0.02\sigma$ and so on to $\varepsilon = 0.1\sigma$, where σ is the standard deviation of the x component. Dashed line is for the linear (Gaussian) cross-redundancy discussed below.

Source: Adapted from Prichard and Theiler (1995) p. 488.

Example 2.1
The **lagged variable dependence**, Eq. 2.14, gives

$$R_q^*(m) \equiv \mathfrak{R}_q(X_1, \ldots, X_{m-1}; X_m, \varepsilon) - \mathfrak{R}_q(X_2, \ldots, X_{m-1}; X_m, \varepsilon)$$

$$\approx -\text{Log} \frac{C_q(X_1, \ldots, X_{m-1}; \varepsilon)C_q(X_2, \ldots, X_m; \varepsilon)}{C_q(X_1, \ldots, X_m; \varepsilon)C_q(X_2, \ldots, X_{m-1}; \varepsilon)}$$

$$R_q^*(2) \approx -\text{Log} \frac{C_q(X_1; \varepsilon)C_q(X_2; \varepsilon)}{C_q(X_1, X_2; \varepsilon)}$$

$$R_q^*(1) \approx \text{Log} \, C_q(X_1; \varepsilon)$$

Which for a stationary process reduces to

$$R_q^*(m) = -\text{Log}\,\frac{[C_q(X_1, \ldots, X_{m-1}; \varepsilon)]^2}{C_q(X_1, \ldots, X_m; \varepsilon)C_q(X_1, \ldots, X_{m-2}; \varepsilon)}$$

The **_Kolmogorov–Sinai entropy_** of order q, Eq. 2.17, gives

$$
\begin{aligned}
K_q & \equiv \lim_{\substack{m \to \infty \\ \varepsilon \to 0}} \frac{1}{\tau}[H_q(X_1, \ldots, X_m; \varepsilon) - H_q(X_1, \ldots, X_{m-1}; \varepsilon] \\
& \equiv \lim_{\substack{m \to \infty \\ \varepsilon \to 0}} \frac{1}{\tau}[H_q(X_m; \varepsilon) - \mathfrak{R}_q(X_1, \ldots, X_m; \varepsilon)] \\
& \equiv \lim_{\substack{m \to \infty \\ \varepsilon \to 0}} \frac{-1}{\tau}\text{Log}\,\frac{C_q(X_1, \ldots, X_m; \varepsilon)}{C_q(X_1, \ldots, X_{m-1}; \varepsilon)}
\end{aligned}
$$

which is not very helpful as it requires passing to a double limit. However, in principle, K_q can be estimated from the graph of the marginal redundancy $\mathfrak{R}_q$ as follows. Recall from Eq. 2.13 that

$$\mathfrak{R}_q(X_1, \ldots, X_{m-1}; X_m, \varepsilon) = H_q(X_m; \varepsilon) - H_q(X_m|X_1, \ldots, X_{m-1}; \varepsilon)$$

Thus, for a stationary process with inter-measurement time τ, we can use Eq. 2.17 to write

$$\lim_{\substack{m \to \infty \\ \varepsilon \to 0}} \mathfrak{R}_q(X_1, \ldots, X_{m-1}; X_m, \varepsilon) = \lim_{\varepsilon \to 0} H_q(X_1; \varepsilon) - \tau K_q$$

The $\lim_{\varepsilon \to 0} H(X_1; \varepsilon)$ does not depend on τ, so for large m and small ε, $\mathfrak{R}_q$ should be linear in τ with slope K_q.

Example 2.2
Based on clean data from the Rössler equations, the second order marginal redundancy plotted in panel (b) of Figure 2.15 yields the stable estimate $K_2 = 0.03$ bits per time step. However, it is clear from panel (d) of the figure that K_1 cannot be reliably estimated with this method. For comparison, panels (a) and (c) show 'normalized' redundancy, and panels (e) and (f) show the linear statistics discussed below under the heading 'Entropy for a Gaussian process'

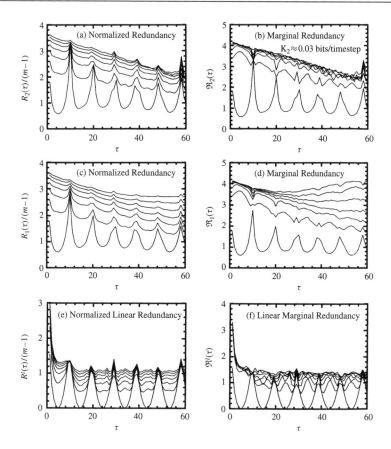

Figure 2.15 The redundancy analysis for data from the Rössler attractor, for $m = 2 - 8$ delay variables with delay times $\tau = 1 - 60$ times the sampling time τ_s. Higher curves are for larger values of m. All curves are based on $\varepsilon = 0.1\sigma$, where σ is the standard deviation of the time series. (a) C_2 based normalized redundancies ($R_2(\tau)/(m - 1)$). (b) C_2 based marginal redundancies $\mathfrak{R}_2(\tau)$. (c) C_1 based normalized redundancies ($R_1(\tau)/(m - 1)$). (d) C_1 based marginal redundancies $\mathfrak{R}_1(\tau)$. (e) Normalized linear redundancies ($R^l(\tau)/(m - 1)$). (f) Linear marginal redundancies $\mathfrak{R}^l(\tau)$. R^l and $\mathfrak{R}^l$ are defined below.

Source: Adapted from Pritchard and Theiler (1995) p. 484.

Example 2.3

The data for this example came from a chaotic laser experiment[13]. As in Example 2.2, we can get a reliable estimate for K_2, but not for K_1. For comparison, panels (a) and (c) show 'normalized' redundancy and panels (e) and (f) in Figure 2.16 show the linear statistics discussed below under the heading 'Entropy for a Gaussian process'.

---

[13]U. Hubner *et al.* (1993).

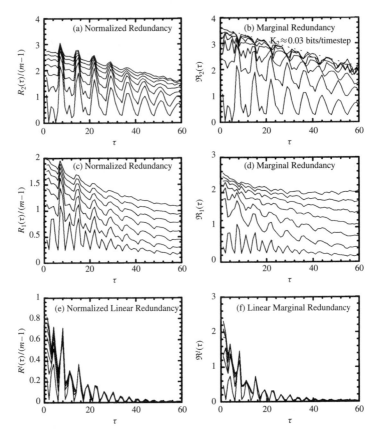

Figure 2.16 Redundancy analysis for the experimental data set for $m = 2 - 8$ delay variables with delay times $\tau = 1 - 60$ times the sampling time τ_s. Higher curves are for larger values of m. Curves (a) and (b) are based on $\varepsilon = 0.1\sigma$, curves (c) and (d) are based on $\varepsilon = 0.25\sigma$, where σ is the standard deviation of the time series. (a) C_2 based normalized redundancies $(R_2(\tau)/(m-1))$. (b) C_2 based marginal redundancies $\Re_2(\tau)$. (c) C_1 based normalized redundancies $(R_1(\tau)/(m-1))$. (d) C_1 based marginal redundancies $\Re_1(\tau)$. (e) Normalized linear redundancies $(R^l(\tau)/(m-1))$. (f) Linear marginal redundancies $\Re^l(\tau)$. R^l and $\Re^l$ are defined below.

Source: Adapted from Pritchard and Theiler (1995) p. 485.

4 Green–Savit statistics[14]

Consider the $m + 1$-length contiguous subset of a stationary measurement time series $x(t)$:[15]

$$[x_0(t), x_1(t), \ldots, x_m(t)] \equiv [x(t), x(t-\tau), \ldots, x(t-m\tau)]$$

For a deterministic system, we expect that $[x_1(t), x_2(t), \ldots, x_m(t)]$ will contain some information about $x_0(t)$. That is, there will be some dependence of $x_0(t)$ on

---

[14]Savit and Green (1991) and Green and Savit (1991).

[15]The time t and the delay τ are necessarily a multiple of τ_s, the sampling time between observations. Hence, if there are N_d observations, there will be $N_r = N_d - m$ sequences of length $m + 1$.

previous measurements $x_k(t)$, $k > 0$. In that case, $|x_k(t) - x_k(s)|$, $t \neq s$, will contain information about $|x_0(t) - x_0(s)|$. In particular, the fact that $|x_k(t) - x_k(s)| < \varepsilon_k$, says something about the likelihood that $|x_0(t) - x_0(s)| < \varepsilon_0$. With these thoughts as motivation, we can explore the dependencies, or determinism, in the data by examining estimates of the conditional probability[16]

$$P\big[|x_0(t) - x_0(s)| < \varepsilon_0 \big| |x_k(t) - x_k(s)| < \varepsilon_k\big]$$

or more generally, by analyzing conditional probabilities of the form

$$P(t_0|t_1, t_2, \ldots, t_m)$$

where t_k stands for the event

$$|x_k(t) - x_k(s)| < \varepsilon_k$$

Now let $\varepsilon = \varepsilon_k$, and observe that, for a fixed ε, $C_2(m; \varepsilon)$, the correlation integral of order 2, approximates the probability that two consecutive m-length sequences of observations are within ε of each other. Hence, if we use the max norm, $C_2(m; \varepsilon)$ yields an estimate of the joint distrubution of the t_k. That is,

$$C_2(m; \varepsilon) \quad = \quad \frac{1}{[N_r]^2} \sum_{t,s} \left[\prod_{k=0}^{m-1} \Theta(\varepsilon - |x_k(t) - x_k(s)|) \right]$$

$$\xrightarrow{\quad N_r \to \infty \quad} P[t_0, t_1, \ldots, t_{m-1}]$$

and in particular, for any k

$$P[t_k] = C_2(1; \varepsilon)$$

Now, let $C_m \equiv C_2(m; \varepsilon)$. Then by definition

$$P(t_0|t_1) = \frac{P(t_0, t_1)}{P(t_1)} = \frac{C_2}{C_1} \qquad\qquad\qquad \textbf{Eq. 2.22}$$

But, if the $x(t)$ are independent, then so are $[x(t) - x(s)]$, and

$$P(t_0|t_1) = P(t_0) = C_1 \qquad\qquad\qquad \textbf{Eq. 2.23}$$

From Eq. 2.22 and Eq. 2.23, $C_2 = [C_1]^2$, and it follows that the statistic

$$\delta_1 \quad = \quad 1 - \frac{[C_1]^2}{C_2}$$

$$= \quad 1 - \frac{P^2(t_1)}{P(t_0, t_1)}$$

$$= \quad 1 - \frac{P(t_0)P(t_1)}{P(t_0, t_1)}$$

[16]The approach seems to have been first advocated by Packard and Crutchfield (1980), who conjectured that if one forms the conditional probability

$$P\big[x(t)|x(t - \tau), x(t - 2\tau), \ldots, x(t - m\tau)\big]$$

then, for reasonable values of the delay time τ, the system's dimensionality coincides with the value of m corresponding to $P(x|\ldots)$ assuming a sharp spike-like form.

measures the extent to which t_0 depends on t_1. If t_1 implies t_0, then $\delta_1 = 1 - C_1$. If t_0 and t_1 are independent, then statistically $\delta_1 = 0$.

To generalize these ideas, consider

$$
\begin{aligned}
P(t_0 | t_1, \ldots, t_{n-1}) &= \frac{P(t_0, \ldots, t_{n-1})}{P(t_1, \ldots, t_{n-1})} \\[2mm]
&= \frac{P(t_0, \ldots, t_{n-1})}{P(t_0, \ldots, t_{n-2})} \qquad \textbf{Eq. 2.24} \\[2mm]
&= \frac{C_n}{C_{n-1}}
\end{aligned}
$$

If $x(t)$ depends only on the previous $n - 2$ observations, that is, $x(t)$ depends only on $x(t - k)$, $k = 1, \ldots, n - 2$, but not on $x(t - (n - 1))$, then it is also true that

$$
\begin{aligned}
P(t_0 | t_1, \ldots, t_{n-1}) &= P(t_0 |, t_1, \ldots, t_{n-2}) \\[2mm]
&= \frac{P(t_0, t_1, \ldots, t_{n-2})}{P(t_1, t_2, \ldots, t_{n-2})} \\[2mm]
&= \frac{P(t_0, t_1, \ldots, t_{n-2})}{P(t_0, t_1, \ldots, t_{n-3})} \qquad \textbf{Eq. 2.25} \\[2mm]
&= \frac{C_{n-1}}{C_{n-2}}
\end{aligned}
$$

That is, if $x(t)$ depends only on the previous $n - 2$ values, then from Eq. 2.24 and Eq. 2.25

$$
\frac{[C_{n-1}]^2}{C_n C_{n-2}} = 1
$$

Conversely, if

$$
\frac{[C_{n-1}]^2}{C_n C_{n-2}} \neq 1
$$

then $x(t)$ depends on more than the previous $n - 2$ values. Specifically, there is some information about $x(t)$ in $x(t - (n - 1))$, in addition to the information conveyed by $x(t - 1), \ldots, x(t - (n - 2))$.

Define the **dependency index** by

$$
\delta_j = 1 - \frac{[C_j]^2}{C_{j-1} C_{j+1}}
$$

It follows from the above discussion that the degree to which δ_j differs from 0 is a measure of the dependence between $x(t)$ and $x(t - j)$, above that already accounted for in $x(t - 1), \ldots, x(t - (j - 1))$.

Suppose we want to predict the next value in the time series, $x(m + 1)$. If we use only the present value $x(m)$, and $\delta_1 = 0$, then the best we can do is embodied in the whole distribution $P(t_{m+1}) = C_1$. If $x(m + 1)$ is completely predictable from

$x(m)$, then $\delta_1 = 1 - C_1$. So the quantity $S_1 = C_1/(1 - \delta_1)$ ranges from C_1 for a random process to 1 for a process that is completely predictable from one previous measurement.

Since δ_2 will range from 0 to $1 - C_1/(1 - \delta_1)$, the quantity $S_2 = C_1/(1 - \delta_1)(1 - \delta_2)$ measures the predictability when two previous observations are used, and so on. The generalization defined by

$$S_j = \frac{C_1}{\prod_1^j (1 - \delta_i)}; \quad S_0 \equiv C_1$$

$$S = \lim_{j \to \infty} S_j = \frac{C_1}{\prod_1^\infty (1 - \delta_i)}$$

is called the **predictablity index**. If the sequence $x(t)$ is no more predictable than a sequence of independent indentically distributed random variables, then $S = C_1$. If for some $j, x(t)$ is completely predictable from j previous measurements, then $S_n = 1$, for $n \geq j$, and so $S = 1$.

Since

$$1 - \delta_j = \frac{[C_j]^2}{C_{j-1} C_{j+1}}$$

$$\prod_1^m (1 - \delta_j) = \frac{C_1 C_m}{C_{m+1}}$$

and we get the intuitive result that

$$S_m = \frac{C_{m+1}}{C_m} = P(t_0 | t_1, \dots, t_m)$$

Thus, S_m measures the likelihood that a prediction, $x(s)$, will fall within ε of the true value, $x(t)$, given the two series have been within ε over the past m observations.

Remarks

The predictability and dependency indices are related to the lagged variable dependence R^* (discussed above) by

$$\text{Log}(1 - \delta_j) = -R^*(j + 1)$$

and

$$\text{Log } S_m = -\sum_{j=1}^{m+1} R^*(j)$$

Examples 2.4, 2.5 and 2.6

For the random processes (Table 2.1) $\delta_j \cong 0$ and $S_j \cong S_0$ for $j > 0$. In contrast, for the logistic map (Table 2.2), $\delta_1 \gg 0$ and $S_1 \gg S_0$, while $\delta_j \cong 0$ and $S_j \cong S_1$ for $j > 1$, which confirms the expected 1-lag dependence and one degree of freedom in this system. Finally, for the Hénon map (Table 2.3), $\delta_1, \delta_2 \gg 0$ and $S_1, S_2 \gg S_0$, while $\delta_j \cong 0$ and $S_j \cong S_2$ for $j > 2$, which confirms the expected dependence on two lagged variables and two degrees of freedom in this system.

j'	Uniform random numbers		Gaussian random numbers	
	δ_j	S_j	δ_j	S_j
0		0.265		0.264
1	−0.00296	0.264	0.0495	0.278
2	−0.00882	0.262	−0.00314	0.277
3	0.00809	0.264	0.0158	0.281
4	0.0671	0.283	0.00738	0.283
5	0.0228	0.290	−0.00344	0.282

Table 2.1 δ_j and the predictabilty index S_j for a random time series.
Source: Adapted from Savit and Green (1991) p. 101.

j	($\sigma = 0.0$)		($\sigma = 0.1$)		($\sigma = 0.3$)		($\sigma = 0.6$)	
	δ_j	S_j	δ_j	S_j	δ_j	S_j	δ_j	Sj
0		0.296		0.290		0.283		0.284
1	0.452	0.540	0.390	0.475	0.364	0.445	0.193	0.352
2	0.0377	0.561	0.0726	0.513	0.0474	0.467	0.00841	0.355
3	−0.0166	0.552	−0.00315	0.511	0.00630	0.470	−0.00284	0.354
4	−0.0134	0.545	−0.00457	0.509	0.000818	0.470	0.0140	0.359
5	−0.0117	0.539	0.000965	0.509	−0.00296	0.469	0.0194	0.366
6	−0.000480	0.539	0.0180	0.519	0.00119	0.469	0.0143	0.371
7	−0.0165	0.530	−0.0337	0.502	−0.0142	0.463	0.0204	0.379

Table 2.2 δ_j and the predictabilty index S_j for a time series generated by the logistic map corrupted by adding a noise variable σh, where h is uniformily distributed on $[-.5, .5]$.
Source: Adapted from Savit and Green (1991) p. 103.

j	δ_j	S_j
0		0.290
1	0.442	0.520
2	0.193	0.644
3	−0.0343	0.623
4	0.0397	0.648
5	0.0225	0.663
6	0.00574	0.667
7	0.0295	0.687
8	−0.00171	0.686
9	0.0435	0.717

Table 2.3 δ_j and the predictabilty index S_j for a time series generated by the Hénon map.
Source: Adapted from Savit and Green (1991) p. 104.

Remarks

- The fecundity of the δ_j and the S_j is greater than it may first appear. In the above discussion, the $x_k(t)$ were delay variables $x(t - k\tau)$, but $x_k(t)$ could represent, for example, a derivative.
- Note that, the significance of a collection of independent variables in predicting $x(t)$ depends on the particular permutation of the $x_k(t)$, $k > 0$, chosen. A different permutation $x_{\pi(k)}(t)$ may give a simpler form requiring fewer variables to capture the dynamical motion.

5 Estimating local statistics

Other interesting variations come from defining local versions of the entropies and redundancies. Since these are local quantities, we will only need $W_V[v : \varepsilon]$ to estimate them.

For example, a *local Shannon entropy* can be defined by

$$
\begin{aligned}
h_1(v; \varepsilon) &\equiv -\text{Log}\, P_V[v; \varepsilon] \\
&\approx -\text{Log}\, W_V[v; \varepsilon]
\end{aligned}
\qquad \textbf{Eq. 2.26}
$$

which is just the unaveraged H_1.

A local information theoretic measure of the interdependence of two coordinates, given a third coordinate in the state space, is given by the *local conditional redundancy*

$$
\begin{aligned}
r_1(z_j, y_j | x_j; \varepsilon) &\equiv h_1(z_j | x_j; \varepsilon) + h_1(y_j | x_j; \varepsilon) - h_1(z_j, y_j | x_j; \varepsilon) \\
&= h_1[z_j, x_j; \varepsilon] - h_1[x_j; \varepsilon] + h_1[y_j, x_j; \varepsilon] - h_1[x_j; \varepsilon] \\
&\quad - h_1[z_j, y_j, x_j; \varepsilon] + h_1[x_j; \varepsilon] \\
&= h_1[z_j, x_j; \varepsilon] + h_1[y_1, x_j; \varepsilon] - h_1[z_j, y_j, x_j; \varepsilon] - h_1[x_j; \varepsilon] \\
&\approx -\text{Log}\, \frac{W_{Z,X}[z_j, x_j; \varepsilon] W_{Y,X}[y_j, x_j; \varepsilon]}{W_{Z,Y,X}[z_j, y_j, x_j; \varepsilon]\, W_X[x_j; \varepsilon]}
\end{aligned}
\qquad \textbf{Eq. 2.27}
$$

Example 2.7

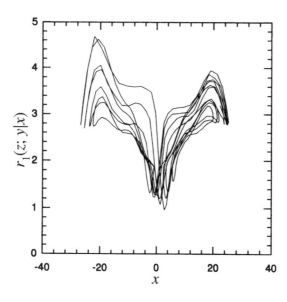

Figure 2.17 The local mutual information $r_1(z; y|x)$–between z and y given x–plotted against x for a short trajectory of the Lorenz equations. When x is small the coupling between y and z is weak, therefore, $r_1(z; y|x)$ is small.

Source: Adapted from Prichard and Theiler (1995) p. 491.

Similarly, a *local version of the Kolmogorov–Sinai entropy* can be defined from the local entropy h_1.

$$k_1(x_j; m, \varepsilon) \equiv \frac{h_1(x_j, \ldots, x_{j+m-1}; \varepsilon) - h_1(x_j, \ldots, x_{j+m-2}; \varepsilon)}{\tau}$$

$$= \frac{1}{\tau} \text{Log} \frac{W_{X_j, \ldots, X_{j+m-1}}[x_j, \ldots, x_{j+m-1}; \varepsilon]}{W_{X_j, \ldots, X_{j+m-2}}[x_j, \ldots, x_{j+m-2}; \varepsilon]} \qquad \textbf{Eq. 2.28}$$

The local Kolmogorov–Sinai entropy gives a qualitative measure of predictability as it varies over the state space.

Example 2.8

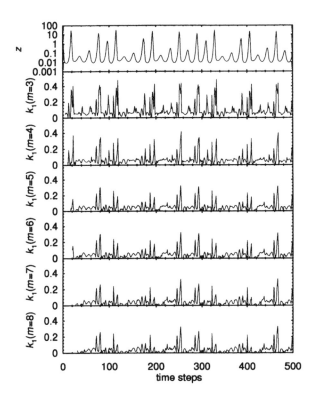

Figure 2.18 Top panel: z component of the Rössler equation. Next six panels are the approximate local Kolmogorov–Sinai entropy calculated with m = 3–8 lagged values of the x component.

Source: Adapted from Prichard and Theiler (1995) p. 490.

6 Entropy for a Gaussian process

If $X = X_1, \ldots, X_m$ is Gaussian, then

$$P_X(x) = \frac{|\xi|^{\frac{1}{2}}}{(2\pi)^{\frac{m}{2}}} \exp\left[-\frac{1}{2} \sum_{i,j}^m \xi_{ij} x_i x_j\right]$$

where $|\xi|$ is the determinant of ξ,

$$\xi = [\xi_{ij}] = \Xi^{-1}$$

$$\Xi = Y^T Y = \left[\frac{\langle x_{t+i}, x_{t+j} \rangle_t}{N_d} \right]$$

and $\langle x_t \rangle = 0$

By inserting this $P_X(x)$ into the definition of H_q, we get the *linearized* entropies H_q^l, redundancies R_q^l, marginal redundancies $\mathfrak{R}_q^l$ and correlation integrals C_q^l of order q:

$$H_q^l(X_1, \ldots, X_m) = \frac{1}{2} m \operatorname{Log} 2\pi + \frac{m \operatorname{Log}_2 q}{2(q-1)} + \frac{1}{2} \operatorname{Log} |\Xi|$$

$$= -\operatorname{Log} C_q^l(x_1, \ldots, X_m) \qquad \textbf{Eq. 2.29}$$

$$C_q^l(X_1, \ldots, X_m) = \left[\frac{|\xi|^{\frac{q-1}{2}}}{q^{\frac{m}{2}} (2\pi)^{\frac{m(q-1)}{2}}} \right]^{\frac{1}{q-1}} \qquad \textbf{Eq. 2.30}$$

$$R_q^l(X_1, \ldots, X_m) \equiv \sum_1^m H_q^l(X_i) - H_q^l(X_1, \ldots, X_m)$$

$$= -\operatorname{Log} \frac{\prod_1^m C_q^l(x_i)}{C_q^l(x)} \qquad \textbf{Eq. 2.31}$$

$$= \frac{1}{2} \sum_1^m \operatorname{Log}(\Xi_{ii}) - \frac{1}{2} \operatorname{Log} |\Xi|$$

Note that $R_q^l = R^l$ is independent of q, so the linear marginal redundancy

$$\mathfrak{R}^l(X_1, \ldots, X_m) = R^l(X_1, \ldots, X_m) - R^l(X_1, \ldots, X_{m-1}) \qquad \textbf{Eq. 2.32}$$

is also independent of q.

Remark
Note that the linear quantities H_q^l, C_q^l, R_q^l, and $\mathfrak{R}_q^l$, given by Eq. 2.29 through Eq. 2.32, are defined even when the system is not Gaussian.

Examples 2.9 and 2.10
See panels (e) and (f) in Figure 2.15 and Figure 2.16 above.

7 BDS statistic[17]

The Brock, Dechert, and Scheinkman (BDS) statistic $W_{m,N_r}(\varepsilon)$ is defined by

$$W_{m,N_r}(\varepsilon) = \frac{\sqrt{N_r}[C_2(m; \varepsilon, N_r) - (C_2(1; \varepsilon, N_r))^m]}{S_{m,N_r}(\varepsilon)}$$

---

[17]Brock, Dechert and Scheinkman (1987); or see Hsieh (1991).

where $C_2(m; \varepsilon, N_r)$ is the correlation integral defined by

$$C_2(m; \varepsilon, N_r) = \frac{2}{N_r(N_r - 1)} \sum_{i>j} \Theta(\varepsilon - \|\underline{y}(i) - \underline{y}(j)\|)$$

where

$$\underline{y}(n) = [x(n), x(n+1), \ldots, x(n+m-1)] \in \mathfrak{R}^m \quad 1 \le n \le N_r$$

and $S_{m,N_r}(\varepsilon)$ is the standard error of $C_2(m; \varepsilon, N_r) - [C_2(1; \varepsilon, N_r)]^m$ which can be estimated using Monte Carlo techniques[18].

It can be shown that asymptotically, as $N_r \to \infty$, the distribution of $W_{m,N_r}(\varepsilon)$ is $\mathfrak{N}(0, 1)$[19] when the $\underline{y}(n)$ are iid[20]. Knowing the asymptotic distribution of $W_{m,N_r}(\varepsilon)$ enables one to formulate a test of the iid hypothesis.

Remark

Note, however, that the alternative to iid behavior includes linear dependence, non-stationarity, chaotic dynamics, and nonlinear stochastic processes[21].

REFERENCES AND FURTHER READING

Auerbach, D., Cvitanovic, P., Eckmann, J., Gunaratne, G. and Procaccia, I. (1987) 'Exploring chaotic motion through periodic orbits', *Physical Review Letters* 58.23, pp. 2387–2389.

Brock, W., Dechert, W. and Scheinkman, J. (1987) 'A test for independence based on the correlation dimension', Working Paper, Univ. Wisconsin, Univ. Houston, and Univ. Chicago.

Buzug, Th., Pawelzik, K., von Stamm, J. and Pfister, G. (1994) 'Mutual information and global strange attractors in Taylor–Couette flow', *Physica* D72, pp. 343–350.

Eckmann, J. and Ruelle, D. (1985) 'Ergodic theory of chaos and strange attractors', *Reviews of Modern Physics* 57.3, pp. 617–656.

Fraser, A. and Swinney, H. (1986) 'Independent coordinates for strange attractors from mutual information', *Physical Review* A33.02, pp. 1134–1140.

Grassberger, P. and Procaccia, I. (1983) 'Measuring the Strangeness of Strange Attractors', *Physica* D9, pp. 189–208.

Green, M. and Savit, R. (1991) 'Dependent variables in broad band continuous time series', *Physica* D50, pp. 521–544.

Hsieh, D. (1991) 'Chaos and Nonlinear Dynamics: Application to Financial Markets', *J. of Finance* 46.05, pp. 1839–1877.

Hubner, U. *et al.* (1993) 'Lorenz-like chaos in NH-FIR lasers (data set A)', in Weigend and Gershenfeld (1993) pp. 73–104.

Ott, E. (1993) *Chaos in Dynamical Systems*, Cambridge University Press, Cambridge.

Packard, N. and Crutchfield, J. (1980) 'Geometry from a time series', *Physical Review Letters* 45.09, pp. 712–716.

Prichard, D. and Theiler, J. (1995) 'Generalized redundancies for time series analysis', *Physica* D84, pp. 476–493.

[18]In fact, the techniques discussed in Appendix 7, on surrogate data, can be used to compute the BDS statistic for a large class of null hypotheses on the underlying process, not just iid as assumed here.

[19]Standard Normal distribution with zero mean and unit variance.

[20]Independent and identically distributed.

[21]For a related statistic that has more discriminating power, see Wu *et al.* (1993).

Savit, R. and Green, M. (1991) 'Time series and dependent variables', *Physica* D50, pp. 95–116.

Shaw, R. (1981) 'Strange attractors, chaotic behavior and information flow', Z. *Naturforsch* A36, pp. 80–112.

Weigend, A. and Gershenfeld, N. (eds), (1993) 'Time series prediction: forecasting the future and understanding the past', *SFI Studies in the Sciences of Complexity*, Vol. XV, Addison-Wesley.

Wu, K. *et al.* (1993) 'Statistical tests for deterministic effects in broad band time series', *Physica* D69, pp. 172–188.

3

Phase space reconstruction

INTUITION

If we wish to understand or predict the behavior of a system with unknown dynamics, then we need to model the system's dynamics either locally or globally. In the experimental situation the only information we have to build such a model consists of a sequence of scalar measurements on specific aspects of the system taken at various points in time. The idea that we could model a general multidimensional dynamical system using only a one-dimensional time series of observations at first blush seems daunting if not absurd. In fact, under rather robust assumptions, retrieving a model of the system's dynamics from such experimental data is feasible, and in this chapter we investigate both the theoretical foundations and the practical techniques for doing so.

If we had a sample trajectory from the system, then we could estimate the system's invariants as well as its dynamics. We have assumed that not only do we not know the equations governing the dynamics of the system, but also that we cannot directly observe the original state space M in which those dynamics take place. Hence, we cannot observe an actual trajectory of the system. Thus, our only hope is that we can use the available information in the time series of observations to reconstruct a sample trajectory in some real p-dimensional Euclidean space $\Re^p$.

Suppose that a system with dynamics f has an attractor $Q \subset M$[1]. Recall that for a system with dynamics f, the attractor Q is where the trajectories $s_t = f^t[s_0], s_0 \in M$[2], reside as $t \to \infty$. As already mentioned, if we want to model the long term behavior of the system in order to be able to retrieve system invariants and dynamics, then we will need a representation in $\Re^p$ of a sample trajectory from the original attractor Q. Is that possible? To begin with, we have to be sure, at least in principle, that Q can be represented, or transformed, into a set $A \subset \Re^p$ such that the trajectories in A retain all the important geometric characteristics of the trajectories in Q. That is, that the spatial and temporal relationship among the trajectory points of A must be no more than a smooth deformation of the spatial and temporal relationship among the trajectory points in Q.

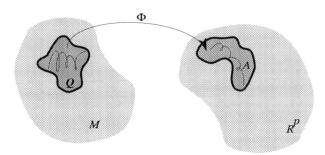

Figure 3.1 The transformation Φ maps the original unobservable attractor Q to the representation A in such a way that a trajectory in A is no more than a smooth deformation of the corresponding trajectory in Q. The original and transformed trajectories are symbolized by the curly lines in Q and A respectively.

---

[1] $Q \subset M$ means that Q is a subset of M.

[2] $s_0 \in M$ means that s_0 is a member of the set M.

If such a transformation Φ exists, and it is invertible, then the dynamics f on Q imply a dynamics φ on A as follows

$$A = \Phi[Q]$$
$$z_u \in A \subset \mathfrak{R}^p; \quad u \geq 0 \qquad\qquad \text{Eq. 3.1}$$
$$z_t = \varphi^t[z_0] \equiv \Phi \cdot f^t \cdot \Phi^{-1}[z_0]$$

where u and t denote the time. The function Φ is said to **embed** Q into $\mathfrak{R}^p$, and $A = \Phi[Q]$ is said to be an **embedding** of Q (see Figure 3.2).

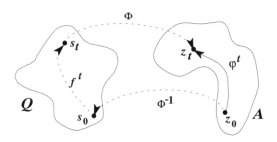

Figure 3.2 The dynamics φ on A (solid line from z_0 to z_t) are obtained, in principle, by first mapping an initial point $z_0 \in A$ back to $s_0 \in Q$ using $\Phi^{-1}[z_0]$, then applying the true dynamics f to the point s_0 to get the evolved point $s_t = f^t[s_0]$ also in Q, and finally mapping the evolved point $s_t \in Q$ back to $z_t \in A$ using $z_t = \Phi[s_t]$.

The existence of the invertible function Φ was established in a celebrated and difficult theorem due to H. Whitney[3]. As intuitive motivation, we point out that one of the assumptions of this theorem is that Q is a manifold, a mathematical object that looks like Euclidean space locally (i.e. in small neighborhoods), so a smooth global transformation from $Q \subset M$ to $A \subset \mathfrak{R}^p$ seems plausible. That result allows us to talk in principle about a dynamical system defined on a space M with dynamics f, as if it were a dynamical system defined on $\mathfrak{R}^p$ with dynamics φ as defined above. However, neither the existence of Φ, nor the implied dynamics of Equation 3.1, is particularly useful when all we have to work with is an observed time series of measurements. To model the system in $\mathfrak{R}^p$ we need to know something about the attractor A, namely, we still need a sample trajectory on A.

An ordered sequence $\{x_t\}$ of observations, that is a time series, can be interpreted as being the result of an unknown mapping $h : Q \rightarrow X$ from states s_t on the unobservable attractor Q to real valued measurements x_t in an observable set X. In general, the set X and the set A are not comparable: a point z_t in A represents the state of the system at time t, but in many experimental situations z_t cannot be determined from the observation x_t. For example, it is not possible in general to identify the state of a multidimensional system from a scalar observation x_t.

[3]Whitney, H., 1936.

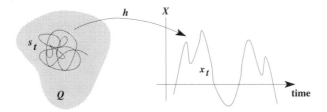

Figure 3.3 To the left, a trajectory $s_t, t \geq 0$, on the original unobservable attractor Q. While we cannot observe the state s_t directly, we can use a measuring device to observe some aspect of it. The output of the measuring device that we use to observe the system can be thought of as the result of the application of a function h to the state s_t. Observing the system at times t_i gives rise to a time series $x_t = h[s_t], t = t_1, t_2, \ldots$.

Because we only get partial information on the actual state of the system from each measurement, it seems that the transformation h from the unobservable states s_t to observable measurements x_t has resulted in the loss of critical information needed to reconstruct the attractor A, and hence the dynamics of the system. In other words, it appears that we just do not have enough information in the observation set X to reconstruct the attractor A.

The key to overcoming this dilemma is to argue as follows. For any realistic measurement function h, each observation x_t must contain some information about s_t, and x_t and x_{t+1} will typically contain different information if $s_t \neq s_{t+1}$. Consider the ordered subsequence $\underline{y}_t = [x_{t-n\tau}, \ldots, x_{t-\tau}, x_t, x_{t+\tau}, \ldots, x_{t+m\tau}]$ of observations around time t. Because the system is assumed to be deterministic, only certain such ordered subsequences are possible. That is, such subsequences can only arise as a result of the system having been in certain states near time t. As the number of observations around the point in time t grows large, the number of possible states that the system could have been in, which could give rise to those subsequences, diminishes. To see why this should be so, we consider the construction $\underline{y}_t$ in terms of the constraint each of its component measurements, $x_{t\pm i\tau}$, puts on the location of the state of the system in the original state space. Define the ***measurement surface***

$S(t) = \{$ all states s, such that $h[s] = x_t\}$

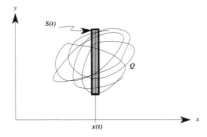

Figure 3.4 Illustration of the concept of a measurement surface in two dimensions where the states are $s(t) = [x(t), y(t)]$ and the time series is generated by $h[s(t)] = x(t)$. The measurement surface $S(t)$ is, in principle, the intersection of the vertical line $x = x(t)$ with the attractor Q. In practice, measurement errors cause $S(t)$ to have a thickness in the horizontal direction.

The measurement surface $S(t)$ is just the set of all states s in the original state space which are consistent with the measurement x_t. Thus, the actual state of the system at time t must be in $S(t)$ for any legitimate value of t. In particular, the actual state of the system at time $t + u$ must be in $S(t + u)$. For a deterministic system, the state of the system at time $t + u$ is determined by the state of the system at time t, that is, $s_{t+u} = f^u[s_t]$. It follows that s_t is in $f^{-u}[S(t + u)]$, the set of all states that evolve to a state consistent with the observation x_{t+u} after exactly u units of time. Clearly, the same argument holds for negative values of u. Now, consider again the subsequence of measurements $\underline{y}(t) = [x_{t-n\tau}, \dots, x_{t-\tau}, x_t, x_{t+\tau}, \dots, x_{t+m\tau}]$. The actual state s_t in the original state space must be consistent with *all* these measurements and so must be in

$$f^{n\tau}[S(t - n\tau)] \quad \cap \dots \cap \quad f^{\tau}[S(t - \tau)] \cap S(t) \cap f^{-\tau}[S(t + \tau)]$$
$$\cap \dots \cap \quad f^{-m\tau}[S(t + m\tau)]$$

the intersection of the time-t measurement surface $S(t)$ with its adjacent $n + m$ measurement surfaces translated to time t (see Figure 3.5).

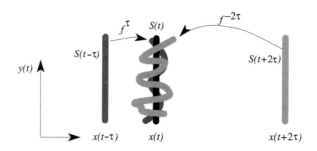

Figure 3.5 Illustration of intersecting measurement surfaces in two dimensions. At the center of the figure is the intersection of the measurement surfaces $S(t)$, $S(t - \tau)$ mapped forward to its image at time t, and $S(t + 2\tau)$ mapped backward to its image at time t. Each of these contains $s_t = [x(t), y(t)]$, the actual state of the system in the original state space at time t. Hence, their intersection contains s_t.

Source: Adapted from Casdagli *et al.* (1991) p. 74.

Hence, it would not be unreasonable to conjecture that a sufficiently large subsequence $\underline{y}_t$ would be enough to identify a state of the true system near time t. Now, the subsequence $\underline{y}_t$ can also be thought of as a point in $(n + m + 1)$-dimensional Euclidean space. If the set $\{\underline{y}_t\}$, treated as points in $\mathfrak{R}^{n+m+1}$, retains approximately the same spatial and temporal relationship among its elements as the set of points $\{s_t\}$ in M, then we could use the points $\{\underline{y}_t\}$ to simulate a real valued representation of an actual trajectory of the system. We emphasize that the subsequence $\underline{y}_t$ is not the actual state of the system but, rather, a real valued representation of the state of the system at a particular point in time t.

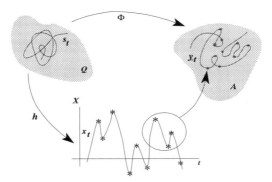

Figure 3.6 Measurements on s_t, the original trajectory of the system, create a time series $x_t = h[s_t]$ (starred (*) points). $\underline{y}_t$, subsequences of the time series, create a representation of s_t in $\mathfrak{R}^p$. Φ is the combination of the two steps, that is, $\underline{y}_t = \Phi[s_t]$. The points $\underline{y}_t$ trace out a segment of a trajectory on an attractor A that is a partial reconstruction of a representation of the original attractor Q.

The plausibility of that conjecture becomes compelling if we think of the $\underline{y}_t$ as a mapping Φ from the true states of the system in M to representations of these states in a real Euclidean space $\mathfrak{R}^p$. That is, symbolically

$$\Phi(s_t) = \underline{y}_t = [x_{t-n\tau}, \dots, x_{t+m\tau}] = [hf^{-n\tau}(s_t), \dots, hf^{+m\tau}(s_t)]$$

If the $\underline{y}_t$, viewed as a mapping, can be shown to be an embedding of the system's true attractor, then the subsequences $\underline{y}_t = \Phi[s_t]$ trace out a ***reconstructed trajectory*** which emulates the important geometric characteristics of a true trajectory, and hence, those of the true attractor. The conjecture was proposed by several researchers[4] at about the same time, and proven to be true by F. Takens (1981).

THEORY

Arguably, reconstructing the experimental trajectory is the most important issue in this book since estimating most of the invariants and dynamics cannot proceed without it. At the heart of a successful state space reconstruction is the choice of parameters for the transformation Φ. Homing in on the optimal parameters will require some work. We start with theorems establishing the existence and form of Φ. That leads to alternative state space coordinate systems. Analysis of the relationship between those coordinate systems yields some clues about the location of the best parameters. A bit more effort then produces the boundaries defining the region containing good parameter choices.

---

[4]D. Ruelle, unpublished communication with N. Packard; N. Packard *et al.* (1980).

Embeddings

We will need the following definitions:

- A set Q is called **compact** if every covering of Q by open sets contains a finite subset which also covers Q. In Euclidean space a set is compact if it is closed and bounded.
- Φ, a function, is **one-to-one** if it does not collapse points. That is, $\Phi[s] = \Phi[s']$ implies $s = s'$.
- Φ is **smooth** if it has a continuous derivative.
- Φ is in C^k if it has continuous partial derivatives of order up to and including k.
- $\Phi \in C^1$ is an **immersion** if $D_s\Phi$ (the Jacobian matrix of Φ at s) is one-to-one. That is, Φ is an immersion if no differential structure is lost going from M to $\Phi[M]$. Note that Φ may be an immersion without being one-to-one.
- Φ is a **homeomorphism** if it is continuous and one-to-one in both directions.
- Φ is a **diffeomorphism** if it has a continuous derivative and an inverse with a continuous derivative.
- M is an m-dimensional **manifold** if for each $s \in M$ there is an open neighborhood U_s of s in M and a homeomorphism g_s from U_s to an open set in $\mathfrak{R}^n$, the composite function $g_x(g_y^{-1})$ for overlapping neighborhoods U_x and U_y also being a homeomorphism on open sets in $\mathfrak{R}^n$. The manifold is called differentiable if the g_s are differentiable. Alternatively (and equivalently) a (differentiable) manifold is a subset of $\mathfrak{R}^n$ defined at each point by expressing m of the point's coordinates as (differentiable) functions of the other $n-m$ coordinates; such as, for example, the surface of a sphere is a two-dimensional manifold.
- $\Phi[M]$ is an **embedding** of M if Φ is a diffeomorphism from M onto its image $\Phi[M]$, that is, a smooth one-to-one map which has a smooth inverse. For a compact manifold M, $\Phi[M]$ is an embedding if and only if Φ is a one-to-one immersion. We will occasionally use an abuse of language and call Φ an embedding if $\Phi[M]$ is an embedding of M.
  An embedding takes an m-dimensional manifold M to a representation as a surface in $\mathfrak{R}^p$ (or a smooth manifold in another space). As $s(t)$ roams over M, $\Phi[s(t)]$ roams over the surface in $\mathfrak{R}^p$ in such a way that the orbit $s(t)$ is only smoothly distorted and chronological order of the trajectory points is maintained. Embeddings preserve differential structures; in particular, they preserve the structure of the surface tangent to the manifold M (the **tangent space**) at every point of the domain of Φ.
- A property P of a function (system) is **generic** if, for randomly chosen parameters of the function (system), the function (system) exhibits the property P with probability 1.

A fundamental result is given by:

Theorem 1 (H. Whitney)
If M is an m-dimensional manifold, and $\Phi : M \to \mathfrak{R}^{2m+1}$ is a smooth map, then generically $\Phi[M]$ is an embedding, that is, generically Φ maps diffeomorphically M into some m-dimensional submanifold M' of $\mathfrak{R}^{2m+1}$. Moreover, the set of maps into $\mathfrak{R}^{2m+1}$ that form embeddings of M is an open and dense set in the C^1 topology of maps.

In spite of the elegence of H. Whitney's Theorem, the genericity of an embedding is not much help in the experimental situation because such maps may have vanishing probability of being realized. What we need is a result that says that the map that we actually determine experimentally is almost always an embedding[5]. By defining a property similar to 'almost always' that works for finite and infinite dimensional vector spaces, Sauer, Yorke and Casdagli (1991) proved the following theorem.

Theorem 2 (T. Sauer, J. Yorke and M. Casdagli)
If $M \subset \mathfrak{R}^k$ is a compact smooth manifold of dimension m, then almost every smooth map $\Phi : \mathfrak{R}^k \to \mathfrak{R}^{2m+1}$ is an embedding of M. In particular, given any smooth map Φ, not only are there maps arbitrarily near Φ that are embeddings, but almost all of the maps near Φ are embeddings.

Reconstruction coordinate systems

What we need now is a way to construct Φ from experimental data. As mentioned in the previous section, Packard *et al.* (1980) conjectured that *any m independent observables* from an m-dimensional dynamical system should be sufficient to reconstruct the asymptotic behavior (attractor) of the system simply by forming a **reconstructed** or **experimental trajectory** consisting of an m-tuple of the observables. Specifically, a reconstructed trajectory $\underline{y}_t$ could be defined in terms of m independent observables $x_t^{(k)}, k = 0, \ldots, m - 1$, as follows,

$$\underline{y}_t \equiv [x_t^{(0)}, x_t^{(1)}, \ldots, x_t^{(m-1)}]^T \in \mathfrak{R}^m$$

F. Takens (1981) provided a rigorous proof of the conjecture in several forms embodied in the following theorem.

Theorem 3 (F. Takens)
Let M be a compact m-dimensional manifold, $F : M \to M$ a C^2 vector field, and $h : M \to \mathfrak{R}$ a smooth function. Let f be the flow of F, that is, $s_t = f^t[s_0], s_t \in M, t \in \mathfrak{R}$. Then it is a generic property that

1. $\quad \Phi_{f,h}(s) : M \to \mathfrak{R}^{d_r}$
    defined by

    $$\Phi_{f,h}(s) = (h[s], hf^\tau[s], hf^{2\tau}[s], \ldots, hf^{(d_r-1)\tau}[s])^T$$

    is an embedding for $d_r \geq 2m + 1$, and non-zero $\tau \in \mathfrak{R}$.
2. Point 1, above, also holds if we replace the vector field (flow) with a map f^n that is a diffeomorphism, and require τ to be a non-zero integer.
3. If there is an attractor Q for the system defined by the dynamics $f^t[s]$, then there is an attractor A for $\varphi^t[s]$[6] which is diffeomorphic with Q. That is, A is an embedding of Q in $\mathfrak{R}^{d_r}$.
4. Point 1, above, continues to hold if we replace $\Phi_{f,h}$ with

    $$\Psi_{f,h}(s) : M \to \mathfrak{R}^{d_r}$$

---

[5]The terms 'almost always,' 'almost every,' and 'almost all' mean 'with probability 1' or 'except on a set of measure zero.'

[6]See Eq. 3.1.

defined by

$$\Psi_{f,h}(s) = (h[s], D^1 h f^t[s], D^2 h f^t[s], \dots, D^{d_r-1} h f^t[s])^T$$

where

$$D^n = \left. \frac{d^n}{dt^n} \right|_{t=0}$$

In the above theorem, the vector function $\Phi_{f,h}$ is called a **delay coordinate map** because its components (coordinates) $h f^t[s]$ are often identified with a time series x_t. In that case, the delay coordinate map defines the experimental trajectory. That is,

$$
\begin{aligned}
\underline{y}_t &= [x_t, x_{t+\tau}, \dots, x_{t+(d_r-1)\tau}]^T \\
&= (h[s(t)], h f^\tau[s(t)], h f^{2\tau}[s(t)], \dots, h f^{(d_r-1)\tau}[s(t)])^T \\
&= \Phi_{f,h}[s_t] \qquad\qquad\qquad\qquad\qquad\qquad\qquad \textbf{Eq. 3.2}
\end{aligned}
$$

In practice the observations x_t are obtained at discrete points in time t_i, $i = 1, \dots,$ N_d, and it will often be the case that the times between observations $t_i - t_{i-1}$ are constant, that is, $t_i = t_o + i\tau_s$ for some constant τ_s called **the sampling time**. When that is the case, perforce, the delay time $\tau = p\tau_s$ for some positive integer p. In the sequel, except where noted otherwise, we will assume that $t_i = i\tau_s$, so that $\tau = p\tau_s$ and

$$t_i + n\tau = (i + np)\tau_s$$

Under that assumption, no confusion arises if we further simplify the notation as follows

$$
\begin{aligned}
x_i &= x(i) = x(t_i) \qquad i = 1, \dots, N_d \\
\underline{y}_j &= \underline{y}(j) = \underline{y}(t_j) \\
&= [x(t_j), x(t_j + \tau), \dots, x(t_j + (d_r - 1)\tau)]^T \\
&= [x(j), x(j + p), \dots, x(j + (d_r - 1)p)]^T
\end{aligned}
$$

where the index $j = 1, \dots, N_r$ and

$$N_r = N_d - (d_r - 1)p$$

We will use the term (d_r, τ)-**reconstruction** to mean a state space reconstruction based on a delay coordinate map with parameters d_r and τ. Note that part 4 of theorem 3 says that derivatives, as well as iterates of the dynamics f, can be used as the coordinates of the reconstructed state space, and hence form alternative coordinate systems. An earlier remark suggested that almost any vector of data containing independent information should work. For example,

$$y(i) = \left\{ \sum_{j<i} x(j) e^{-(i-j)\tau_s}, x(i), x(i+1) - x(i-1) \right\} \qquad \textbf{Eq. 3.3}$$

The above theorem does not establish that rigorously and it is an open question as to how well constructs like Eq. 3.3 work in practice. However, the theorems of Sauer, Yorke and Casdagli below, do extend the range of embedding functions considerably.

Theorem 3 states that $d_r \geq 2m + 1$ is *sufficient* to obtain an embedding. But this is not a necessary condition. In fact, an embedding can sometimes be obtained for $d_r \geq m$. When $\Phi_{f,h}$ is an embedding, the space $\Phi_{f,h}(M)$ is called the **embedding space** and its dimension d_E is called the **embedding dimension**. We will also use d_E to denote any value of d_r sufficiently large to form an embedding, while d_e will denote the smallest value of d_r necessary to achieve an embedding. For arbitrary d_r, $\Phi_{f,h}(M)$ is called the **reconstructed state** (or **phase**) **space**, and $A = \Phi_{f,h}(Q)$ is called the **reconstructed** or **experimental attractor** of the true attractor Q. A given delay coordinate map, Eq. 3.2, is defined in terms of a fixed number d_r of observations, which, for a fixed **delay** or **lag time** τ, defines a fixed interval of time $\tau_w = (d_r - 1)\tau$ called the **reconstruction window**. While the results of Takens's theorem, and those of Sauer, Yorke and Casdagli, which follow, depend on the dimension d_r, the value of the delay time τ is theoretically arbitrary. In practice the value of τ, and by implication τ_w, is critical to obtaining a representation of the trajectories (attractor) which will faithfully reproduce the system's invariants.

As with Whitney's theorem, Takens's theorem gives a generic result as opposed to a more useful result which is true almost always. Moreover, as the true state space M is unobservable, its (topological) dimension m is generally unknown so we do not know in practice how to form a lower bound on d_r in order to guarantee that $\Phi_{f,h}$ is an embedding. The following theorems from Sauer, Yorke and Casdagli (1991) give practical conditions under which $\Phi_{f,h}$ is almost always an embedding.

Theorem 4 (T. Sauer, J. Yorke and M. Casdagli)
Assume $Q \subset \mathfrak{R}^k$ is compact and let $m > 2D_0(Q)$ [7] be an integer, then almost every smooth map $\Phi : \mathfrak{R}^k \to \mathfrak{R}^m$ is,

1. one-to-one on Q
2. an immersion on each compact subset of a smooth manifold contained in Q.

Theorem 5 (T. Sauer, J. Yorke and M. Casdagli)
Let f be a diffeomorphism on an open subset U of $\mathfrak{R}^k$, assume $Q \subset U$ is compact and let $m > 2D_0(Q)$ be an integer. Assume that for every positive integer $p \leq m$, the set Q_p of periodic points of period p has $D_0(Q_p) < p/2$, and that the linearization (Jacobian) $D_x f^p$ for each of these orbits has distinct eigenvalues. Then for almost every smooth function $h : U \to \mathfrak{R}$, the delay coordinate map $\Phi_{f,h} : U \to \mathfrak{R}^m$ is

1. one-to-one on Q
2. an immersion on each compact subset of a smooth manifold contained in Q.

Corollaries (T. Sauer, J. Yorke and M. Casdagli)

1. Let $h_1, h_2, \ldots, h_t$ be any set of polynomials in k variables up to order $2m$. Then given any smooth function h_0 on $U \in \mathfrak{R}^k$, for almost all choices of $\alpha = (\alpha_1, \ldots, \alpha_t) \in \mathfrak{R}^t$ the function

$$h_\alpha = h_0 + \sum_1^t \alpha_i h_i$$

satisfies properties 1 and 2 in Theorem 5.

[7] $D_0(A)$ is the capacity or box-counting dimension of A. See Chapter 4, Fractal dimension.

2. The more general delay coordinate map

$$\Phi[s] = [h_1(s), \ldots, h_1 f^{n_1-1}[s], \ldots, h_q[s], \ldots, h_q f^{n_q-1}[s]]$$

also satisfies Theorem 5 as long as h_i is smooth, $n_1 + \ldots + n_q > 2D_0(Q)$, and $D_0(Q_p) < p/2$ for $p \leq \max\{n_1, \ldots, n_q\}$.

3. Theorem 5 continues to hold if $\Phi_{f,h} : \mathfrak{R}^k \rightarrow \mathfrak{R}^w$ is transformed by an $m \times w$ matrix B of rank $m > 2D_0(Q)$, and if f has no periodic orbits of period less than or equal to w.

The reader may be wondering why replacing the topological dimension of the attractor Q with the box counting dimension D_0 of Q yields a tractable lower bound on the embedding dimension. Since Q is unobservable, neither the topological nor the box counting dimension of Q is computable. However, recall that the primary reason for focusing on representations of Q in Euclidean space which are a result of a transformation which is an embedding, is that an embedding is a smooth transformation of Q which preserves the important dynamical and geometric properties of Q. That is, such properties are invariant under a transformation which is an embedding. Among the invariant properties of Q is its spectrum of fractal dimensions, and in particular its box counting dimension $D_0(Q)$. Hence, if we have a transformation Φ which is an embedding, then the box counting dimension $D_0(A)$ of the reconstucted attractor $A = \Phi(Q)$ is equal to the box counting dimension $D_0(Q)$ of the original attractor Q. Now, A is observable and there are a number of efficient algorithms for computing its fractal dimension[8]. So, if we have an embedding, we can compute $D_0(A)$ which gives a lower bound on the dimension parameter d_r of the delay coordinate map that yields an embedding. Clearly that reasoning is circular, but it does suggest the solution. If the reconstruction dimension d_r of the delay coordinate map is large enough to form an embedding, then any reconstruction dimension greater than d_r will also form an embedding. Moreover, as long as the reconstruction is an embedding, $D_0(A)$ is a constant. Hence, as the state space is reconstructed in higher and higher dimensions, eventually $D_0(A)$ converges to its true value $D_0(Q)$. That procedure is the basis of an algorithm discussed in Chapter 4, on fractal dimension, for obtaining simultaneously the dimension of the embedding space and the fractal dimension of the attractor. Other methods for estimating the dimensions of the embedding space and the attractor are discussed later in this chapter.

Principal components

Define the $N_r \times d_r$ *trajectory matrix* by

$$Y = \frac{1}{\sqrt{N_r}} \begin{bmatrix} \underline{y}_1^T \\ \underline{y}_2^T \\ \vdots \\ \underline{y}_{N_r}^T \end{bmatrix}$$

---

[8] See Chapter 4, Fractal dimension.

The well known *singular value decomposition*[9] (SVD) of the (trajectory) matrix Y is given by

$$Y = S \Sigma C^T \qquad \text{Eq. 3.4}$$

The right singular vectors $\{c_i\}$ of Y, that is the columns of C, form an orthonormal basis for the embedding space. The projection YC of the trajectory matrix onto that orthonormal basis of the embedding space is called the *matrix of principal components* and in that context the right singular vectors $\{c_i\}$ are called *principal axes*. Principal components are an alternative coordinate system for phase space reconstruction with some important properties. First note that since C is orthonormal, it is a pure rotation. Thus, the projection YC does not change distances between the trajectory points. In particular, it does not change the dimension of the attractor. Clearly, by Eq. 3.4, the c_i and σ_i^2 are respectively the eigenvectors and eigenvalues of the covariance matrix $\Xi_Y = Y^T Y$. Hence, the c_i and σ_i are the directions and lengths of the principal axes of the d_r-dimensional volume explored (on average) by the trajectory $\underline{y}_i \in \Re^{d_r}, i = 1, \ldots, N_r$.

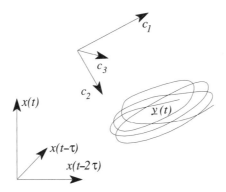

Figure 3.7 Delay coordinate system, bottom left, and principal component coordinate system, top center, for the experimental trajectory $\underline{y}(t)$.

Since

$$C^T Y^T Y C = \Sigma^2 \qquad \text{Eq. 3.5}$$

in this coordinate system the components of the trajectory points are *linearly* uncorrelated, and the variance of the ith principal component is σ_i^2. Hence, principal components have maximum signal-to-noise ratios[10] of any coordinate system for a given pair of reconstruction parameters d_r and τ. If $\sigma_i^2 = 0$ for some i, then

---

[9] See Appendix 4, Matrix decomposition.

[10] If noise is additive (see Chapter 6), SVD is optimal in the sense that principal components have maximum variance and, therefore, maximum signal-to-noise ratios. But, the principal components with small singular values may have signal-to-noise ratios worse than delays. Moreover, principal components are optimal only for a fixed d_E and τ.

the reconstructed trajectory does not visit principal axis i and the number of non-zero principal values equals the dimension of the embedding space d_E when the reconstruction is an embedding. Even in principle though, measurement error, as well as dynamical nonlinearities, will produce a **noise floor** under the singular values which divides the singular value spectrum into significant and non-significant values rather than zero and non-zero values. In practice the noise levels in the data can be so large as to mask the true dynamics, e.g., embedding dimension. In effect the noise raises the noise floor under the singular value spectrum so that both significant and insignificant singular values get absorbed into the floor. If the singular values satisfy

$$\sigma_1^2 \geq \sigma_2^2 \geq \cdots \geq \sigma_q^2 \gg \langle \eta^2 \rangle \approx \sigma_j^2 \quad j = q + 1, \ldots, d_r$$

where $< \eta^2 >$ is the noise variance, then the reconstruction does not contain enough independent information to identify the state when $q < d_A$[11] even if $d_r > 2d_A$ theoretically implies an embedding.

Quasi-equivalence of coordinates

Thus far, we have discussed the theoretical basis for delay coordinates, derivatives, and principal components as alternative coordinate systems for phase space reconstruction. We might well ask which of these reconstructions is best? The answer is not at all clear. At present it seems that we have to adopt a trial and error approach to finding the best resonstruction in any given situation. Theoretically, the choice of coordinate system (and delay time) should not matter, but in practice the best choice appears to depend on things like the accuracy and quantity of the data, the characteristics of the true dynamics, and the objectives of the analysis. However, it will be shown below that if the delay coordinate reconstruction window size is small and the reconstruction is an embedding, then the above reconstructions are all similar in the sense that each of the reconstructions is related to the others by a linear coordinate transformation. In order to show that, we will introduce another coordinate system as the basis for comparison.

For the moment shift the reconstruction window defined by[12],

$$\underline{y}(t) = [x(t), \ldots, x(t + (d_E - 1)\tau)]$$

to

$$\underline{y}(t) = [x(t - p\tau), \ldots, x(t), \ldots, x(t + p\tau)]$$

where p satisfies $d_E = 2p + 1$. Note that the trajectory matrix Y and the covariance matrix $Y^T Y$ are invariant with respect to this shift. For a fixed p let

$$\{r_{j,p}(-p), r_{j,p}(-p + 1), \ldots, r_{j,p}(p)\}$$

be a d_E-tuple of weights and for $0 \leq j \leq 2p$ define the weighted sum

$$w_j(t) = \sum_{n=-p}^{p} r_{j,p}(n)x(t + n\tau) \qquad \text{Eq. 3.6}$$

---

[11]d_A is the topological dimension of the manifold containing A.
[12]Recall that $x(t)$ and $\underline{y}(t)$ are alternative notation for x_t and $\underline{y}_t$ respectively.

Assume $x(t)$ is analytic and in general well behaved. Then $x(t)$ has a Taylor expansion and we can write

$$w_j(t) = \sum_{n=-p}^{p} r_{j,p}(n) \left(\sum_{i=0}^{\infty} \frac{(n\tau)^i}{i!} D^i x(t) \right)$$

$$= \sum_{i=0}^{\infty} \frac{\tau^i}{i!} D^i x(t) \left(\sum_{n=-p}^{p} n^i r_{j,p}(n) \right) \qquad \text{Eq. 3.7}$$

where

$$D^i x(t) = \frac{d^i x(t)}{dt^i}$$

By choosing

$$\sum_{n=-p}^{p} n^i r_{j,p}(n) = 0 \quad \text{for} \quad i < j$$

$w_j(t)$ becomes proportional to $D^j x(t)$ to leading order in Eq. 3.7. By adding the orthogonality constraints

$$\sum_{n=-p}^{p} r_{i,p}(n) r_{j,p}(n) = \delta_{ij} \quad 0 \le i, j \le 2p \qquad \text{Eq. 3.8}$$

we get a unique set of functions $\{r_{j,p}(n) | 0 \le j \le 2p\}$ the jth member of which is a polynomial of degree j in n that is even (odd) as j is even (odd)[13].

Since $r_{j,p}(n)$ is even (odd) in j,

$$\sum_{n=-p}^{p} n^{j+1} r_{j,p}(n) = 0$$

and from Eq. 3.7 we get

$$w_j(t) = c_j(p) \frac{\tau_w^j}{2^j j!} D^j x(t) + O(\tau_w^{j+2}) \qquad \text{Eq. 3.9}$$

where

$$c_j(p) = p^{-j} \sum_{n=-p}^{p} n^j r_{j,p}(n)$$

and

$$\tau_w = (d_E - 1)\tau = 2p\tau$$

By construction (Eq. 3.8) the vectors $\underline{r}_j$ defined by

$$\underline{r}_j = [r_{j,p}(-p), \dots, r_{j,p}(0), \dots, r_{j,p}(p)]^T$$

[13]As $p \to \infty$, the $r_{j,p}(n)$ approach the Legendre polynomials. For p finite the $r_{j,p}(n)$ are called **discrete Legendre polynomials**. Note that $r_{2p,p}(n)$ is the standard finite difference filter for estimating the derivative $D^p x(t)$. Hence, in light of Eq. 3.6, $w_j(t)$ is not a standard finite difference estimate of $D^j x(t)$.

form an orthonormal basis in $\Re^{d_E=2p+1}$. Since by definition (Eq. 3.6)

$$w_j(t) = [\underline{y}(t)]^T \cdot \underline{r}_j$$

the $w_j(t)$ are the projections or coordinates of the $\underline{y}(t)$ in the $\{\underline{r}_j\}$ system. They are called **Legendre coordinates**. The full set of Legendre coordinates is given by

$$\underline{w}^T(t) = \underline{y}^T(t) \cdot R$$

where

$$\underline{w}^T(t) = [w_0(t), w_1(t), \ldots, w_{2p}(t)]$$

and $R : \Re^{d_E} \to \Re^{d_E}$ is the orthonormal transformation defined by

$$R = (\underline{r}_0, \ldots, \underline{r}_{d_E-1})$$

Since R is orthonormal, Legendre coordinates are a simple rotation of delay coordinates. When τ_w is small the Legendre coordinates are proportional to the derivatives of $\underline{y}(t)$, so the derivatives are a rotation and rescaling of delays. Now, to see the connection with principal components let $W = YR$ and $\Xi_W = W^T W$. Then

$$\Xi_W = R^T Y^T Y R = R^T \Xi_Y R \qquad \qquad \text{Eq. 3.10}$$

Let V diagonalize (be the eigenvectors of) Ξ_W, then

$$V^T \Xi_W V = \Sigma_W^2$$

so by Eq. 3.10

$$V^T R^T \Xi_Y R V = \Sigma_W^2$$

Hence, RV is the eigenvector matrix of Ξ_Y. That is, $C = RV$ by Eq. 3.5.

Now, it can be shown[14] that if the window size τ_w is small, that is τ_w is something less than

$$\tau_w^* = 2\sqrt{\frac{3\kappa_0}{\kappa_1}} \quad \text{where} \quad \kappa_i = \langle [D^i x(t)]^2 \rangle_t \qquad \qquad \text{Eq. 3.11}$$

then[15]

$$V = \Im + O(\tau_w^2)$$

so

$$C = R + R \cdot [O(\tau_w^2)] = R + O(\tau_w^2)$$

and the eigenvectors of Ξ_Y are related to the discrete Legendre polynomials by

$$\underline{c}_j = \underline{r}_j + O(\tau_w^2)$$

The above does not imply that R diagonalizes Ξ_Y to leading order, because the

[14] J. Gibson et al., (1992).
[15] $\Im$ is the identity matrix.

correction terms to $R^T \Xi_Y R$ are of the same order or greater than some of the eigen-values of Ξ_Y. But the above does imply that the rotation from Legendre coordinates to principal components is small for small windows τ_w. In fact, the principal components $P = YC$ are nearly a Gram-Schmidt orthogonalization of the Legendre coordinates $W = YR$:

$$p_j(t) = w_j(t) - \sum_{i=0}^{j-1} p_i(t) \frac{\langle p_i w_i \rangle}{\langle p_i^2 \rangle} + O(\tau_w^{j+2})$$

While the introduction of Legendre polynomials facilitated the demonstration of the similarity between the different coordinate systems, they can also have some important computational advantages. Attractors reconstructed from principal components are similar to attractors reconstructed from Legendre coordinates when $\tau_w \ll \tau_w^*$. Since principal components have to be estimated numerically, there may be substantial error for small data sets. In contrast, the discrete Legendre polynomials do not have to be estimated and provide a ready-made set of basis vectors.

Multidimensional time series

Suppose we are able to observe not one but several (L) simultaneous measurements of a dynamical system at time t, that is

$$\underline{x}(t) = [x_1(t), \; x_2(t), \ldots, x_L(t)] = hf^t[s_0]$$

Note that the general embedding theorem does *not* require h to be single valued; it can be vector valued. Hence, we can form the ***multi-channel delay vectors***

$$
\begin{aligned}
\underline{y}'(t) &= [\underline{x}(t), \ldots, \underline{x}(t + (d_r' - 1)\tau)]^T \\
&= [hf^t[s_0], \ldots, hf^{t+(d_r'-1)\tau}[s_0]]^T \\
&= \Phi'[s_t]
\end{aligned}
$$

and the embedding theorem still holds if $d_r > 2d_A$ where $d_r = L \times d_r'$.

Reconstruction parameters

Keep in mind in what follows that d_r is the dimension of a given state space recon-struction, d_E is any value of d_r that yields an embedding, and d_e is the smallest value of d_r that yields an embedding.

Since we are in practice unlikely to be in posession of a vector time series where each multi-valued observation contains enough information to identify the state of the system[16], we will need more than one observation from the time series to reconstruct a point on the system's trajectory, and that leads directly to a reconstruction based on delay coordinates. From the theory, a delay coordinate map will yield an em-bedding for *any* sufficiently large reconstruction dimension and *any* non-zero delay time. Hence, there is a vast set of parameter values which produce embeddings. But the theory assumes a world in which there is an endless amount of clean[17]

---

[16]For example, data that can be used directly to form derivative coordinates is excluded.

[17]Clean data is free from noise and measurement error, and dirty data is contaminated by these.

data, and in that world all embeddings result in equivalent attractors. In the real world, attractors are reconstructed from a finite amount of dirty data and, as shown below, in that world different reconstruction parameters can result in very different experimental attractors even though the delay coordinate maps which produce them are all theoretically embeddings.

Some of the problems caused by contaminated data can be mitigated with noise reduction techniques covered in Chapter 6, Noise reduction. Here, we assume that we have done what we can to clean the data and now we have to do the best we can in the presence of any residual contamination. However, we must emphasize that noise reduction and state space reconstruction are not independent processes, and an iterative approach of alternating noise reduction and embedding methods can result in a better phase space reconstruction.

Our objective in reconstructing a representation of the system's attractor from observations is to obtain an experimental trajectory from which we can compute accurate estimates of the system's invariants and dynamics. Among the vast collection of parameter values which theoretically yield embeddings, there will be some that actually result in reconstructions that are close to real embeddings and some that will not. In general, the closer we are to a real embedding, the better we are able to meet our objective. That does not necessarily mean that one set of parameters is optimal for all purposes. For example, it would not be unreasonable to find that one set of parameters yielded a reconstruction that produced accurate dimension estimates, while another set of parameters yielded a reconstruction that produced dynamical models with better prediction statistics. In any case, we still need a method to identify the parameter values which produce reconstructions which are nearly embeddings. Unfortunately, there is no known computation we can make that will generate good reconstruction parameters. For example, consider the following: we need to know the dimension of the attractor to guarantee, even in theory, that a given reconstruction is an embedding, but we need the embedding to calculate the dimension of the attractor. Hence, we will be forced to adopt some iterative or trial and error scheme to obtain the best reconstruction parameters.

There are a variety of effects that arise as a result of poor state space reconstructions which can undermine our efforts to estimate a system's invariants and dynamics. Some of these effects are discussed in the next section in the context of particular algorithms, and others in the remainder of this section where we explore the relationship between geometry and 'information' in state space reconstruction, and develop a rough outline of the area in parameter space where the good reconstruction parameter values lie. We will investigate the roles that d_r and τ play in the delay coordinate reconstruction through a device introduced in an earlier example where the reconstruction process is considered in terms of the constraint each measurement puts on the location of the present state in the original state space. The device was a construct called a measurement surface which we reintroduce here for two purposes. We use it first to reveal a not-so-obvious but fundamental cause of prediction error. Namely, the failure of the delay coordinate map to achieve one-to-oneness in the experimental situation. We then go on to use the same construct to introduce the important concepts of irrelevance and redundancy which provide insight into the relationship between geometry and 'information' in state space reconstruction. Recall[18] that the measurement surface

$$S(t) = \{s : x(t) = h[s]\}$$

is the set of all system states, in the original state space, consistent with the observation $x(t)$. Consider the measurement surfaces $S(t - i\tau), i = 0, 1, \ldots, k$, corresponding to a sequence of past observations on the system. Define the **inverted measurement surfaces**

$$f^{i\tau}[S(t - i\tau)] = \{f^{i\tau}[s] : s \in S(t - i\tau)\}$$

Determinism requires that if the elements of $S(t - i\tau)$ are consistent with the measurement $x(t - i\tau)$, then $s(t)$, the present state in the original state space, must be an element of the inverted measurement surface $f^{i\tau}[S(t - i\tau)]$, that is, since $s(t - i\tau) \in S(t - i\tau)$ and $s(t) = f^{i\tau}[s(t - i\tau)]$, trivially $s(t) \in f^{i\tau}[S(t - i\tau)]$. Since that is true for each i, it follows that for any k

$$s(t) \in I(t) \equiv f^{\tau k}[S(t - k\tau)] \cap f^{\tau(k-1)}[S(t - (k - 1)\tau)] \cap \ldots \cap S(t)$$

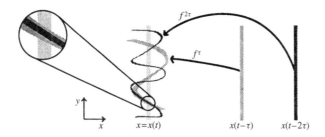

Figure 3.8 A two-dimensional illustration. Lagged measurement surfaces $S(t - \tau)$ and $S(t - 2\tau)$ mapped forward to time t to form what we have called inverted measurement surfaces. The intersection of the inverted measurement surfaces with $S(t)$ is the set $I(t)$, a magnification of which appears on the left of the diagram

Source: Adapted from Casdagli *et al.* (1991) p. 74.

Now, for any reconstructed trajectory point $\underline{y}(t) = [x(t), x(t + \tau), \ldots, x(t + (d_r - 1)\tau)]^T$ we can use the above construct (with the sign of τ reversed) to obtain a set $I(t) = G[\underline{y}(t)]$ where evidently $G[\underline{y}(t)]$ uses the observations $x(t + n\tau), n = 0, \ldots, d_{r-1}$, to locate the true state $s(t)$. Since $\underline{y}(t) = \Phi[s(t)]$, it can be seen that G is trying to reverse the process of taking measurements on the system and reconstructing its trajectory. Thus, G is a pseudo inverse of the delay coordinate map. When the delay coordinate map is an embedding it is fully invertible and the inverted measurement surfaces intersect at the point $s(t)$ in the original state space. It follows from the embedding theorems that the intersection of at most $2d_A + 1$ inverted measurement surfaces[19] will uniquely identify the original state $s(t)$. From the schematic picture

[18]See Figures 3.4 and 3.5.
[19]d_A is the dimension of the manifold containing the attractor A.

in Figure 3.8, we can see that the reconstruction dimension d_r controls the number of intersecting inverted measurement surfaces while the time delay τ influences the angle at which the inverted measurement surfaces intersect. Since $I(t)$ consists of the single point $s(t)$ when the reconstruction is an embedding, the intersection of the inverted measurement surfaces is infinitely sharp and it does not matter what value of $\tau > 0$ is used. But the embedding theorems assume infinitely accurate data. Inaccuracies in the data cause the inverted measurement surfaces to become fat or fuzzy and their intersection is not sharp, especially when the inverted measurement surfaces are nearly parallel at their point of intersection. So even if the delay coordinate map is in theory an embedding, it may in fact not even be one-to-one and the resulting experimental trajectory will be an innaccurate representation of the true dynamics[20]. To see why, suppose that we reconstruct a trajectory $\underline{y}$ with a delay coordinate map Φ from a measurement time series x, and then, use the experimental trajectory $\underline{y}$ to construct a function φ to model the system's dynamics.

When the corresponding sets $I(t)$ are small and well localized φ will be an accurate description of the system's dynamics, but the accuracy of φ can deteriorate substantially if any of the $I(t)$ have a significant spatial extent. The reason is that, in practice, φ is estimated locally from some 'average' of the behavior of a number of trajectory segments passing through a small volume in the reconstructed state space near $\underline{y}(t)$. That makes sense only if those trajectory segments reflect similar dynamics. But if $I(t)$ does have significant spatial extent, then the trajectories of points in $I(t)$ do not necessarily have similar dynamics, and by construction these will all get mapped by Φ to the same small region around $\underline{y}(t)$ in the reconstructed state space. The result is, of course, that no deterministic model will be able to explain or predict the behavior of the system in the viscinity of $\underline{y}(t)$. Thus when the data is dirty, the choice of the delay time τ can be critical to modeling the dynamics of the system or estimating its invariants. A good choice of the delay time will reduce the size of the set $I(t)$ of intersecting inverted measurement surfaces and improve the accuracy of the experimental trajectory.

When the delay time τ is small, successive measurements of the time series $x(t)$ have measurement surfaces of past observations $S(t - i\tau)$ which are nearly vertical near the true state when mapped forward in time. That is, $f^{i\tau}$ is nearly an identity, and the sets $f^{i\tau}[S(t - i\tau)]$ are nearly parallel near $s(t)$ as depicted in insert (b) in Figure 3.9. The interval of time characteristic of this behavior is called the **redundancy time** τ_R. It can be shown (Casdagli *et al.*, 1991) that for a given level η of error in the data, and d_M the dimension of the original state space, the redundancy time τ_R is approximated by

$$\tau_R \approx (\eta^2 d_E^{-1})^{\frac{1}{2(d_M - 1)}} \qquad \textbf{Eq. 3.12}$$

On the other hand, for τ large and $i > 0$, $f^{i\tau}$ will compress the points in $S(t - i\tau)$ along the stable manifold and stretch them along the unstable manifold[21]. The result is that in $I(t)$, the intersection of the inverted measurement surfaces, the location of $s(t)$ is well defined along the stable manifold, but ill-defined along the unstable manifold. Schematically, for large τ and $i > 0$, the inverted measurement surfaces

---

[20] See Chapter 6, Noise reduction.

[21] See Chapters 1, Dynamical systems and 5, Lyapunov exponents.

$f^{i\tau}[S(t-i\tau)]$ are again nearly parallel near $s(t)$ as depicted in insert (a) of Figure 3.9. Since the size of the intersecting set $I(t)$ is not reduced along the unstable direction by adding more of these inverted measurement surfaces when τ is large, measurements made in the distant past of chaotic systems become irrelevant. The interval of time characteristic of this behavior is called the **irrelevance time** τ_I. The irrelevance time τ_I is related to the uncertainty time $Ln(\varepsilon^{-1}/\lambda_1)$, that is, the time after which we lose all information about the initial state and beyond which the system is not predictable[22].

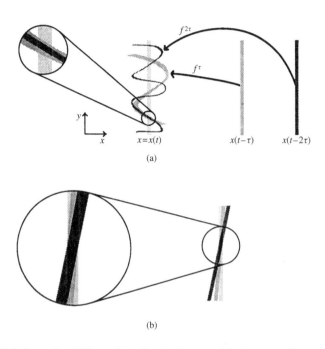

(a)

(b)

Figure 3.9 A 2-dimensional illustration of redundance and irrelevance. Images of measurement strips $S(t - i\tau)$, transported to the same time t. (a) illustrates irrelevance; τ is large, and $f^{i\tau}$ is highly nonlinear. The measurement strips are complicated. Strips from the distant past, with large $i\tau$, are roughly parallel along the unstable manifold near the true state s. Increasing $i\tau$ better determines the state along the stable manifold, but gives no new information about the unstable manifold. Thus at a finite level of coarse-graining imposed by noise or measurement error, measurements from the distant past are irrelevant, since the limiting factor is determination along the unstable manifold. (b) illustrates redundance: When τ is small $f^{i\tau}$ is close to the identity, and is approximately linear, so that the images of the measurement strips are nearly parallel at time t. Their intersection is delocalized, giving $I(t)$ some spatial extent.
Source: Adapted from Casdagli et al., (1991) p. 74.

[22] Although τ_I is positively correlated with the uncertainty time, the exact relationship is a complicated function of the dynamics. λ_1 is the largest Lyapunov exponent. See Chapters 2, Entropy and 5, Lyapunov exponents.

Window size

The total amount of information used to reconstruct each trajectory point is defined by the ***reconstruction window*** $\tau_w = (d_r - 1)\tau$. As will be seen below, given that the reconstruction dimension is theoretically sufficient to yield an embedding, the most important aspect of the reconstruction is the ***embedding window*** $\tau_w = (d_E - 1)\tau$.

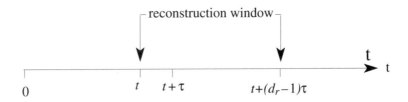

Figure 3.10 Reconstruction window.

If the embedding window size $\tau_w < \tau_R$, then all the information used to define a trajectory point is obtained from measurements made within the redundancy time of the system, and the components of each reconstructed trajectory point are nearly equal. Consequently, the experimental trajectory points $y(t)$ congregate around the main diagonal of the reconstructed phase space and the geometry of the experimental attractor is lost. If the embedding window is of size $\tau_w > \tau_I$, then the initial and final measurements used to define the trajectory point are separated in time by more than the irrelevance time of the system. When a system is chaotic, trajectory points separated in time by more than the irrelevance time become uncorrelated. Consequently, for a chaotic system, the trailing components of the trajectory vectors $y(t)$ start to look random relative to the initial components, and the geometry of the attractor becomes obscure.

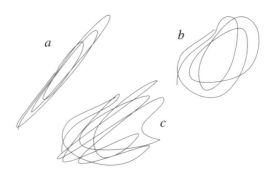

Figure 3.11 Reconstructions using different window widths. (a) $\tau_w < \tau_R$: the reconstruction lies close to the main diagonal of the state space. (b) well formed reconstruction. (c) $\tau_w > \tau_I$: the reconstruction becomes unnecessarily complicated.

When $\tau_R > \tau_I$ the system is random, in any practical sense, because we can no longer reconstruct a faithful representation of the dynamics. That follows from the fact that it is not possible to find an embedding window such that $\tau_w > \tau_R$ and $\tau_w < \tau_I$, so we can never obtain enough relevant, non-redundant measurements to resolve the independent coordinates of the embedding space. In light of Eq. 3.12 and the relationship between τ_I and the uncertainty time of the system, it would not be unreasonable to find that $\tau_R > \tau_I$ when λ_1 and d_M are large, and η is non-zero.

Pulling these comments together, our situation is hopeless if $\tau_R > \tau_I$, but if $\tau_R < \tau_I$, then an optimal window size will lie somewhere between τ_R and τ_I. We would like to relate the optimal window to τ_R and τ_I, but they are inconvenient to work with. However, we can obtain a more precise idea of the size of the optimal window by recalling from Eq. 3.9 that

$$
\begin{aligned}
w_j(t) &= \frac{c_j(p)\tau_w^j}{2^j j!} D^j x(t) + O(\tau_w^{j+2}) \\
&= Q_j(p, \tau) D^j x(t) + O(\tau_w^{j+2})
\end{aligned}
$$

where $p = (d_r - 1)/2$. Now, note that that $D^j x(t)$ is independent of p and τ, and that $Q_j(p, \tau)$ is increasing in both p and τ since $c_j(p)$ is increasing in p. Given that all else is equal, we would like the signal-to-noise ratio in each component of the reconstructed trajectory points to be as large as possible to avoid noise dominating the dynamics in that direction. In the small window regime defined by Eq. 3.11, that is when $\tau_w \ll \tau_w^*$, the signal-to-noise ratio is proportional to the square root of,

$$
\begin{aligned}
\langle w_j^2 \rangle &= Q_j^2(p, \tau)\langle [D^j x(t)]^2 \rangle_t + O(\tau_w^{2j+2}) \\
&= Q_j^2(p, \tau)\kappa_j + O(\tau_w^{2j+2})
\end{aligned}
$$

Increasing $\tau_w \to \tau_w^*$, such that p (equivalently d_r) does not decrease, increases the signal-to-noise ratio for each component while leaving $w_j(t)$ proportional to $D_j x(t)$. So in the small window regime, increasing τ_w in that way maximizes the signal-to-noise ratio without increasing the complexity of the reconstruction. As τ_w passes τ_w^* the $O(\tau_w^{2j+2})$ terms involving higher order derivatives come into play, the reconstruction components are no longer independent, and the reconstructed attractor has a more complicated geometry. Note, however, that although we get better signal-to-noise ratios as $p \to \infty$ (and $\tau \to 0$) for a fixed $\tau_w = 2p\tau$, the fact that τ cannot be less than the sampling time τ_s of the data effectively limits how large p, and therefore d_r, can be.

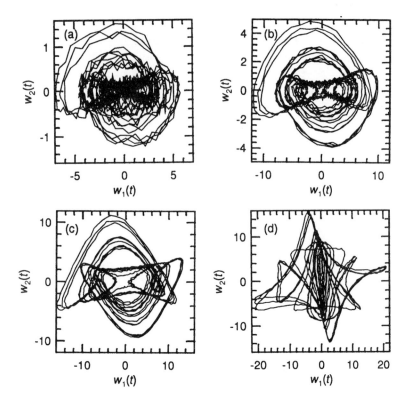

Figure 3.12 Phase portraits in Legendre coordinates for varying window widths. Time series is the $x(t)$ coordinate of the Lorenz I system with 1% additive Guassian IID noise. In all four plots, $p = 1(d_r = 3)$, and we show $w_2(t)$ versus $w_1(t)$. In (a) $\tau_w \approx 1/8\tau_w^*$ (b) $\tau_w \approx 1/4\tau_w^*$ (c) $\tau_w \approx 1/2\tau_w^*$ (d) $\tau_w = 0.64 = \tau_w^*$. As τ_w increases, signal-to-noise ratios increase and the geometry stays constant until τ_w reaches $1/2\tau_w^*$. Past this value signal-to-noise ratios are nearly constant and the geometry becomes increasingly complex.
Source: Gibson *et al.*, (1992) p. 17.

The foregoing analysis suggests choosing reconstruction parameters that result in an embedding window that lies a bit below the critical window level τ_w^*. That is, the optimal window occurs just before the geometry of the attractor starts to become complicated. Intuitively then, the optimal window size is close to τ_I, the irrelevance time of the system. That restricts the parameter space, but may still leave us with many possibilities. Which combination of d_r and τ is better depends on whether our objective is system modeling or estimation of system invariants, and that is a matter for experimentation. That is, locating the best values of d_r and τ will involve some sort of optimization over the restricted parameter space. However, before undertaking any such optimization, we would do well to ask if *any* choice of d_r and τ from the restricted parameter space would result in a deterministic experimental trajectory. By a deterministic trajectory we mean one whose vector components are not overwhelmed by noise or measurement error in the data. To illustrate, consider the principal components of a delay or Legendre coordinate state space reconstruction made from data with a noise variance (or measurement error) equal to $\langle \eta^2 \rangle$. If the qth singular value[23] of the trajectory matrix satisfies

$$\sigma_{q-1}^2 \approx \langle \eta^2 \rangle,$$

then the signal-to-noise ratio for the qth component is on the order of 1, and that coordinate is dominated by noise. Hence, the trajectory defined by the first q principal components cannot be deterministic. Conversely, if q principal components form an embedding and

$$\sigma_{q-1}^2 \gg \langle \eta^2 \rangle, \qquad\qquad \textbf{Eq. 3.13}$$

then the q-dimensional state space defined by the first q singular vectors[24] is said to be ***approximately deterministic***. If $\tau_w \ll \tau_w^*$, then we can approximate σ_{q-1}^2 with $\langle w_{q-1}^2 \rangle$ and substitute Eq. 3.9 into Eq. 3.13 which gives

$$\left(\frac{c_{q-1}(p)\tau_w^{q-1}}{2^{q-1}(q-1)!} \right)^2 \langle [D^{q-1}(t)]^2 \rangle \gg \langle \eta^2 \rangle$$

or

$$c_{q-1}(p)\tau_w^{q-1} \gg 2^{q-1}(q-1)! \sqrt{\frac{\langle \eta^2 \rangle}{\kappa_{q-1}}}$$

The above inequality determines the values of $p = (d_r - 1)/2$ and $\tau_w = 2p\tau$ which result in an approximately deterministic q-dimensional state space for a given noise level η. Now, whichever reconstruction parameters are used, they must produce an approximately deterministic d_e-dimensional state space to accommodate a faithful representation of the true trajectory so that we can reach our modeling or estimation objectives. Hence, those parameters must satisfy the following inequality.

$$c_{d_e-1}(p)\tau_w^{d_e-1} > \gamma = 2^{d_e-1}(d_e - 1)! \sqrt{\frac{\langle \eta^2 \rangle}{\kappa_{d_e-1}}} \qquad\qquad \textbf{Eq. 3.14}$$

Figure 3.13 illustrates the kind of picture we get when we plot the range of admissible values for the reconstruction parameters d_r and τ that yield an approximately deterministic d_e-dimensional state space[25].

---

[23]By convention $\sigma_0^2 \geq \sigma_1^2 \geq \ldots$ where for convenience here the origin of the index is 0. See Appendix 4, Matrix decomposition.

[24]The first singular vectors correspond to the largest singular values.

[25]In Figure 3.13 noise=1% means $\sqrt{\langle \eta^2 \rangle / \langle x^2 \rangle} = 0.01$.

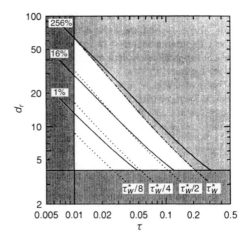

Figure 3.13 The parameter space of a (d_E, τ)-reconstruction of the attractor of the Lorenz system using its x-coordinate as the time series. Dark grey regions represent restricted parameters ($\tau < \tau_s = 0.01$ or $d_r < q = 4$). The light grey region represents parameters in the large window regime ($(d_r - 1)\tau > \tau_w^* \approx 0.63$). The white region represents the set of small-window parameters which form embeddings. Dashed lines indicate constant window widths, and solid lines indicate parameters at which the signal-to-noise ratio of $w_{3=q-1}$ is unity, for three different noise levels, calculated from Eq. 3.14.

Source: Gibson *et al.* (1992) p. 19.

From Figure 3.13 we can pick out the location of the better reconstruction parameters. The reconstruction dimension d_r is bounded below by the horizontal constraint $d_r = d_e$. The delay time τ is bounded below by the vertical constraint $\tau = \tau_s$. Combinations of d_r and τ are constrained to lie below and to the left of the diagonal line defined by the critical window τ_w^*, and to the right and above the diagonal line defining consistency with an approximately deterministic d_e-dimensional state space obtained by equating the two sides of Eq. 3.14. The vertical boundry τ_s is defined by the measurement process which gave rise to the time series data. The remaining three constraints are given in terms of d_e and τ_w^*. Algorithms for computing estimates of d_e and τ_w^* are presented in the next section.

Some final cautionary remarks before moving on to the algorithms. We have argued that, within the area defined by the above constraints, lie the good combinations of the embedding parameters d_E and τ. Signal-to-noise ratio considerations alone suggest using larger values of d_E in this region. However, the signal-to-noise ratio is not our only concern. Higher embedding dimensions increase the influence of noise and can add to the folding of the attractor, and in general add unnecessary complication to the computations. The computational effort required to estimate the system's dynamics and invariants, and the associated accumulation of roundoff errors, increase dramatically for larger values of d_E. The best choice of d_E will involve a compromise between these effects.

More importantly, there are problems with the critical window τ_w^*. First, it is defined in terms of the time derivatives of the system, which do not exist for discrete time systems. Second, the criterion for the optimality of phase space reconstructions based on the critical window, is that they are less complex. Complexity is not necessarily the right criterion for a good reconstruction. For example, the best window size for estimating Lyapunov exponents may not be the best window size for estimating fractal dimension, and still another window size may yield the best reconstruction for modeling and prediction. Since the best window for a particular objective will thus involve iteration, the critical window is a reasonable place to start the search; if it can be computed. Another reasonable starting point for the search is the characteristic cycle time τ_c, that results from the recurrence property of chaotic systems. For simple dynamics, τ_c can be estimated from the number of times the time series $x(t)$ crosses its mean value. For systems with complicated dynamics, some amount of filtering may be required before applying this technique.

ALGORITHMS

In this section we will go through some of the algorithms for deriving estimates of good reconstruction parameters and other important quantities. The present selection, from the references at the end of the chapter, contains a good sample of the techniques used to attack the problem. The algorithms are organized according to which parameter or quantity is the object of the estimation procedure. With few exceptions, the algorithms are iterative in nature and many use some measure of attractor expansion[26] to decide what constitutes a good reconstruction parameter. We assume throughout this section, unless stated otherwise, that we are using a (d_r, τ)-reconstruction, that is, that the experimental trajectory $y(n)$ has been reconstructed from a d_r-dimensional delay coordinate map with delay time τ.

First, a word on an inappropriate use of singular value decomposition, namely as a means of esimating d_e. Recall the singular value decomposition (SVD) of the trajectory matrix

$$Y = S\Sigma C^T$$

where

$$(\Sigma)_{ij} = \delta_{ij}\sigma_i; \quad \sigma_1 \geq \sigma_2 \geq \cdots \geq \sigma_{d_r} \geq 0$$

In practice, noise and non-linearity mean that all singular values are bounded away from zero. But if a particular reconstruction yielded a set of singular values that satisfied

$$\sigma_1 \geq \sigma_2 \geq \cdots \geq \sigma_k \gg \sigma_{k+1} \approx \cdots \approx \sigma_{d_E} \approx 0$$

then we might be tempted to assume $d_e = k$ was a lower bound on the embedding dimension, reasoning that since the system has almost no projection in singular directions corresponding to $\sigma_{k+1} \approx \cdots \approx \sigma_{d_E} \approx 0$, the subspace spanned by the singular vectors corresponding to the significant singular values $\sigma_1, \ldots, \sigma_k$ must contain enough degrees of freedom to accommodate the dynamics of the system. The

---

[26] Also called space filling, it is the extent of phase space volume visited by the reconstructed trajectory.

problem with using SVD in this way is that the number of significant singular values depends on the initial embedding parameters d_E and τ.

If the delay time is greater than the irrelvance time τ_I, then every component of a delay vector is uncorrelated with the corresponding component in every other delay vector. It follows that the covariance matrix $Y^T Y$ has elements approximately equal to $\delta_{ij}\sigma^2$ for large values of N_r. Hence, the number of significant singular values is equal to d_E whatever its value. Conversely, if the embedding window is less than the redundancy time τ_R, then the attractor collapses onto the main diagonal of the embedding space. It follows that there is only one significant singular value; the one that corresponds to the direction of the main diagonal of the embedding space. Even for less extreme window sizes, noise and nonlinearity will affect the size of the singular values (see Figure 3.14).

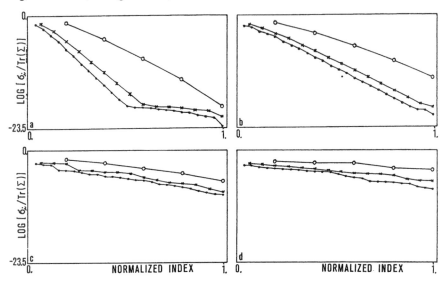

Figure 3.14 Illustration showing that the number of significant singular values increases with the size of the embedding window. The plots are the normalized singular spectra obtained from 4096 data points generated by the x component of the Lorenz I system in the chaotic regime. Embedding dimensions used were 5 (open circles), 15 (crosses), and 25 (full circles). The delay times used were panel (a) $\tau = 0.02$, panel (b) $\tau = 0.04$, panel (c) $\tau = 0.12$, and panel (d) $\tau = 0.40$.

Source: Palus and Dvorak (1992) p. 226.

Having eliminated singular value decomposition as a method of estimating embedding dimension, we proceed with more reliable algorithms.

1 Critical window[27] $\left(\tau_w^*\right)$

Recall (Eq. 3.11) the definition:

$$\tau_w^* = 2\sqrt{\frac{3\kappa_0}{\kappa_1}} = 2\left[\frac{3\langle x^2\rangle}{\langle (dx/dt)^2\rangle}\right]^{\frac{1}{2}}$$

[27]J. Gibson et al. (1992).

The dx/dt term can be computed with a discrete Legendre polynomial filter which can be obtained from any good reference book on numerical methods[28]. Keep in mind that Legendre polynomials must be applied over a window width τ_w, so their use requires iteration. Start with τ_w small and increase it until τ_w^* converges.

2 Global false near neighbors[29] (d_e)

If the reconstruction dimension d_r is too low to form an embedding, then the reconstructed state space will have insufficient degrees of freedom to allow the experimental trajectory to assume its true geometry. In that case, points that are spatially distant on the true trajectory can get projected into the same small neighborhood of the experimental trajectory. Such points are called *false near neighbors*. True near neighbors are a result of the recurrence property of the dynamics which eventually brings the system's trajectory back infinitely often to within an arbitrarily small neighborhood of any previous trajectory point (see Figure 3.15).

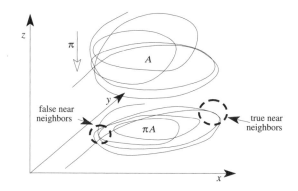

Figure 3.15 π projects A onto the x-y plane creating false near neighbors.

The idea behind the procedure described below is that, as the reconstruction dimension d_r is increased, false near neighbors will separate, but true near neighbors will not. When d_r reaches the minimum embedding dimension d_e, small neighborhoods should contain only true near neighbors. To the extent that real data sets are finite and dirty, the last statement is only approximately true, and some near neighbors will continue to be of the false variety.

Assume a fixed delay time τ. For a given reconstruction dimension d_r, define the

---

[28]See General references.

[29]Kennel, Brown and Abarbanel (1992).

squared distance between the experimental trajectory point $\underline{y}(n) \in \mathfrak{R}^{d_r}$ and its kth nearest neighbor $\underline{y}(n_k)$[30] to be

$$R_{d_r}^2(n, k) = \|\underline{y}(n) - \underline{y}(n_k)\|^2 = \sum_{m=0}^{d_r-1} [x(n + m\tau/\tau_s) - x(n_k + m\tau/\tau_s)]^2$$

and note that

$$R_{d_r+1}^2(n, k) = R_{d_r}^2(n, k) + [x(n + d_r\tau/\tau_s) - x(n_k + d_r\tau/\tau_s)]^2$$

If $y(n_k)$ is a false neighbor, then the change in its distance to $y(n)$, in going from dimension d_r to dimension $d_r + 1$, should be relatively large. Define the relative change in that distance by

$$
\begin{aligned}
\Delta_{d_r}(n, k) &= \left[\frac{R_{d_r+1}^2(n, k) - R_{d_r}^2(n, k)}{R_{d_r}^2(n, k)} \right]^{\frac{1}{2}} \\
&= \frac{|x(n + d_r\tau/\tau_s) - x(n_k + d_r\tau/\tau_s)|}{R_{d_r}(n, k)}
\end{aligned}
$$

Now, for some sensible choice of cutoff level $R_\varepsilon > 0$, most false near neighbors of $y(n)$ will satisfy the condition

$$\Delta_{d_r}(n, k) > R_\varepsilon$$

The smaller the neighborhoods analyzed, the more reliable the test for that condition. The smallest neighborhood that can be analyzed is the one containing only the nearest neighbor. Hence, it is sufficient to consider only $\Delta_{d_r}(n, k = 1)$. Accordingly, we can simplify the notation by defining

$$
\begin{aligned}
R_{d_r}(n) &\equiv R_{d_r}(n, k = 1) \\
\Delta_{d_r}(n) &\equiv \Delta_{d_r}(n, k = 1)
\end{aligned}
$$

Let $P(d_r)$ be the proportion of points on the d_r-dimensional experimental trajectory which have a false near neighbor, that is

$$P(d_r) = \frac{1}{N_r} \sum_{n=1}^{N_r} \Theta(\Delta_{d_r}(n) - R_\varepsilon)$$

where Θ is the Heaviside function. $P(d_r)$ is one measure of the adequacy of the reconstruction dimension d_r. Specifically, as d_r increases to reach the minimum embedding dimension d_e, $P(d_r)$ should saturate at some small value. So our procedure is to choose a tolerance $\delta \geq 0$ and then increase d_r until $P(d_r) \leq \delta$. The smallest d_r which satisfies that condition is then identified with d_e.

Problems

- For a finite noisy data set, while the distance to the nearest neighbor in dimension d_r increases when measured in dimension $d_r + 1$, the relative increase may remain

---

[30] See Appendix 2, Nearest neighbor searches.

small. So a procedure that says increase d_r until $P(d_r) \le \delta$ may have a finite stopping time even when the data set is random. Even with a low noise finite data set, $\Delta_{d_r}(n)$ may not be able to discriminate a false near neighbor if the near neighbor is not actually close. One way to handle that problem is to define 'close' as some fraction A_ε of the attractor size $d(A)$[31], and call a near neighbor false if either the previous condition is satisfied, that is

$$\Delta_{d_r}(n) > R_\varepsilon \qquad\qquad \text{Eq. 3.15}$$

or

$$\frac{R_{d_r+1}(n)}{d(A)} > A_\varepsilon \qquad\qquad \text{Eq. 3.16}$$

Using both criteria

$$P(d_r) = 1 - \frac{1}{N_r} \sum_{n=1}^{N_r} \Theta(R_\varepsilon - \Delta_{d_r}(n)) \cdot \Theta\left(A_\varepsilon - \frac{R_{d_r+1}(n)}{d(A)}\right)$$

That is the recommended definition of $P(d_r)$ and is the one used in the examples below.

- Because of the characteristic stretching and folding of chaotic flows[32], even small neighborhoods can undergo severe stretching. Thus, there may be neighborhoods where the local Lyapunov exponents[33] are so large that even a true near neighbor looks false. In that case, $P(d_r)$ may be significantly different from zero even if the reconstruction is an embedding.
- The tolerance level δ requires some judgement. The minimum of $P(d_r)$ may not be zero, and can be significantly larger than zero, for finite dirty data sets even when they come from observing a deterministic system. In some cases $P(d_r)$ may not be monotonically decreasing in d_r. In practice it makes sense to plot $P(d_r)$ versus d_r and then pick d_e to correspond to the smallest d_r that either gives a minimum in $P(d_r)$ or makes $P(d_r) < \delta$.

Parameter values
- Start with $d_r = 1$.
- The method appears to be robust to the precise value of the delay time τ. For the purposes of the algorithm, τ can be chosen to be something like the first minimum of the mutual information or some fraction of the correlation time, both of which are discussed later in this section. The first minimum of the mutual information is used in the examples below.
- $d(A)$ is just some measure of the spatial extent of the attractor A and could be defined as something like

$$d(A) = \max[x(n)] - \min[x(n)]$$

or

$$d(A) = \left(\frac{1}{N_d} \sum_n [x(n) - \langle x(i)\rangle_i]^2\right)^{1/2}$$

---

[31] See parameters below.
[32] See Chapter 2, Entropy.
[33] See Chapter 5, Lyapunov exponents.

Note that the tolerance level A_ε will depend on the measure $d(A)$ we choose. In the examples below $d(A)$ is the second of the two measures given above, that is, the standard deviation of the $x(n)$.

- The method also appears to be robust to the precise values of both the tolerance levels A_ε and R_ε. For the systems analyzed in the examples below $A_\varepsilon = 2.0$ and $R_\varepsilon = 15$ seem to work well.
- $\delta = 0.01$ is a reasonable level for experimental data.

Example 3.1
This example demonstrates the effectiveness of the algorithm on clean chaotic data and on uniformly distributed pseudo random data. The clean chaotic data was generated using the x coordinate of the flow of the Lorenz II equations

$$
\begin{aligned}
dx/dt &= -y^2 - z^2 - a(x - F) \\
dy/dt &= xy - bxz - y + G \\
dz/dt &= bxy + xz - z
\end{aligned}
$$

The parameter values were set to $a = 0.25$, $b = 4.0$, $F = 8.0$, and $G = 1.0$, where chaotic behavior is encountered. The fractal dimension[34] of this system's attractor is something slightly greater than 2.5.

Figure 3.16 and Figure 3.17 clearly show the difference in the behavior of $P(d_r)$ for chaotic and random systems, and why it is important to use both Eq. 3.15 and Eq. 3.16 to evaluate $P(d_r)$.

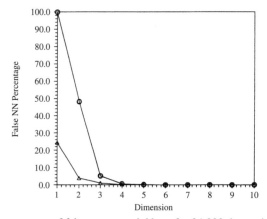

Figure 3.16 The percentage of false nearest neighbors for 24,000 data points from the Lorenz II equations. The data were sampled at $\tau_s = 0.05$. The time delay used for the delay coordinate map was $\tau = 11\tau_s = 0.55$, which is the first minimum of the average mutual information for this system. Three different criteria are compared for detecting false near neighbors. First is the change in distance to the nearest neighbors in dimension d_r when the component $x(n + d_r\tau/\tau_s)$ is added to the vectors, Eq. 3.15. These points are marked with squares. Second is the criterion which compares R_{d_r+1} to the size of the attractor $d(A)$, Eq. 3.16. These points are marked with triangles. The third criterion applies both tests to the data and a point which fails either test is declared false. These points are marked with circles; they overlap the squares in this example.
Source: Adapted from Kennel *et al.* (1992) p. 3405.

[34] See Chapter 4, Fractal dimension.

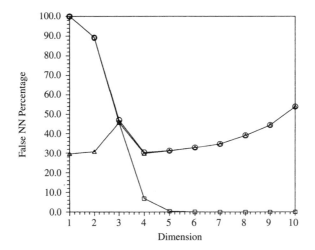

Figure 3.17 Same as Figure 3.16 but the data come from a set of pseudo random numbers uniformly distributed in the interval $[-1.0, 1.0]$. The comparison between this figure and Figure 3.16 is quite striking and points out clearly the difference between a low dimensional chaotic signal and a pseudo random signal (a high dimensional chaotic signal).

Source: Adapted from Kennel *et al.* (1992) p. 3406.

Figures 3.18 through 3.21 show the sensitivity of $P(d_r)$ to R_ε and N_r for the chaotic and pseudo random data at two different values of d_r. We see from these figures that the behavior of $P(d_r)$, illustrated in figures 3.16 and 3.17, rapidly becomes independent of the number of trajectory points N_r, and the tolerance levels R_ε and A_ε.

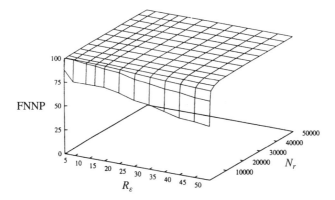

Figure 3.18 Data from the Lorenz II system in $d_r = 1$. The false-near-neighbor percentage (FNNP) is evaluated as a function of R_ε and the length of the data set N_r. The delay time for the reconstruction was $\tau = 0.85$. $d_r = 1$ is clearly not a good choice for embedding this system.

Source: Kennel *et al.*, (1992) p. 3407.

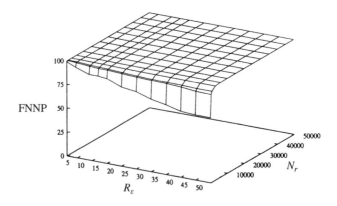

Figure 3.19 Data from a uniform noise distribution on $[-1.0, 1.0]$ in $d_r = 1$. The false-near-neighbor percentage (FNNP) is evaluated as a function of R_ε and the length of the data set N_r.
Source: Kennel *et al.*, (1992) p. 3407.

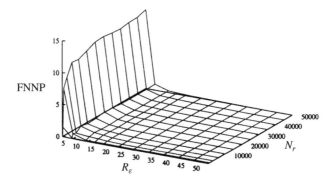

Figure 3.20 Data from the Lorenz II system in $d_r = 5$. The false-near-neighbor percentage (FNNP) is evaluated as a function of R_ε and the length of the data set N_r. The delay time for the reconstruction was $\tau = 0.85$. $d_r = 5$ is clearly a good choice for embedding this system.
Source: Kennel *et al.* (1992) p. 3407.

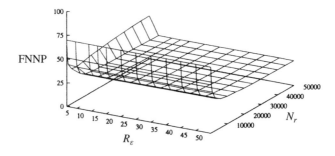

Figure 3.21 Data from a uniform noise distribution on $[-1.0, 1.0]$ in $d_r = 8$. The false-near-neighbor percentage (FNNP) is evaluated as a function of R_ε and the length of the data set N_r.
Source: Kennel *et al.*, (1992) p. 3407.

Example 3.2

This example demonstrates the effectiveness of the algorithm in the presence of additive noise. More precisely random numbers uniformly distributed on $[-L, L]$ were added to data generated from the x coordinate of the flow of the Lorenz I equations

$$\begin{aligned}
dx/dt &= \sigma(y - x) \\
xy/dt &= xz + rx - y \\
dz/dx &= xy - bz
\end{aligned}$$

The parameter values were set to $\sigma = 16.0$, $b = 4.0$, and $r = 45.92$, where chaotic dynamics are encountered. The fractal dimension of this system's attractor is about 2.09.

From Figure 3.22 we can see that the false nearest neighbour approach is quite robust against noise contamination. For example, with $L/d(A) = 10\%$, $P(d_r)$ is only 0.27 when $d_r = 4$. Even with $L/d(A) = 50\%$, $P(d_r)$ is only 2.5 when $d_r = 4$.

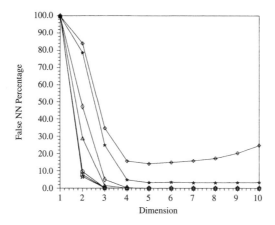

Figure 3.22 Data from the Lorenz I model which have been contaminated with uniformly distributed noise in the interval $[-L, L]$. 41,000 points from the attactor are used here. The percentage of false nearest neighbors is shown for $L/d(A) = 0.0, 0.005, 0.01, 0.05, 0.1, 0.5$, and 1.0. For this system $d(A) \approx 12.7$. The result for $L/d(A) = 0$ is marked with an open star; for $L/d(A) = 0.005$, it is marked with a square; for $L/d(A) = 0.01$, it is marked with a circle; for $L/d(A) = 0.05$, it is marked with a triangle; for $L/d(A) = 0.1$, it is marked with a diamond; for $L/d(A) = 0.5$, it is marked with a filled star; and for $L/d(A) = 1.0$, it is marked with an open cross.

Source: Kennel *et al.* (1992) p. 3409.

3 Mutual information and autocorrelation[35] (τ)

Recall[36] the concept of a measurement strip $S(t - \tau)$ corresponding to an observable $x(t - \tau)$, and $f^\tau[S(t - \tau)]$, the image of it transported to time t under the dynamics f. A measurement strip localizes the state of the system in the original state space.

---

[35] Fraser and Swinney (1986).
[36] See Figures 3.4 and 3.5.

The intersection of the image $f^\tau[S(t-\tau)]$ with the measurement strip $S(t)$ gives a more precise localization of the state of the system in the original state space at time t. It is easy to see that the intersection, and therefore the localization, is sharpest when the image of the measurement strip from $t-\tau$ is both narrow and orthogonal to the measurement strip at t.

The idea can be demonstrated quite simply in two dimensions. Thus, we let $s(t) = [x(t), y(t)]$, $h[s(t)] = x(t)$, and $y(t) = [x(t), x(t-\tau)]$. In this setting, a measurement surface has the simple form $S(t) = \{(x,y)|x = x(t)\}$, that is, a set of points along a vertical line in the (x,y) plane (see Figure 3.23). For large values of τ, the image $f^\tau[S(t-\tau)]$ looks like the full attractor, but for small values of τ, the image tends to be simply a rotation of the vertical measurement strip $S(t-\tau)$. Since the image $f^\tau[S(t-\tau)]$ looks like the whole attractor for large τ, it tells us nothing about the state of the system at time t except that it lies somewhere on the attractor. However, for small values of τ, as the image gets rotated toward the horizontal, we gain more and more information about $y(t)$, and less about $x(t)$. Since that image locates $y(t)$ well, and obviously $S(t)$ locates $x(t)$ well, their intersection locates $s(t) = [x(t), y(t)]$ well.

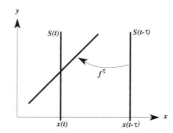

Figure 3.23 For small values of τ, the image $f^\tau[S(t-\tau)]$ tends to be a simple rotation. The intersection of the image with $S(t)$ contains $s(t)$ and is sharpest when the image is rotated to the horizontal, which tends to be when $x(t-\tau)$ is independent of $x(t)$.

The image $f^\tau[S(t-\tau)]$ tends to be horizontal to $S(t)$ when $x(t)$ is independent of $x(t-\tau)$, and that is precisely when the mutual information[37] among the two is lowest. The above is basically the motivation for using the minimum of the mutual information statistic for choosing a time delay. The same reasoning is employed to justify the use of the autocorrelation time[38] or a zero of the autocorrelation function of the time series $x(t)$.

Problems

- The above argument for mutual information is based on a two-dimensional reconstruction and we would have to use a generalization of mutual information[39] among more than two variables to get a comparable time delay for higher dimensional reconstructions.

[37]See Chapter 2, Entropy (theory and algorithms) with Q and S systems replaced by the delay coordinate sets $\{x(t)\}$ and $\{x(t-\tau)\}$, and $Q \otimes S$ replaced by $\{[x(t), x(t-\tau)]\}$.

[38]Time required for the autocorrelation function to reach $1/e$ of its original value, where e is the natural number $2.71\ldots$

[39]The generalization of mutual information is called redundancy. See Chapter 2, Entropy.

- Mutual information may have no minimum or several minimums. If there is no minimum the procedure does not work. The reason for choosing the first minimum is that the longer the time span, the more the image $f^\tau[S(t - \tau)]$ resembles the whole attractor, and the less information it contains about the state at time t. Moreover, large values of τ risk pushing the trailing components of the reconstructed trajectory points past the irrelevance time of the system.

Example 3.3

Figure 3.24 shows the rotation and spreading of the measurement strip image $f^\tau[S(t - \tau)]$ for various values of τ in a two-dimensional projection of the Rössler attractor.

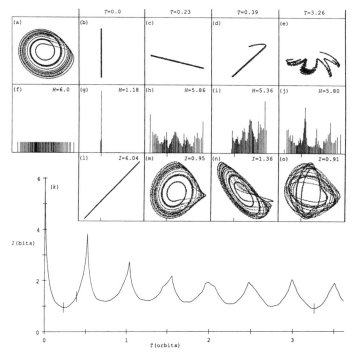

Figure 3.24 Rotation and spreading in the Rössler attractor. The entire attractor projected onto the original x, y coordinates is shown in (a), and (f) shows a histogram of the x measurements for this attractor for 64 equiprobable bins. The measurements corresponding to bin 20 are shown in (b), and the corresponding histogram is (g); figures (c), (d), and (e) show these points at later times T, and the corresponding histograms are (h), (i), and (j), respectively. (Histograms (g)–(j) are given by the probability densities of the original 64 equiprobable bins.) For short times the points are essentially only rotated, as (c) illustrates, while for long times there is spreading as well as rotation, as (e) illustrates. Phase portraits constructed by the time delay method for the delays $\tau = T$ in (b)–(e) are shown in (l)–(o), respectively, where the tick marks indicate bin 20. The mutual information (calculated over 65 536 points) is shown as a function of T in (k). Clearly, the best reconstruction is the one in panel (m) using a delay time of $\tau = 0.23$ which corresponds to the first minimum of the mutual information shown in panel (k).

Source: Fraser and Swinney (1986) p. 1136.

103

Example 3.4

Figure 3.25 shows a comparison of the phase portraits of a reconstructed chaotic Roux[40] attractor first using the first zero of the autocorrelation function, and then using the first minimum of the mutual information, as the delay time of a delay coordinate map. The trajectories in the left hand phase portrait are crushed together and impossible to discern, while those in the right hand phase portrait are spread out and fairly easy to follow.

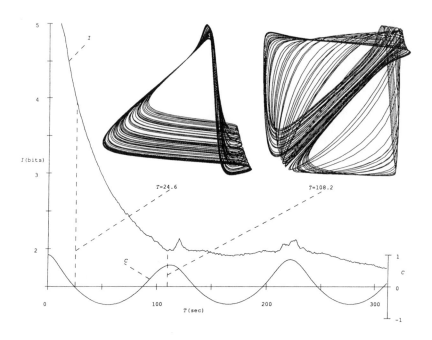

Figure 3.25 Phase portraits of the Roux attractor in the Belousov–Zhabotinskii chemical reaction. The dependence of the mutual information I and the autocorrelation function C on T are shown for calculations over 32 768 points. The coordinates used in constructing the phase portrait on the left are linearly independent ($\tau = 24.6$ = first zero of the autocorrelation), while the coordinates used in the phase portrait on the right are more generally independent ($\tau = 108.2$ = first minimum of the mutual information).

Source: Fraser and Swinney (1986) p. 1135.

4 Singular value fraction[41] (τ given d_E)

Suppose we reconstruct a trajectory in $\Re^{d_E}$ from observations using a delay time τ. From the singular value decomposition[42] of the trajectory matrix Y we get the singular values

$$\sigma_1(\tau) \geq \sigma_2(\tau) \geq \cdots \geq \sigma_{d_E}(\tau) \geq 0 \qquad \textbf{Eq. 3.17}$$

[40]Balousov–Zhabotinskii chemical reaction. Roux *et al.* (1983).
[41]Kember and Fowler (1993).
[42]See Appendix 4, Matrix decomposition.

The singular values measure the average amount that the experimental trajectory (attractor) probes $\Re^{d_E}$ in the direction of the corresponding right singular vectors. If the embedded attractor is crumpled in some directions relative to others it is going to be more difficult to extract the dynamics of the system. So intuitively it makes sense to find an embedding which spreads the attractor out as evenly as possible. Ideally that means $\sigma_1(\tau) = \sigma_2(\tau) = \cdots = \sigma_{d_E}(\tau)$.

Assume that the observations $x(i)$ are normalized to have zero mean and unit variance. Also assume that $N_d \gg 1$ so that, regardless of the value of τ, we can set[43]

$$\sum_1^{d_E} \sigma_i^2(\tau) = d_E \qquad \text{Eq. 3.18}$$

Consider

$$F(k, \tau) = \frac{\sum_1^k \sigma_i^2(\tau)}{d_E} \qquad \text{Eq. 3.19}$$

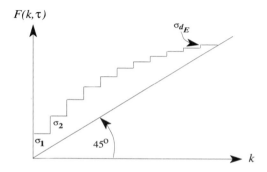

Figure 3.26 Eq. 3.17 and Eq. 3.19 imply $F(k, \tau)$ is monotonically increasing and concave in k. If $F(k, \tau)$ touches the 45° line for any $k < d_E$, then Eq. 3.17 implies that it touches the line for every $k \leq d_E$.

By inspection (see Figure 3.26)

$$\min_\tau F(k, \tau) = k/d_E \qquad k = 1, 2, \ldots, d_E \qquad \text{Eq. 3.20}$$

Furthermore, if for some $k = k_0 < d_E$, and $\tau = \Delta_m$

$$F(k_0, \Delta_m) = k_0/d_E$$

then

$$F(k, \Delta_m) = k/d_E \qquad k = 1, 2, \ldots, d_E$$

and

$$\sigma_1(\Delta_m) = \sigma_2(\Delta_m) = \ldots = \sigma_{d_E}(\Delta_m) = 1, \qquad \text{Eq. 3.21}$$

[43] The assumptions imply $Trc[C^T Y^T Y C] = Trc[\Sigma^2] = \Sigma_1^{d_E} \sigma_i^2(\tau)$, and $Y^T Y \approx$ autocorrelation matrix whose trace is d_E.

that is, the same value Δ_m will minimize $F(k, \tau)$ for every k, and all the singular values are equal to one. Hence, finding a value of τ such that $F(k, \tau) = k/d_E$ for some value of $k < d_E$, is equivalent to finding a value of τ that spreads the attractor uniformly in $\mathfrak{R}^{d_E}$.

Evidently,

$$\frac{k}{d_E} \le F(k, \tau) \le 1$$

On rearranging we get

$$0 \le d_E F(k, \tau) - k \le d_E - k$$

For convenience we define the **singular value fraction (SVF)**

$$f_{sv}(k, \tau) \equiv \frac{d_E F_{sv}(k, \tau) - k}{d_E - k} = \frac{\Sigma_1^k \sigma_i^2(\tau) - k}{d_E - k}$$

which takes on values in the interval [0,1]. Clearly, $f_{sv}(k, \tau)$ takes on its minimum value 0 for the same value of τ that minimizes $F(k, t)$, and since it does not matter what value of $k < d_E$ we use, choosing a delay time τ which minimizes $f_{sv}(\tau) \equiv f_{sv}(k = 1, \tau)$ is equivalent to maximizing the spread of the attractor for a given embedding dimension d_E. In general, $f_{sv}(\tau)$ will not have a unique minimum. In practice, it may not acheive its absolute minimum or even have any minimum at all. $f_{sv}(\tau)$ not obtaining its absolute minimum is not fatal since just minimizing it will yield the most uniformly spread attactor we can get with the given data. If $f_{sv}(\tau)$ does have at least one minimum, then experimental evidence suggests choosing the delay time to be Δ_m, the value of τ corresponding to the first local minimum of $f_{sv}(\tau)$. If $f_{sv}(\tau)$ has no minimum, then the procedure fails.

Parameter values
None.

Example 3.5

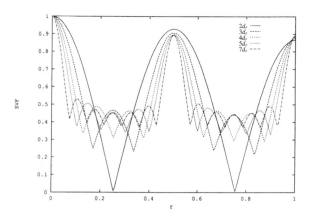

Figure 3.27 $f_{sv}(\tau)$ for the Rössler equations using a time series of the y component and reconstruction dimensions $d_r = 2, 3, 4, 5,$ and 7.

Source: Kember and Fowler (1993) p. 75.

Problems

The method does not work if the attractor has more than one unstable equilibrium point. If $x(t)$ oscillates between several foci c_i it can usually be seen easily in a histogram of the time series. For two symmetric foci, using $x^2(t)$ instead of $x(t)$ as the time series can rehabilitate the algorithm. In the more general situation the problem can sometimes be fixed, for the purposes of the algorithm, by transforming $x(t)$ to $u(t) = \prod[x(t) - c_i]$.

Example 3.6

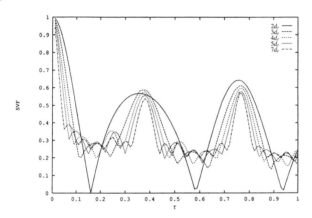

Figure 3.28 $f_{sv}(\tau)$ for the Lorenz I equations, using x^2 as the variable.

Source: Kember and Fowler (1993) p. 78.

Remarks

- If for some value of $k = k_0$, there is a Δ_M such that $f_{sv}(k_0, \Delta_M) = 1$, then $F(k_0, \Delta_M) = 1$ and[44]

$$F(k_0, \Delta_M) = 1 \quad \Rightarrow \quad \sum_1^{k_0} \sigma_i^2 = d_E$$

$$\Rightarrow \quad \sum_{k_0+1}^{d_E} \sigma_i^2 = 0$$

$$\Rightarrow \quad \sigma_i^2 = 0 \quad \text{for} \quad i > k_0$$

In that case, the reconstructed attractor $\Phi[A]$ lies in a subspace $\mathfrak{R}^{k_0}$. Specifically, for $k_0 = 1$, $F(1, \Delta_M) = 1$ implies $\sigma_1^2(\Delta_M) = d_E$. That implies that all relevent information is captured in the first component of the reconstructed trajectory vector. Hence, Δ_M must be close to the irrelevance time τ_I of the system, and as such should be independent of the reconstruction. Although $f_{sv}(\Delta_M)$ may not actually reach the value of 1 in practice, the plots in Figures 3.27 and 3.28 suggest that Δ_M is located at the first *major* maximum of $f_{sv}(\tau > 0)$ which, as expected, occurs after the minimum corresponding to Δ_m.

[44]Noisy data and nonlinear dynamics imply that no singular value is identically zero, so the relationship is only approximate.

- Note from Tables 3.1 and 3.2 that

$$\Delta_m \approx \frac{\Delta_M}{d_E} \qquad\qquad \textbf{Eq. 3.22}$$

so that

$$\tau_w \equiv (d_E - 1)\Delta_m \approx \frac{d_E - 1}{d_E}\Delta_M \approx \Delta_M$$

Hence, the optimal window, using Δ_m, is a bit below the irrelevance time of the system. In fact, for d_E large, the embedding window for the maximally spread attractor converges to the irrelevance time of the system. Based on the discussion at the end of the theoretical section, that would put Δ_m in the optimal region of Figure 3.13.

Δ_m	Δ_M	Δ_M/d_e	Δ_w	d_E
0.25	0.5	0.25	0.25	2
0.17	0.5	0.17	0.34	3
0.13	0.5	0.13	0.39	4
0.1	0.5	0.1	0.4	5
0.07	0.5	0.07	0.42	7

Table 3.1 Delay times corresponding to Rössler data in Figure 3.27.
Source: Kember and Fowler (1993) p. 75.

Δ_m	Δ_M	Δ_M/d_E	Δ_w	d_E
0.16	0.36	0.18	0.18	2
0.12	0.38	0.13	0.26	3
0.09	0.38	0.095	0.29	4
0.07	0.38	0.076	0.30	5
0.05	0.38	0.054	0.32	7

Table 3.2 Delay times corresponding to Lorenz data in Figure 3.28.
Source: Kember and Fowler (1993) p. 77.

- The plot in Figure 3.29 gives the prediction error[45] as a function of the reconstruction dimension and delay time. If predictor error is used to optimize τ, then we might expect its optimal value τ_p to be close to Δ_m. In fact, that is not the case. From the plot we can see

$$d_E \tau_p \approx 0.2$$

---

[45]The mean absolute error of a linear predictor, $\underline{y}(t + T) = b + A\underline{y}(t)$, estimated from the forward images of $d_E + 1$ nearest neighbors.

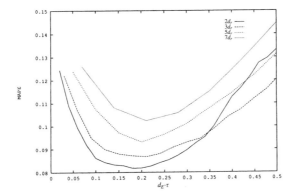

Figure 3.29 Mean average prediction error (MAPE) for the Lorenz I equations as a function of $d_E \tau$. There is a weak minimum near $d_E \tau \approx 0.2$.

Source: Kember and Fowler (1993) p. 78.

From Table 3.2 for the same system, we see that, independent of d_E,

$$\Delta_M \approx 0.38$$

Hence,

$$d_E \tau_p \approx 0.53 \Delta_M$$

Finally, from Eq. 3.22

$$\tau_p \approx 0.53 \Delta_m$$

Hence, τ_p and Δ_m are not equal, and that means that a uniformly spread attractor is not necessarily optimal for prediction.

- Nonetheless, as expected, and the plot in Figure 3.30 illustrates, predictability is lost when $\tau \approx \Delta_M$, which is about 0.5 for the Rössler system (see Table 3.1).

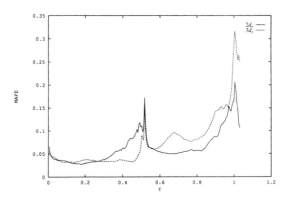

Figure 3.30 Mean average prediction error (MAPE) for the Rössler equations as a function of τ.

Source: Kember and Fowler (1993) p. 79.

5 Average displacement[46] (τ given d_E)

The problem with approaches to optimizing the delay time τ by maximizing some measure of the space filling of the reconstruction is that such a measure may not have a maximum. The algorithm discussed below adopts a heuristic approach to developing a rule for choosing delay times which produce estimates of system invariants that are invariant to moderate changes in the chosen delay time.

If τ is very small, then the experimental trajectory will lie close to the main diagonal of the reconstructed state space, and the relationship between the trajectory points is masked by noise and measurement error. Clearly, we cannot expect to obtain reliable estimates of the system's invariants from such a reconstruction. As τ is increased, the trajectory begins to unfold from the main diagonal into the surrounding reconstructed state space. The unfolding of the trajectory is accompanied by an increase in the signal-to-noise ratio of the components of the trajectory points, and the relationship between the points becomes clearer. We can measure the degree of unfolding as follows. Let $\{y(n; d_r, \tau) : n = 1, 2, \ldots, N_r\}$ denote an experimental trajectory reconstructed in d_r dimensions with a time delay equal to τ. The degenerate trajectory $\{\underline{y}(n; d_r, 0)\}$ lies squarely on the main diagonal of $\Re^{d_r}$. Hence,

$$
\begin{aligned}
S(d_r, \tau) &= \frac{1}{N_r} \sum_{i=1}^{N_r} \| \underline{y}(i; d_r, \tau) - \underline{y}(i; d_r, 0) \| \\
&= \frac{1}{N_r} \sum_{i=1}^{N_r} \left[\sum_{j=0}^{d_r-1} [x(i + j\tau/\tau_s) - x(i)]^2 \right]^{\frac{1}{2}}
\end{aligned}
$$

measures how much $\{y(n; d_r, \tau)\}$ has moved away from the main diagonal. $S(d_r, \tau)$ is called the **average displacement**. The behavior of $S(d_r, \tau)$ is illustrated for several attractors in the plots in Figures 3.31–3.34. Since the dimensions of these systems are known, we can fix the embedding dimension at $d_E = 7$, which is the criterion given in Takens's theorem for three-dimensional attractors. As can be seen from Figures 3.31–3.34, $S(d_E, \tau)$ tends to plateau for larger values of τ.

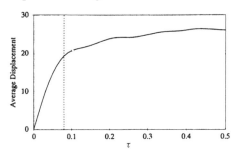

Figure 3.31 $S(d_r, \tau)$ versus τ for the Lorenz attractor with $d_r = 7$ and $N_d = 2500$. The value of τ_x (the delay corresponding to the 40% slope threshold) is distinguished using a dashed vertical line. The point corresponding to τ_w^* (the critical window width) is marked with an open circle.

Source: Rosenstein *et al.* (1994) p. 90.

[46]Rosenstein, Collins and De Luca (1994).

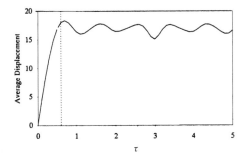

Figure 3.32 $S(d_r, \tau)$ versus τ for the Rössler attractor with $d_r = 7$ and $N_d = 2500$. The value of τ_x is distinguished using a dashed vertical line. The point corresponding to τ_w^* is marked with an open circle.

Source: Rosenstein *et al.* (1994) p. 90.

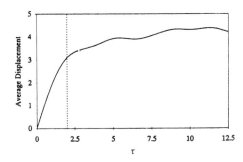

Figure 3.33 $S(d_r, \tau)$ versus τ for the three-torus with $d_r = 7$ and $N_d = 2500$. The value of τ_x is distinguished using a dashed vertical line. The point corresponding to τ_w^* is marked with an open circle.

Source: Rosenstein *et al.* (1994) p. 91.

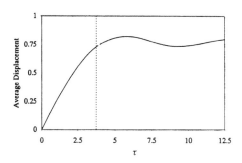

Figure 3.34 $S(d_r, \tau)$ versus τ for the Mackay–Glass system with $d_r = 7$ and $N_d = 2500$. The value of τ_x is distinguished using a dashed vertical line. The point corresponding to τ_w^* is marked with an open circle.

Source: Rosenstein *et al.* (1994) p. 90.

111

The procedure for choosing a good delay time is to compute $S(d_E, \tau)$ for a wide range of values of τ, say $0 < \tau_1, \tau_2, \ldots$, and set the delay time to that value τ_i where the *rate of change* in $S(d_E, \tau)$ has declined sufficiently from its initial value[47]. That is, the best delay time is chosen to be the value τ_x which satisfies

$$\tau_x(\alpha) = \min_{i>2}\left\{\tau_i \left| \frac{\Delta S(d_E, \tau_i)}{\Delta \tau_i} \le \alpha \frac{\Delta S(d_E, \tau_2)}{\Delta \tau_2} \right.\right\}$$

where

$$\frac{\Delta S(d_E, \tau_i)}{\Delta \tau_i} = \frac{S(d_E, \tau_i) - S(d_E, \tau_{i-1})}{\tau_i - \tau_{i-1}}$$

and $0 < \alpha < 1$ is a threshold parameter of the algorithm. Note in Figures 3.31–3.34, the proximity of the delay time corresponding to the critical window τ_w^* size, that is, $\tau_w^*/(d_E - 1)$, and the point $\tau_x(\alpha = 0.4)$ where $S(d_E, \tau)$'s slope drops to 40% of its initial value.

Parameter values
$\alpha = 0.40$ as used in the examples below.

Example 3.7
For the chaotic systems studied, the plots in Figures 3.35 and 3.36 indicate that using the 40% threshold point $\tau_x(0.4)$ as the delay time gives consistently wide plateaus for the correlation dimension D_2[48]. Moreover, at least for the Lorenz system, the estimated D_2 is robust in τ around the level of $\tau_x(0.4)$.

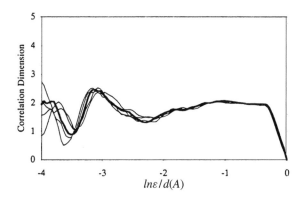

Figure 3.35 D_2 versus $\ln \varepsilon / d(A)$ for the Lorenz attractor with $d_r = 7$, $N_d = 2500$ and $\tau = 0.06, 0.07, 0.08, 0.09$ and 0.10s. (The heavy line corresponds to $\tau = \tau_x = 0.08$s. Delays of 0.06s and 0.10s coincided with slope thresholds of 60% and 20% respectively).

Source: Rosenstein *et al.* (1994) p. 91.

---

[47]Because $S(d_E, \tau)$ is in general not smooth at the origin, we only look at the rate of change in $S(d_E, \tau)$ for $\tau > 0$. Specifically, $S(d_r, 0)$ is excluded from the slope calculation.

[48]See Chapter 4, Fractal dimension.

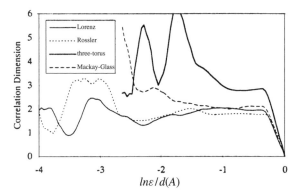

Figure 3.36 D_2 versus $\ln \varepsilon / d(A)$ for the dynamical systems of Figures 3.31 through 3.34. The respective values for τ correspond to τ_x as marked in Figures 3.31 through 3.34.

Source: Rosenstein *et al.* (1994) p. 91.

Remarks

Considering the window size

$$\tau_w = (d_E - 1)\tau_x$$

as a function of d_E, we can see from Figure 3.37 that, as expected, τ_w is approximately independent of d_E for the systems analyzed.

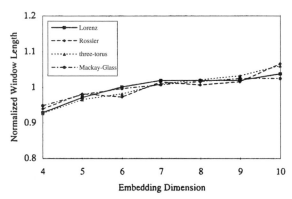

Figure 3.37 Normalized window length versus embedding dimension for the dynamical systems of Figures 3.31–3.34. In order to calculate a more precise estimate of τ_w, N_d was increased to 10,000. For each system, the values of τ_w were divided by the mean window length. Group means were as follows: (1) Lorenz attractor. $\tau_w = 0.41 \pm 0.02$s. (2) Rössler attractor, $\tau_w = 2.95 \pm 0.12$s. (3) three-torus, $\tau_w = 11.1 \pm 0.5$s, and (4) Mackay–Glass system, $\tau_w = 21.9 \pm 0.6$s.

Source: Rosenstein *et al.* (1994) p. 93.

The slight upward bias is consistent with the estimate $\tau_w \approx (1 - 1/d_E)\Delta_M$, given earlier in the remarks section of the Singular Value Fraction algorithm, where Δ_M is close to the irrelevence time τ_I.

6 Fill factor[49] $(d_E$ and $\tau)$

This algorithm, and the next, simultaneously estimate d_E and τ. This, the first algorithm, uses a measure of the space filling of the experimental attractor.

Let $\{r_j(k) : 0 \leq j \leq d_r\}$ be the kth sample of $d_r + 1$ points from the experimental trajectory $\{y(n)\}$. Assume there are N_{ref} sample sets, that is, $1 \leq k \leq N_{\text{ref}}$. For each of the sample sets we form the difference vectors $d_j(k) = r_0(k) - r_j(k)$ and the volume element

$$V_{d_r,k}(\tau) = |\det[d_1(k), \ldots, d_{d_r}(k)]|$$

where the $d_j(k)$ are treated as column vectors. Now define

$$V_{d_r}(\tau) = \frac{\langle V_{d_r,k}(\tau)\rangle_k}{[\max\{x_k\} - \min\{x_k\}]^{d_r}}$$

$V_{d_r}(\tau)$ measures the average space filling over the attractor relative to the volume of a cube with equal fixed sides containing the attractor. In general $V_{d_r}(\tau)$ is non-increasing in d_r for a fixed τ, and decreases exponentially in d_r for $d_r > d_e$. So we work with the logarithm and define the *fill factor* as

$$\Upsilon_{d_r}(\tau) = \text{Log} \, V_{d_r}(\tau)$$

As a function of τ, $\Upsilon_{d_r}(\tau)$ will increase to a maximum, and then go into some complex pattern reflecting the geometry of the attractor. For $d_r < d_e$, $\Upsilon_{d_r}(\tau)$ has very little structure in τ. For $d_r \geq d_e$, $\Upsilon_{d_E}(\tau)$ has structure in τ, and it maintains that same structure in τ as d_r increases. As a function of d_r, $\Upsilon_{d_r}(\tau)$ is non-increasing until $d_r \geq d_e$, and then it decreases linearly in d_r. An example here would greatly help clarify the above description of $\Upsilon_{d_r}(\tau)$. Hence, unusually, we will inject an example here, in the middle of the discussion, instead of the end. The equation of motion for this system is the non-autonomous[50] Duffing forced oscillator[51]

$$\frac{dy}{dt} + ay + (x^3 - x) = 6\cos(t)$$

$$\frac{dx}{dt} = y$$

The plots in Figure 3.38 clearly show the eventual linear decrease in $\Upsilon_{d_r}(\tau)$ and convergence to a stable pattern as a function of τ.

---

[49] Buzug and Pfister (1992a).

[50] This variety of non-autonomous equation can be converted to the autonomous type by converting time to an angle variable.

[51] See Holmes (1979).

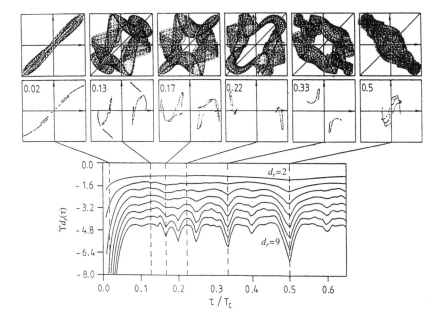

Figure 3.38 Fill factor $\Upsilon_{d_r}(\tau)$ is plotted vs delay time τ/τ_c for the Duffing attractor from $d_r = 2$ to 9. The reconstructed Duffing attractor reveals its intrinsic structure for $d_r > 3$. For six special values of τ/τ_c the reconstruction of the embedding space and the corresponding Poincaré sections are shown. Except for $\tau/\tau_c = 0.02$ and $\tau/\tau_c = 0.5$ no qualitative difference is seen.

Source: Buzug and Pfister (1992a).

Since $\Upsilon_{d_r}(\tau)$ decreases linearly for $d_r \geq d_e$, the following statistic should plateau for a reconstruction that forms an embedding

$$\Delta\Upsilon(d_r) = \frac{\tau_s}{\tau_f - \tau_i} \sum_{\tau=\tau_i}^{\tau_f} [\Upsilon_{d_r+1}(\tau) - \Upsilon_{d_r}(\tau)]$$

where τ_s is the sampling time and τ_i and τ_f, the initial and final delay times in the sum, are parameters. Hence, our procedure is to increase d_r until $\Delta\Upsilon(d_r)$ saturates, then pick τ such that $\Upsilon_{d_E}(\tau)$ is maximized for $\tau_s \leq \tau \leq 0.5\tau_c$[52]. The reason for the lower bound on τ is obvious. The upper bound on τ is supposed to keep the delay time below the irrelevance time of the system. Choosing τ to maximize $\Upsilon_{d_E}(\tau)$ is an experimentally determined heuristic.

Problems
The problem with the fill factor is that it does not work for attractors with more than one unstable focus. For these attractors $\Upsilon_{d_r}(\tau)$ seems to have no structure in τ for any value of d_r, and it is not known if a transformation similar to the one used for the Singular Value Fraction, to handle the same problem, will work here.

---

[52] τ_c is the average cycle or reccurence time of the system.

Parameter values

- Set $\tau_i = \tau_s$ and $\tau_f = \tau_c/2$.
- Set $N_{\text{ref}} = \max\,[100,\,2\%\ \text{of}\ N_r]$.

Remarks

- Note that additive noise lowers $\Upsilon_{d_r}(\tau)$, as noise tends to fill whatever space it is put into. Noise also complicates $\Upsilon_{d_r}(\tau)$, but experiments with noisy non-chaotic data indicate that neither change diminishes the effectiveness of the algorithm.

- $\Upsilon_{d_r}(\tau)$ is a blunt tool for estimating d_e.

7 Deformation integral[53] (d_E and τ)

If the reconstruction is an embedding, then from first principles, nearby points on the attractor should remain nearby points over short periods of time, separating on average at an exponential rate equal to the largest Lyapunov exponent[54].

Let $y(t_j) \in \{y(n)\}$ be a set of N_{ref} reference points along the experimental trajectory. Let $B_j = \{y(t_{j_k})\}$ be the set of $K_\varepsilon(j)$ neighbors of the reference point $y(t_j)$ contained in $B_\varepsilon[y(t_j)]$, a hypersphere with radius ε centered at $y(t_j)$. Let t_{ev} be the minimum evolution time over which we observe the system, and let $1 \le Q$ be the maximum number of such steps we go forward in time. Let the initial distance between the jth reference point $y(t_j)$ and the center of mass of B_j be given by $\delta^0_{d_r,j}(\tau)$, and denote by $\delta^t_{d_r,j}(\tau)$ that distance after the system has evolved for t units of time. Finally, define the absolute change in that distance after $q t_{ev}$ units of time by

$$D^q_{d_r,j}(\tau) = \delta^{q t_{ev}}_{d_r,j}(\tau) - \delta^0_{d_r,j}(\tau)$$

In terms of the raw data, the distances to the center of mass at time $q t_{ev}$ are given by

$$\delta^{q t_{ev}}_{d_E,j}(\tau) = \left[\sum_{m=0}^{d_E-1} [x[t_j + m\tau + q t_{ev}] - C^{q t_{ev}}_{m,j}(\tau)]^2\right]^{\frac{1}{2}}$$

and the center of mass at time $q t_{ev}$ is given by

$$C^{q t_{ev}}_{m,j}(\tau) = \frac{1}{K_\varepsilon(j)} \sum_{k=1}^{K_\varepsilon(j)} x[t_{j_k} + m\tau + q t_{ev}]$$

for $t_j < N_r\tau - Q t_{ev}$.

---

[53] Buzug and Pfister (1992a).
[54] See Chapter 5, Lyapunov exponents.

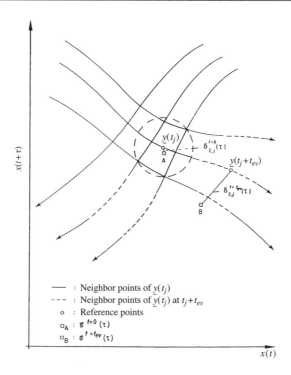

Figure 3.39 Illustration of the method based on the local dynamics of the flow. Trajectories intersect in this two-dimensional projection. So a number of the trajectories neighbouring the reference point $y(t_j)$ have a different direction from the reference trajectory when proceeding in time. The distance between the reference point and the center of mass of neighboring points is a measure of the noncausal behavior of the dynamics. The growth in that distance should be minimal for a proper embedding.

Source: Buzug and Pfister (1992a) p. 7079.

The minimum growth in $\delta^t_{d_r,j}(\tau)$ is roughly $e^{\lambda_1 t}$ for a properly reconstructed chaotic system[55]. However, in a poorly reconstructed phase space, trajectory points may be close because the attractor is crumpled in some directions, so the growth in $\delta^t_{d_r,j}(\tau)$ will be larger than the $e^{\lambda_1 t}$ minimum[56]. Hence, our estimates for the reconstruction parameters are those values of d_r and τ which, in some sense, minimize the growth in $D^q_{d_r,j}(\tau)$ on average over the neighborhoods of the reference points.

Accordingly, we smooth the $D^q_{d_r,j}(\tau)$ statistic by forming a discrete integral,

---

[55]λ_1 is the largest Lyapunov exponent.

[56]It is logical to wonder if the log difference

$$Ln \frac{\delta^{qt_{ev}}_{d_r,j}(\tau)}{\delta^0_{d_r,j}(\tau)} = \alpha t$$

would not be a better statistic than $D^q_{d_r,j}(\tau)$ for the actual growth rate α. But it turns out that that statistic is not robust with respect to noise; a small error in $\delta^0_{d_r,j}(\tau)$ grows rapidly in t, and $\delta^0_{d_r,j}(\tau)$ tends to be small when the data is noisy.

normalized relative to the spatial extent of the attractor and the sampling time of the data. That gives the ***normalized integral deformation***

$$\Delta_{d_r,j}(\tau) = \frac{\sum_{q=1}^{Q} \frac{D_{d_r,j}^{q-1}(\tau)+D_{d_r,j}^{q}(\tau)}{2} \tau_{ev}}{\tau_s \{\max[x(k)] - \min[x(k)]\}}$$

Viewing the local expansion of the attractor through $\Delta_{d_r,j}(\tau)$ instead of $D_{d_r,j}^{q}(\tau)$ fundamentally alters nothing. If the average growth in $D_{d_r,j}^{q}(\tau)$ is small for some choice of τ, then the integral under $D_{d_r,j}^{q}(\tau)$ is small, and so is $\Delta_{d_r,j}(\tau)$.

Finally, we average the normalized local deformation over all N_{ref} reference points to obtain the global statistic

$$\langle \Delta_{d_r,j}(\tau) \rangle_j = \frac{1}{N_{\text{ref}}} \sum_{j=1}^{N_{\text{ref}}} \Delta_{d_r,j}(\tau)$$

For reconstructions that are embeddings $\Delta_{d_r}(\tau) = \langle \Delta_{d_r,j}(\tau) \rangle_j$ should be minimal in τ and convergent in d_r.

Example 3.8
Figure 3.40 illustrates the effectiveness of the algorithm in locating good reconstruction parameters for the Lorenz I attractor.

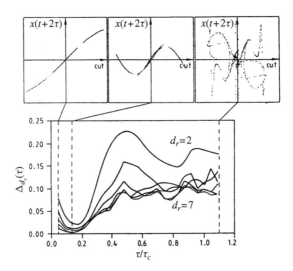

Figure 3.40 Averaged integral local deformation vs delay time for $d_r = 2 - 7$ for the reconstructed Lorenz attractor. For three values of τ/τ_c (indicated by the dashed lines) the corresponding Poincaré sections are shown. The direction of the Poincaré cut is the diagonal in the $x(t) - x(t + \tau)$ plane. According to the meaning of $\Delta_{d_r}(\tau)$ we expect a proper delay time at its minima. A proper delay time in this case is $\tau/\tau_c \approx 0.13$ and the estimated embedding dimension is $d_r = 3$.

Source: Buzug and Pfister (1992a) p. 7080.

Example 3.9

This example compares the Deformation Integral and the Fill Factor statistics for laboratory data from a chaotic Couette–Taylor experiment[57]. Based on this example, when the Fill Factor works, it gives the same optimal reconstruction parameters as those obtained with the Deformation Integral.

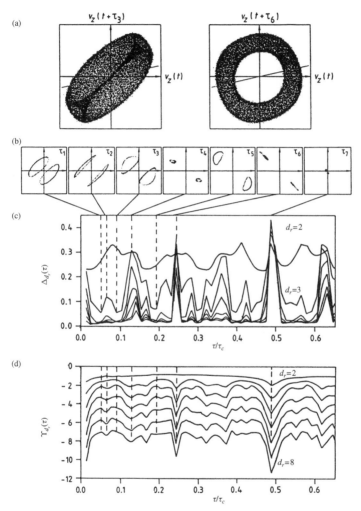

Figure 3.41 Fill Factor $\Upsilon_{d_r}(\tau)$ (d) and Normalized Integral Deformation $\Delta_{d_r}(\tau)$ (c) versus normalized delay time from $d_r = 2 - 8$. For seven values of τ/τ_c, Poincaré sections are shown in (b). Shown in (a) are two-dimensional projections of the experimental trajectory, and the direction of the Poincaré cut used to obtain the Poincaré sections at τ_3 and τ_6 in (b). The time series $v_z(t)$ is the axial component of the local velocity of a Couette–Taylor flow.

Source: Buzug and Pfister (1992a) p. 7082.

[57] An example of a physical system with unknown dynamics. See Brandstater and Swinney (1987).

Parameter values

- The choice of t_{ev} is not arbitrary. If the evolution time is equal to the delay time, the distance between successive $y(qt_{ev})$ will differ only by a factor relating to their first and last components. Since the delay time is an unknown parameter that we are trying to estimate, and we want t_{ev} to be small, it is reasonable to set $t_{ev} = \tau_s$
- Let τ_c be the characteristic cycle time of the attractor. Then choose

$$1 < Q < \frac{\tau_c/4}{t_{ev}}$$

- Set $N_{ref} = \max\{100, 2\%\ \text{of}\ N_r\}$
- Set the neighborhood radius $\varepsilon = 4\%$ of $d(A) = \max\{x(n)\} - \min\{x(n)\}$.

8 Trajectory alignment (d_E and τ)

Below we discuss two approaches to measuring the amount of determinism reflected in a given reconstruction, rather than explicit methods for obtaining the reconstruction parameters. Both approaches are motivated by the following. Recall the equations of motion for a continuous time dynamical system

$$ds(t)/dt = F[s(t)]$$

and the equations of motion for a discrete time dynamical system

$$\Delta s_n = s_{n+1} - s_n = f[s_n] - s_n$$

In either case, since F and f are assumed differentiable in the state variable, $F[s]$ (Δs) will be close to $F[u]$ (Δu) when s is close to u. Hence, the tangents to the trajectories of continuous time systems – or the interpolated vectors between trajectory points in the case of discrete time systems – should align in small regions of the state space (see Figure 3.42).

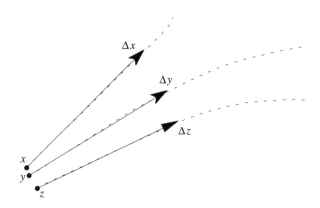

Figure 3.42 The tangent vectors Δx, Δy, and Δz at, respectively, the trajectory points x, y, and z, should be nearly identical when x, y, and z are close together.

Method A[58]

For a given state space reconstruction of dimension d_r, we partition the state space with d_r-dimensional hypercubes c_j of size ε^{d_r}. We then form d_r-dimensional vectors $\Delta y(n) = \underline{y}(n+1) - \underline{y}(n)$, and record the location where $\Delta y(n)$ entered cube c_j and where the first subsequent $\Delta y(n+i)$ left it (see Figure 3.43). We thus obtain a vector $v_{k,j}$ each time some contiguous sequence of $\Delta y(n)$ pass through a given hypercube c_j, where the subscript k is a sequential, but arbitrary ordering of the collection of tangent vector segments that pass through a given hypercube. Finally, we normalize the $v_{k,j}$ to have unit length, that is, $\|v_{k,j}\| = 1$ for all k and all j.

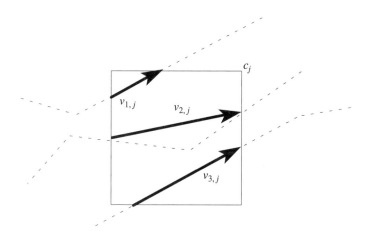

Figure 3.43 Partitioning cube c_j in a two-dimensional reconstruction with intercepting trajectory segments and unnormalized $v_{k,j}$. Note that we had to use more than one $\Delta y(n)$ to define $v_{2,j}$ because the single tangent vector did not completely traverse c_j.

Now, consider the normalized vector addition

$$V_j = \frac{\sum_{k=1}^{n_j} v_{k,j}}{n_j}$$

where n_j denotes the number of tangent vectors that pass through the hypercube c_j. If the v_{kj} are nearly parallel for all k, then $\|V_j\| \approx 1$. To the extent that the v_{kj} are not aligned, $\|V_j\|$ is reduced. In particular, if the $v_{k,j}$ were randomly oriented in some hypercube c_j, then the vector sum $\sum_k v_{k,j}$ would look like a d_r-dimensional random walk consisting of n_j steps of unit length, each of which is taken in a random direction (see Figure 3.44).

---

[58]Kaplan and Glass (1992).

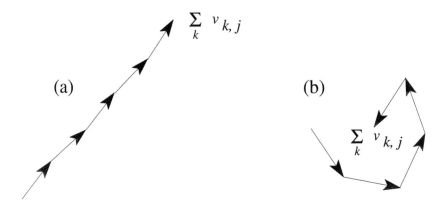

Figure 3.44 In panel (a) the $v_{k,j}$ from a deterministic system are nearly parallel and their vector sum is almost a straight line. In panel (b) the $v_{k,j}$ from a stochastic system form a random walk.

It can be shown that the average displacement for the n-step random walk is approximately

$$R_n^{d_r} = \left(\frac{2}{nd_r}\right)^{\frac{1}{2}} \frac{\Gamma[(d_r + 1)/2]}{\Gamma[d_r/2]} \qquad \text{Eq. 3.23}$$

where Γ is the Gamma function[59].

To characterize the alignment over the whole phase space (attractor) for a given reconstruction, we form the weighted average of the lengths of the V_j relative to that of a random walk:

$$\Lambda(\tau, d_r) = \left\langle \frac{\|V_j\|^2 - (R_{n_j}^{d_r})^2}{1 - (R_{n_j}^{d_r})^2} \right\rangle_j$$

For a deterministic system $\Lambda \approx 1$, while for a random walk $\Lambda \approx 0$. Hence, Λ is a measure of the amount of determinism in a given reconstruction. Λ should be used to distinguish between the system of interest and a random alternative, rather than to optimize the reconstruction by maximizing it with respect to the parameters d_r and τ.

Remarks
When data is abundant, we can also obtain a good deal of information, with less computation, from the unweighted statistic

$$L_n^{d_r} = \langle \|V_j\| \rangle_{j:n_j=n} \qquad \text{Eq. 3.24}$$

where each $L_n^{d_r}$ is the average taken over all cubes with the same number n of trajectory passes. Again $L_n^{d_r}$ should be close to 1 for well aligned trajectories. Note, however, that some systems have an autocorrelation function ψ that oscillates while falling off to zero slowly. For such systems, $L_n^{d_r}$ may look like $R_n^{d_r}$ when the time

---

[59]Eq. 3.23 is valid asymptotically as $n \to \infty$, but is within 3% even for $n = 2$.

delay τ of the reconstruction is chosen such that $\psi(\tau) \approx 0$, or when τ is at a maximum or minimum of ψ.

Parameter values

In principle, for a deterministic system, $L_n^{d_r} \to 1$ as $\varepsilon \to 0$. But to observe that result we would need huge amounts of data. In practice, ε must be set high enough to ensure statistical significance of $L_n^{d_r}$. In the examples below, except where noted otherwise, the phase space was partitioned with a 16^{d_r} grid[60]. In the case of the Lorenz attractor, that would yield hypercubes with edge size $\varepsilon = 2.5$.

Example 3.10

The raw data for this example (Figures 3.45 and 3.46) came from the x-component of the Lorenz system[61]. The random time series are Fourier-phase-randomized versions of the original data[62]. The data span 655 time units (apparently yielding a time series on the order of 10^5 points).

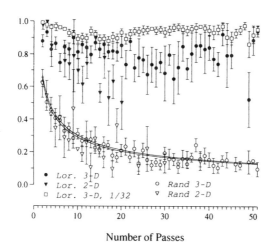

Figure 3.45 $L_n^{d_r}$ versus n for $d_r = 2$ and 3, and time delay $\tau = 0.75$. The grid for the phase space partition was 16^{d_r}, except for the open squares which represent a 3-dimensional embedding with a 32^3 grid. Error bars show the standard error for the estimate of the mean in Eq. 3.24. The solid lines show the theoretical values computed from Eq. 3.23 for random flights in two and three dimensions. The two processes are distinguishable, but note the expected decline in $L_n^{d_r}$ for the deterministic Lorenz system for smaller d_r or larger ε.

Source: Kaplan and Glass (1992) p. 428.

---

[60]The notation 16^3, for example, means a three-dimensional 16 by 16 by 16 grid.

[61]See Chapter 1, Dynamical systems, for the equations of motion.

[62]See Appendix 7, on surrogate data, for the methodology.

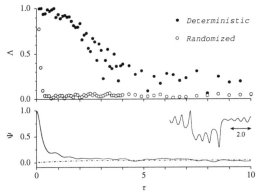

Figure 3.46 Upper panel shows Λ versus the delay time τ for a three-dimensional embedding with a 32^3 grid partition of the phase space. The lower panel shows the autocorrelation function and an insert showing a small section of the original time series. The two processes are distinguishable, but note the expected decline in Λ for the deterministic Lorenz system as τ gets large.

Source: Adapted from Kaplan and Glass (1992) p. 429.

Example 3.11

The data for this example (see Figure 3.47) came from the Mackay–Glass equation[63]. The random time series are again Fourier-phase-randomized versions of the original data. The data span 1.32×10^5 time units (apparently yielding a time series with on the order of 10^5 points).

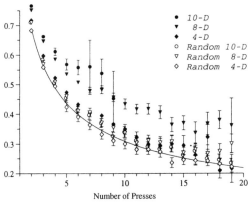

Figure 3.47 $L_n^{d_r}$ versus n for $d_r = 4$, 8 and 10, and time delay $\tau = 75$. The grid for the phase space partition was 4^{d_r} for $d_r = 8$ and 10, and 16^4 for $d_r = 4$. Error bars show the standard error for the estimate of the mean in Eq. 3.24. The solid line shows the theoretical values computed from Eq. 3.23 for random flights in four dimensions. The two processes are distinguishable, but note the expected decline in $L_n^{d_r}$ for the deterministic system for decreasing d_r, and also the surprising decline as n gets large, corresponding to high density regions of the attractor.

Source: Adapted from Kaplan and Glass (1992) p. 429.

[63] See Chapter 1, Dynamical systems, for the equation of motion.

Method B[64]

Although similar in spirit, this method differs substantially from the previous method in execution. Instead of partitioning the phase space around the experimental attractor, we take a uniform sample of the reconstructed trajectory points $y(t)$ – which is the same as sampling from the natural measure of the experimental attractor – and from these we construct local tangent vectors for the vector alignment analysis.

Accordingly, let $\underline{y}(c_i), i = 1, \ldots, N_c$ be the sampled points, and $\underline{y}(n_{ij}), j = 1, \ldots, K$, be the K nearest neighbors[65] of $\underline{y}(c_i)$. Now, let $n_{i0} = c_i$ and form the tangent vectors

$$v_{ij} = \underline{y}(n_{ij} + 1) - \underline{y}(n_{ij}) \qquad j = 0, \ldots, K; \quad i = 1, \ldots, N_c$$

If the neighborhood of each sampled point is indeed small, then differentiable dynamics implies that the v_{ij} should be nearly identical, and close to the normalized vector sum

$$v_{i\cdot} = \frac{1}{K+1} \sum_{j=0}^{K} v_{ij}$$

When that is the case, the neighborhood *translation error*

$$e_{\text{trans}}(i) = \frac{1}{K+1} \sum_{j=0}^{K} \frac{\|v_{ij} - v_{i\cdot}\|^2}{\|v_{i\cdot}\|^2}$$

will be 'small,' and so the *median* translation error over all the sampled neighborhoods will also be small. Hence, the median translation error measures the extent to which the phase space reconstruction is consistent with smooth deterministic dynamics.

Remarks

- Use the median translation error to distinguish between the system of interest and a random alternative, rather than to optimize the reconstruction by minimizing it with respect to the reconstruction parameters.
- When the data sets are small, care should be taken to ensure that the neighbors used to construct the tangent vectors v_{ij} are a result of close returns of the system's experimental trajectory, rather than the proximity of trajectory points caused by observations that are close in time.

Parameter values

In the examples below the neighborhood size K is 4, and the number of sample points N_c is about equal to 10% of the number of observations N_d.

Example 3.12

The raw data for this example came from 2048 measurements of the x-component of the Lorenz[66] system sampled at $\tau_s = 0.0314$. The random time series used for comparison is a Fourier-phase-randomized[67] version of the original time series. The two processes are clearly distinguishable in Figure 3.48.

---

[64] Wayland *et al.* (1993) p. 580.

[65] See Appendix 2, Nearest neighbor searches.

[66] See Chapter 1, Dynamical systems, for the equations of motion.

[67] See Appendix 7, on surrogate data.

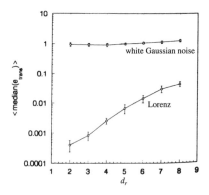

Figure 3.48 Plot of the median translation error vs reconstruction dimension d_r for the Lorenz data and a Gaussian stochastic process constructed from a phase-randomized version of the Lorenz data. Each point on the plot is the average median translation error for thirty 2048 point realizations of the Lorenz and random systems. The error bars represent ± 1 standard deviation of the median translation error for the thirty realizations. The time delay for the phase space reconstruction was fixed at $4\tau_s \approx 0.125$, the number of neighborhoods sampled N_c was 200, and the neighborhood size K was 4.

Source: Adapted from Wayland *et al.* (1993) p. 580.

Example 3.13

The raw data for this example came from 1024 measurements of the x-component of the Hénon[68] system. The plot in Figure 3.49 shows the degradation in determinism as measured by an increase in the median translation error with decreasing signal-to-noise (SNR) ratios[69] created by adding increasing amounts of white Gaussian noise.

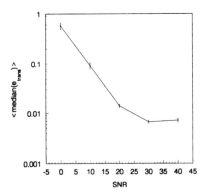

Figure 3.49 Plot of the median translation error vs signal-to-noise ratio (SNR) for the Hénon data contaminated by adding white Gaussian noise. Each point on the plot is the average median translation error for thirty 1024 point realizations of the noisy Hénon data. The error bars represent ± 1 standard deviation of the median translation error for the thirty realizations. The dimension and time delay for the phase space reconstruction were fixed at 5 and $2\tau_s$, respectively, the number of neighborhoods sampled N_c was 100, and the neighborhood size K was 4.

Source: Adapted from Wayland *et al.* (1993) p. 581.

[68] See Chapter 1, Dynamical systems, for the equations of motion.
[69] SNR = 10 Log [Signal Variance/Noise Variance].

Example 3.14

The raw data for this example came from 2048 measurements of ambient sea noise sampled at 820 Hz. The random time series used for comparison is a Fourier-phase-randomized[70] version of the original time series. The two processes are clearly not distinguishable in Figure 3.50.

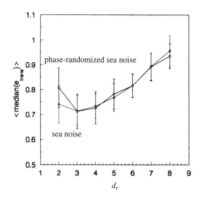

Figure 3.50 Plot of the median translation error vs reconstruction dimension d_r for the sea noise data and a Gaussian stochastic process constructed from a phase-randomized version of the sea noise data. Each point on the plot is the average median translation error for thirty 2048 point realizations of the sea noise and Gaussian systems. The error bars represent ± 1 standard deviation of the median translation error for the thirty realizations. The time delay for the phase space reconstruction was fixed at $3\tau_s$, the number of neighborhoods sampled N_c was 200, and the neighborhood size K was 4.

Source: Adapted from Wayland *et al.* (1993) p. 582.

9 Prediction error[71] $\left(d_E \text{ and } \tau\right)$

Assume that the experimental trajectory $\underline{y}(n)$ is the result of a (d_r, τ)-reconstruction, and let

$$\delta_{jk} = \|\underline{y}(j) - \underline{y}(k)\|$$

and

$$\varepsilon_{jk} = \|\underline{y}(j+1) - \underline{y}(k+1)\|$$

If there is a smooth dynamical mapping φ generating the $\underline{y}(n)$, then $\underline{y}(j)$ close to $\underline{y}(k)$ implies

$$\underline{y}(j+1) - \underline{y}(k+1) \approx J_k \cdot [\underline{y}(j) - \underline{y}(k)]$$

where J_k is the Jacobian of φ at $\underline{y}(k)$. Hence, to first order

$$
\begin{aligned}
\varepsilon_{jk} &= \|\underline{y}(j+1) - \underline{y}(k+1)\| \\
&= |J_k| \cdot \|\underline{y}(j) - \underline{y}(k)\| \\
&= |J_k| \cdot \delta_{jk}
\end{aligned}
$$

[70] See Appendix 7, on surrogate data.
[71] Kaplan (1994).

If we now average ε_{jk} over all (j, k) for which $r \leq \delta_{jk} < r + \Delta r$, we get the conditional average

$$
\begin{aligned}
\varepsilon(r) &= \frac{1}{n_r} \sum_{\{j,k|r \leq \delta_{jk} < r+\Delta r\}} \varepsilon_{jk} \\
&= \langle |J_k| \cdot \delta_{jk} \rangle \\
&< \langle |J_k| \rangle (r + \Delta r) \\
&\equiv \lambda(r + \Delta r)
\end{aligned}
$$

where

$$
n_r = \text{cardinality}^{[72]} \text{ of } \{(j, k)|r \leq \delta_{jk} \leq r + \Delta r\}
$$

Clearly, for $\lambda \leq K < \infty$

$$
\lim_{r+\Delta r \to 0} \varepsilon(r) = 0
$$

In contrast, consider the auto-regressive process

$$
x_{i+1} = \sum_{k=0}^{p-1} \alpha_k x_{i-k} - \eta_i
$$

where $x_i \in \mathfrak{R}$, and η_i is random noise. For a one-dimensional reconstruction

$$
\varepsilon_{ij} = \left| \sum_{k=0}^{p-1} \alpha_k (x_{i-k} - x_{j-k}) + (\eta_i - \eta_j) \right|
$$

and, conditioning on the proximity of x_{i-k} and x_{j-k}

$$
\epsilon'(r) = \left\langle \left| \sum_{k=0}^{p-1} \alpha_k \beta_{ij}(k) + (\eta_i - \eta_j) \right| \right\rangle
$$

where

$$
\beta_{ij}(k) \in (-r - \Delta r, -r] \cup [r, r + \Delta r)
$$

Thus,

$$
\Upsilon \equiv \lim_{r+\Delta r \to 0} \epsilon'(r) = \langle |\eta_i - \eta_j| \rangle
$$

and this value of Υ will be smaller than the true Υ for this process computed in any dimension. Hence, if

$$
\Upsilon - \lim_{r+\Delta r \to 0} \varepsilon(r)
$$

is significantly larger than zero in a statistical sense, then we can rule out a linear stochastic model as the process generating the data. More general stochastic processes can be obtained by the methods discussed in Appendix 7, on surrogate data, but whatever stochastic model is used, if we can find an r^* such that $\varepsilon(r) < \Upsilon$ for $r \leq r^{*[73]}$, then we can conclude that the process generating the measurements $x(t)$ is not the postulated stochastic process.

[72] number of elements in the set.

[73] Here, Υ denotes $\lim_{r+\Delta r \to 0} \varepsilon(r)$ for the postulated stochastic model.

128

Remarks

- We can avoid the need to set Δr, and reduce the size of confidence intervals, by looking at the cumulative version of $\varepsilon(r)$ defined by

$$E(r) = \frac{1}{\upsilon_r} \sum_{\{i,j|\delta_{ij}<r\}} \varepsilon_{ij}$$

where

$$\upsilon_r = \text{cardinality of } \{(j,k) \mid \delta_{jk} < r\}$$

- The relationship between $E(r)$ and the predictability index $S_m(r)$ of Green and Savit[74] can be established as follows. Without loss of generality, we can define the ε_{jk} for higher dimensional (> 1) reconstructions to be $|x(j+\tau/\tau_s) - x(k+\tau/\tau_s)|$, as the other components are bounded in an interval of size Δr by the conditioning of ε_{jk} on δ_{jk}. Now, recall that for $\underline{y}(n) \in \Re^m$

$$S_m(r) = P(\varepsilon < r \mid \delta < r)$$

and it can be seen that

$$E(r) = \int_0^r u \, dP(\varepsilon < u \mid \delta < r)$$

- The method relies on the statistics $E(r)$ and $\varepsilon(r)$ which depend on the exceptional events $\delta_{jk} < r$. On the one hand, that means that we can establish determinism with relatively few data points. On the other hand, the method can be sensitive to noise. Specifically, in a statistical sense, noise puts a lower bound ϑ on interpoint distances, so it may not be possible to find enough exceptional points (with $\delta_{jk} < r^*$) to make a test significant[75]. In addition, the fact that $\varepsilon(r) \geq \vartheta$ statistically means that, if $\vartheta > \Upsilon$, then it will not be possible to distinguish between a deterministic and a stochastic process.
- The method, and others for that matter, can be spoofed by the occurrence of a few very nearly identical sequences that are indeed exceptions (a.k.a. outliers), and do not reflect any deterministic dynamics. In any case, it is worth taking a close look at the $\underline{y}(j)$ and $\underline{y}(k)$ corresponding to the δ_{jk} that are included in the computation of conditional expectations $\varepsilon(r)$, for $r < r^*$.

Example 3.15
The data for this example was obtained from the Mackay–Glass[76] delay differential equation sampled at $\tau_s = 19$. The resulting time series reflects a high dimensional discrete time dynamical system, but the system's attractor has a dimension of about 7.

---

[74]See the algorithm section in Chapter 2, Entropy.

[75]If we need N^* data points to make the test significant, then we need

$$N_r > \sqrt{\frac{2N^*}{C_2(d_r; r^*)}}$$

where N_r is the length of the trajectory, and $C_2(d_r; r^*)$ is the correlation integral. Contrast this with the estimate based on *typical* points given in Appendix 1, Data requirements.

[76]See Chapter 1, Dynamical systems, for the equation of motion.

The random time series were obtained from Fourier-phase-randomized[77] versions of the original data (see Figure 3.51). Using reconstruction parameters of $d_r = 10$ and $\tau = 3\tau_s$, we obtain a value for $\Upsilon \approx 0.3$, and a value for $r^* = 0.75$. From Figure 3.52, it can be seen that, with this reconstruction, as few as 200 data points give a clear indication of determinism.

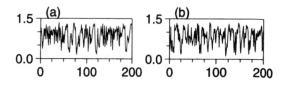

Figure 3.51 In panel (a), the time series from the Mackay–Glass system, and in panel (b) a Fourier-phase-randomization of the time series in panel (a).
Source: Adapted from Kaplan (1994) p. 44.

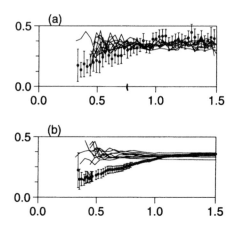

Figure 3.52 Plotted in panel (a) is $\varepsilon(r)$ versus r for $(d_r = 10, \tau = 3\tau_s)$-reconstructions based on the original Mackay–Glass data (black dots), and 10 phase-randomized versions of the data (thin lines). The error bars for $\varepsilon(r)$ are equal to the standard error of the mean in each bin $(r, r + \Delta r)$. Plotted in panel (b) is the cumulative statistic $E(r)$ versus r for the same data.
Source: Adapted from Kaplan (1994) p. 44.

10 Test for determinism[78] $\left(d_E \text{ and } \tau\right)$

A distinguishing characteristic between (low dimensional) chaos and noise is some short term predictability for the former. Suppose we have a (d_r, τ)-delay reconstruction and a set of prediction errors $A = \{e_{d_r,\tau}(n), n = 1, \ldots, N_A\}$ for each point

---

[77]See Appendix 7, on surrogate data.
[78]Kennel and Isabelle (1992).

on the experimental trajectory for which an error can be computed. If these errors are significantly smaller than the errors computed in a similar fashion from delay reconstructions of random time series, then we could say with some confidence that the first reconstruction represented a deterministic system.

To start, we need a definition of prediction error. Let $\underline{y}(n^*)$ be the nearest neighbor of $\underline{y}(n)$ on the experimental trajectory. A predictor for the state of the system τ_f time units in the future, starting from $\underline{y}(n)$, is $\underline{y}(n^* + \tau_f/\tau_s)$, the image of its nearest neighbor. Thus, the error in the prediction can be measured by

$$\|\underline{y}(n^* + \tau_f/\tau_s) - \underline{y}(n + \tau_f/\tau_s)\|$$

But we make a further simplification and define the prediction error to be the absolute error in the last component, that is

$$e_{d_r,\tau}(n) = |x[n^* + (d_r - 1) + \tau_f/\tau_s] - x[n + (d_r - 1) + \tau_f/\tau_s]| \qquad \textbf{Eq. 3.25}$$

Admittedly, that is about as simple a predictor, and predictor error, as one can devise, but it turns out that more sophisticated models do not improve the effectiveness of the algorithm.

Next, we need some random time series for comparison. These have to be chosen with some sophistication if the algorithm is to be useful. Comparing the prediction error of the original time series with that of any-old-random time series, and concluding that the original was deterministic because it had lower prediction errors, would not be convincing. For a convincing conclusion, we need random time series that mimic as many of the statistical properties of the original time series as possible. Generating such random time series, also known as **surrogate data**, is the subject of Appendix 7. Here, we assume that we have obtained M (Gaussian) surrogate time series, and reconstructed an experimental trajectory for each using a delay coordinate map with the same parameters as were used to reconstruct the experimental trajectory from the original time series. We then use Eq. 3.25 to compute $e^{(k)}{}_{d_r,\tau}(n)$, the prediction error for the nth point on the experimental trajectory reconstructed from the kth surrogate time series. Let $B = \{e^{(k)}{}_{d_r,\tau}(n), n = 1, \ldots, N_A, k = 1, \ldots, M\}$, the combined set of *all* prediction errors from *all* the surrogate time series.

What we need now is a way to quantify what we mean by 'the prediction errors for the original time series are significantly smaller than the prediction errors for the surrogate time series.' In other words, we need a statistic, with a known distribution, that compares the errors in the set A with the errors in the set B. For that, we use the Mann–Whitney rank-sum

$$U_{d_r,\tau} = \sum_{i=1}^{N_A} \sum_{j=1}^{N_s} \Theta(A_i - B_j)$$

where A_i is the ith element of A, B_j is the jth element of B, Θ is the Heaviside function, and $N_S = M N_A$. Assuming that all the errors are pairwise independent

$$z_{d_r,\tau} = \frac{U_{d_r,\tau} - N_A N_S/2}{\sqrt{\frac{1}{12} N_A N_S (N_A + N_S + 1)}}$$

131

is $\aleph(0, 1)$[79] under the null hypothesis that the samples A and B are from the same distribution. From here it is easy to see the way forward. We will conclude that the original time series comes from a deterministic system if we can reject the null hypothesis confidently. So we choose a level of confidence α, and look up the value of β_α that satisfies

$$\alpha = \text{Probability that } z_{d_r,\tau} < \beta_\alpha$$

from a table of the cumulative standard normal distribution. Then, if we conclude that the original time series comes from a deterministic system whenever $z_{d_r,\tau} < \beta_\alpha$, we will be wrong with probability α.

To turn this into a mechanism for determining good reconstruction parameters, we need to test the hypothesis for different values of d_r and τ. If a given (d_r, τ)-reconstruction passes the above test for determinism with flying colors, then reconstructions in the same dimension, but with slightly different delay times, should also pass the test. Hence, a reconstruction which is strongly deterministic in the sense defined by the above test, will remain deterministic for small changes in the delay time.

Let τ_i, $i = 1, \ldots, K$, be a set of delay times and let z_{d_r,τ_i} be the corresponding test statistics. All $K(d_r, \tau_i)$-reconstructions jointly (and conservatively) pass the test for determinism, at the confidence level α, if $z_{d_r,\tau_i} < \beta_{\alpha/K}$ for all i. To find a set of optimal reconstruction parameters, we increase the dimension d_r until all the z_{d_r,τ_i}, or a significant contiguous subset of them, converge on a level below $\beta_{\alpha/K}$. When that occurs, we set $d_E = d_r$, and we set the delay time τ to any value τ_i in the range of the minimum plateau.

Parameter values
- Choose the number of surrogate data sets M between 10 and 100.
- Choose K between 5 and 10.
- Choose K values τ_i, $i = 1, \ldots, K$, near τ_{mi}, the first minimum of the mutual information[80].
- $\alpha = 0.01$ is reasonable, but remember the test is based on α/K.
- Choose the forecast horizon τ_f on the order of $1/2$ the mutual information time τ_{mi}.

Problems
- To insure the independence among the prediction errors, the sets $e_{d_r,\tau}^{(k)}(j)$ and $e_{d_r,\tau}(j)$ should be decimated as follows. Compute the correlation time in units of τ_s for each set of prediction errors and set q equal to the least integer upper bound on the maximum of those correlation times. Now, for $0 \leq n < N_r/q$, replace the set $e_{d_r,\tau}(j)$ with the set $e_{d_r,\tau}(1+nq)$ and the set $e_{d_r,\tau}^{(k)}(j)$ with the set $e_{d_r,\tau}^{(k)}(1+nq)$.

- To prevent the algorithm from giving a false positive reading for non-Gaussian random processes, it is sufficient to replace the original time series with a 'histogram transform'[81] of itself. The same maneuver can be performed to get the input for each new surrogate data set.

---

[79]Standard normal distribution (mean 0 and variance 1).
[80]The algorithm is discussed earlier in this chapter.
[81]See Appendix 7, on surrogate data.

Example 3.16

Figures 3.53 through Figure 3.56 clearly show that while there are deterministic phase space reconstructions for even a very noisy Lorenz II system, there are no deterministic phase space reconstructions for the colored noise process or the physical system (wind velocity data).

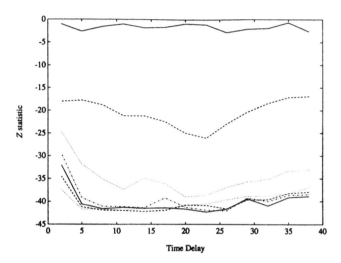

Figure 3.53 Statistic for Lorenz II data set, no added noise. x axis is in multiples of the sampling time $\tau_s = 0.05$. Each curve is for a different reconstruction dimension: $d_r = 1$ is the top curve, with d_r increasing for successively lower curves.

Source: Kennel and Isabelle (1992) p. 3114.

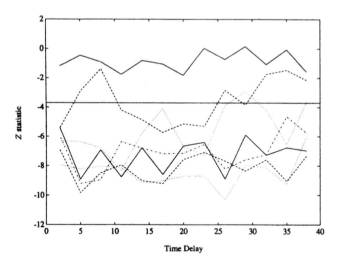

Figure 3.54 Statistic for Lorenz II data set, 50% added noise. Horizontal line is at 99% confidence level for rejection. Lower curves are for increasing values of d_r.

Source: Kennel and Isabelle (1992) p. 3115.

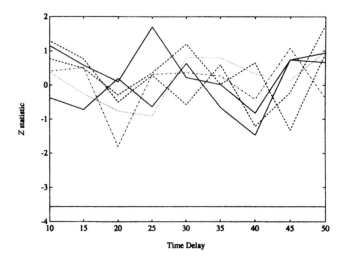

Figure 3.55 Statistic for colored noise data set for various values of d_r. Horizontal line is at 99% confidence level for rejection.

Source: Kennel and Isabelle (1992) p. 3115.

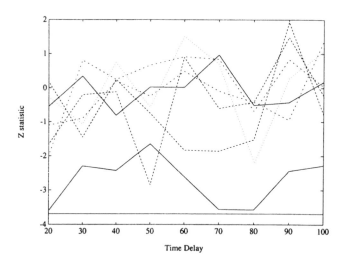

Figure 3.56 Statistic for wind velocity data for various values of d_r. Straight horizontal line is 99% rejection threshold.

Source: Kennel and Isabelle (1992) p. 3116.

11 Tangent space approach[82] $\left(d_A\right)$

The purpose of the next two algorithms is to estimate d_A, the dimension of the manifold containing the attractor. d_A, also called the **topological dimension**, quantifies

---

[82]Broomhead, Jones and King (1987) p. L563.

the minimum degrees of freedom required to capture the dynamics of the system in every neighborhood of the attractor. Unlike d_e, which is sensitive to the embedding process, d_A is an invariant of the system. As such, d_A is useful in classifying the dynamics and may lead to another smooth transformation of the state space that further unfolds the attractor from a set in $\Re^{d_e}$ to one in $\Re^{d_A}$. If that results in a reduction of the dimension of the reconstruction, it will simplify the computations needed to estimate the $d_A \times d_A$ Jacobians for the true d_A Lyapunov exponents[83], and reduce the complexity of forecasting models.

In order to obtain an estimate of d_A we note that, locally, the manifold containing the reconstructed attractor looks like $\Re^{d_A}$ which is closely approximated by the tangent space[84] of the manifold at any point on the attractor. We approximate the tangent space at some trajectory point $y(n)$ by forming the pseudo tangent vectors $z(k) = y(n_k) - y(n)$, $1 \le k \le K_\varepsilon(n)$, where the $y(n_k)$ are the $K_\varepsilon(n)$ points contained in $B_\varepsilon[y(n)]$, an ε-neighborhood about $y(n)$[85].

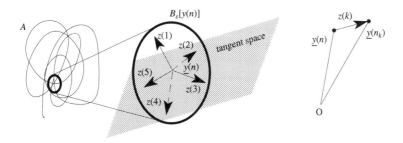

Figure 3.57 Pseudo tangent vectors constructed from points in $B_\varepsilon[y(n)]$, a neighborhood of the experimental trajectory point $y(n)$. Most pseudo tangent vectors will not lie precisely in the tangent space due to noise and/or curvature of the attractor. In the figure, only $z(3)$ and $z(5)$ lie in the tangent space.

The 'trajectory' matrix

$$Z = \frac{1}{\sqrt{K_\varepsilon(n)}} \begin{bmatrix} z(1)^T \\ z(2)^T \\ \vdots \\ z(K_\varepsilon(n))^T \end{bmatrix},$$

or equivalently $Z^T Z$, can be decomposed to yield a set of d_r singular vectors[86], d_A of which span the tangent space. To determine which of the d_r singular vectors define the tangent space, we reason as follows. Assume we project the pseudo tangent vectors onto the singular basis and let z_i be the ith principal component of the projection z.

---

[83] See Chapter 5, Lyapunov exponents.

[84] See Chapter 1, Dynamical systems.

[85] An ε-neighborhood of a state space point y is the set of all state space points u such that $\|u - y\| < \varepsilon$.

[86] The eigenvectors of $Z^T Z$ are the same as the right singular vectors of Z. See Appendix 4, Matrix decomposition.

Since the manifold is of dimension d_A, and the tangent space approximates well the manifold in small neighbohoods, we can write[87] a pseudo tangent vector in the form

$$\underline{z}^T = [z_1, z_2, \dots, z_{d_A}, f_{d_A+1}[\{z_i\}], \dots, f_{d_r}[\{z_i\}]]$$

where for $d_A + 1 \le j \le d_r$ and $1 \le i \le d_A$

$$f_j : \Re^{d_A} \rightarrow \Re$$

$$f_j[\{0\}] = 0 \quad \text{and} \quad \frac{\partial f_j[\{0\}]}{\partial z_i} = 0$$

As $B_\varepsilon[y(n)]$ expands, the f_j soak up the effect of curvature and noise. Let ζ be a d_A-dimensional vector of length ε. Since the functions f_i and their partial derivatives vanish at $y(n)$, the origin of the neighborhood $B_\varepsilon[y(n)]$, we can write for some constant C and $\alpha > 2$

$$f_j(\xi) = C\varepsilon^2 + O(\varepsilon^\alpha) \qquad \qquad \textbf{Eq. 3.26}$$

Now, the non-zero components of a true tangent vector in $B_\varepsilon[y(n)]$ lie in the tangent plane and will expand at a rate proportional to ε as the size of the neighborhood $B_\varepsilon[y(n)]$ is increased. Similarly, the first d_A components of any pseudo tangent vector will increase at a rate proportional to ε, but by Eq. 3.26, the remaining $d_r - d_A$ components will increase at a rate proportional to ε^2. Recall that the singular values of Z measure the average amount that the pseudo tangent vectors probe $\Re^{d_r}$ in the directions of the corresponding singular vectors. It follows that, as $B_\varepsilon[y(n)]$ changes size, the first d_A singular values will scale as ε, while the last $d_r - d_A$ singular values will scale as ε^2. So if we plot the singular values as a function of ε, we can pick out those that correspond to the singular directions of the tangent space by observing their rates of growth. The number of singular values which scale as ε is equal to the topological dimension of the manifold at the reference point $y(n)$.

Remember that d_A is a global measure over the whole attractor. Hence, the above local analysis for the neighborhood of the reference point $y(n)$ must be done at a selection of reference points over the whole attractor. The true topological dimension d_A is then given by the maximum of the local topological dimensions at each reference point.

Problems

- To the extent that the manifold is 'thin' in some singular direction, the linear increase in the corresponding singular value will saturate when ε approaches the thickness of the attractor in that direction.
- Singular values corresponding to singular directions that are normal to the tangent space are dominated by noise for small ε, and so will remain independent of ε until curvature effects dominate. After curvature effects begin to dominate, those singular values will start to scale with ε^2.
- Equating the topological dimension with the maximum of the local topological dimensions is a bit dangerous since one anomalous region can spoil the whole procedure. In practice it makes sense to examine the phase portraits in regions where the local dimension seems uncharacteristically high as those may be caused by suspect data.

---

[87]Follows from one of two equivalent definitions of a manifold given earlier in this chapter.

Parameter values
- Choose $d_r \gg d_e$.
- Choose a neighborhood size that will give $5 \leq K_\varepsilon(n) \leq 250$ for all n.

Example 3.17
The above analysis was carried out on 2.5×10^5 observations from a (chaotic) non-linear electronic oscillator. The reconstruction was carried out with $d_r = 50$, and then projected down to a five-dimensional principal component representation. A two-dimensional projection of the resulting experimental trajectory is shown in Figure 3.58. Neighborhood sizes were chosen in the range $5 \leq K_\varepsilon(n) \leq 250$.

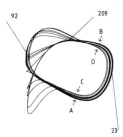

Figure 3.58 The projection of 2000 samples of the trajectory reconstructed from the time series. The numbers indicate the positions $y(23)$, $y(92)$ and $y(209)$ of the origins of the neighborhoods used in the analysis. The direction of the flow is anticlockwise. The point B maps to C, while D maps to a region near A.
Source: Adapted from Broomhead (1987) p. L566.

Analysis of the low curvature area of the attractor near $y(23)$ yields Figure 3.59. By inspection $d_A \geq 2$ locally, but the noise floor is obscuring the growth rates of the small singular values preventing us from drawing any better conclusions. Note that the saturation of one singular value for $\varepsilon > 10^{-1/2}$ indicates 'thinness' in that direction.

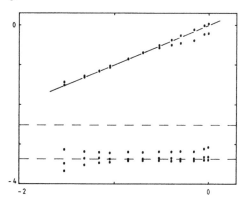

Figure 3.59 A log-log plot of the local singular spectrum (scaled by $2K_\varepsilon(n)^{-1/2}$) against the radius, ε, of the neighborhoods taken about the point $y(23)$. The horizontal broken lines mark the range of the global singular values found in the noise floor. For comparison of the data with the expected growth of singular values corresponding to tangent space directions, the diagonal line is a plot of ε.
Source: Adapted from Broomhead et al. (1987) p. L566.

137

Analysis of the area of the attractor near $y(209)$ yields Figure 3.60. The small singular values appear to scale like ε^2 as curvature sets in at $\varepsilon > 10^{-1/2}$, but the noise in this direction is still obscuring the initial growth rates of the small singular values.

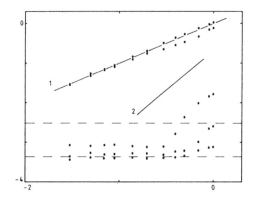

Figure 3.60 A plot similar to Figure 3.59, but for neighborhoods taken about $\underline{y}(209)$. The line marked 1 is a plot of ε. The line marked 2 has slope 2, and is for comparison with the plot of the singular values growing from the noise floor. Here the comparison is inconclusive.

Source: Adapted from Broomhead *et al.* (1987) p. L567.

Analysis of the area of the attractor near $\underline{y}(92)$, where curvature is present at all accessible scales, yields Figure 3.61. Here the curvature is high enough to expose the nonlinear growth of the smaller singular values. The manifold is probably flat for $\varepsilon < 10^{-2}$. Behavior around $\underline{y}(23)$, $\underline{y}(92)$ and $\underline{y}(209)$ looks typical for the attractor, so based on the analysis of the neighborhoods of these points we would conclude that $d_A = 2$.

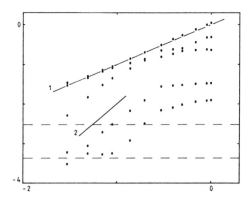

Figure 3.61 A plot similar to Figure 3.59 and Figure 3.60, but for neighborhoods taken about $\underline{y}(92)$. Note that the small singular values begin to scale like ε^2 as they grow out of the noise floor.

Source: Adapted from Broomhead *et al.* (1987) p. L567.

138

12 Local false near neighbors[88] (d_A)

Obtaining an estimate of the local topological dimension d_A by applying the methodology of Global False Near Neighbors[89] to local neighborhoods has problems. However, the problems can be sidestepped by exploiting a defining property of dynamical systems: namely that points that are close together on the attractor stay close over short periods of time.

Assume we have a global embedding. Let N_b be a fixed integer, and let $B_{N_b}[y(n)]$ be the set consisting of $y(n)$ and its N_b nearest neighbors. Because the reconstruction is an embedding, points in $B_{N_b}[y(n)]$ that are close to $y(n)$ are true near neighbors whose proximity is a result of the dynamics of the system rather than a geometrical artifact of the reconstruction. Let $p(n_k)$ be the principal component representation of $y(n_k)$ obtained from a global singular value decomposition of the trajectory matrix Y in $\Re^{d_E}$ (see Figure 3.62). Let $w(n_k)$ be the projection of $p(n_k)$ into $\Re^{d_L}$ obtained by truncating $p(n_k)$ to its first d_L components. Finally, let $w(n^*)$ be $w(n)$'s closest neighbor *as measured in* $\Re^{d_L}$.

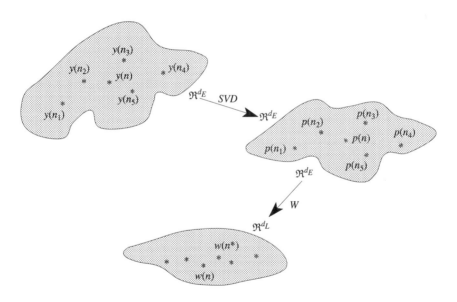

Figure 3.62 Schematic showing the *SVD* transformation from the original delay coordinate system to principal components in $\Re^{d_E}$ and then the projection W down onto $\Re^{d_L}$.

Now, if $\Re^{d_L}$ contains enough degrees of freedom to accommodate the dynamics of the system, then the projection W will not collapse the attractor locally, and $y(n^*)$ will be a true near neighbor of $y(n)$. In that case, the sections of the experimental trajectory containing $y(n)$ and $y(n^*)$ will remain close over short periods of time.

---

[88] Abarbanel and Kennel (1993) p. 3057.

[89] The algorithm is discussed earlier in this chapter.

That is, if $\|y(n^*) - y(n)\|$ is small, then $\|y(n^* + \Delta) - y(n + \Delta)\|$ is also small for a short time interval Δ. For simplicity, we only measure the drift in the last component of $y(n_k) - y(n)$, namely

$$|x(n^* + (d_E - 1)\tau/\tau_s + \Delta) - x(n + (d_E - 1)\tau/\tau_s + \Delta)|$$

as a function of Δ, and determine the smallest value of Δ for which the drift becomes larger than some fraction β of $d(A)$[90], the extent of the attractor. We label that value of Δ as $K(n; d_L)$ for each reference point $y(n)$ and its nearest neighbor $y(n^*)$, where n^* is determined from $w(n^*)$ *locally in dimension* d_L. That is

$$K(n; d_L) \quad = \quad \{\min \Delta | \beta d(A)$$
$$< |x(n^* + (d_E - 1)\tau/\tau_s + \Delta) - x(n + (d_E - 1)\tau/\tau_s + \Delta)|\}$$

If $K(n; d_L)$ is small, then $y(n^*)$ is moving away from $y(n)$ quickly, and it is an indication that they are close for geometric rather than dynamical reasons[91].

As a test statistic we take $P_K(d_L)$ to be the proportion of the total number of points surveyed N_A that have $K(n; d_L) < T$, where T, like β and N_b, is a parameter of the algorithm. That is,

$$P_K(d_L) = \frac{1}{N_A} \sum_{n=1}^{N_A} \Theta(T - K[n; d_L])$$

where $N_A \leq N_d - (d_E - 1)\tau/\tau_s - \Delta$.

If $d_L < d_A$, then $P_K(d_L)$ should be sensitive to N_b because for larger N_b there is a greater chance of including a point that gets projected close to the reference point, but actually lies on some other part of the attractor. For $d_L \geq d_A$, $P_K(d_L)$ should be insensitive to the value of N_b. Except where the largest Lyapunov exponent is very large, true near neighbors should not move a distance that is a significant portion of the attractor over short periods of time. Hence, for a sufficiently large value of d_L, that is for $d_L \geq d_A$, $P_K(d_L)$ should plateau at a level that is independent of the value of N_b.

Parameter values

- Apparently, the sufficient value of d_L does not depend on the precise values of either β or T chosen, as long as $T \approx \tau_{mi}$[92] and $0.05 \leq \beta \leq 0.2$.

- Choose $10 \leq N_b \leq 100$.

Examples 3.18 and 3.19 using clean data

For the Ikeda map we find $P_K(d_L)$ plateaus independently of N_b at $d_L = 2$, hence the procedure recovers the correct $d_A = 2$ (see Figure 3.63).

---

[90] See parameter $d(A)$ of Global False Near Neighbors earlier in this chapter.

[91] This is a sufficient condition, not a necessary one, as it only looks at the last component.

[92] τ_{mi} is the first minimum of the mutual information discussed earlier in this chapter.

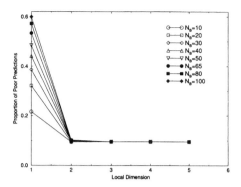

Figure 3.63 $P_k(d_L)$ for 20,000 points of the real part of $z(n)$ from the Ikeda map for various $10 \leq N_b \leq 100$; $d_E = 5$. $d_L = 2$ is clearly the dimension where $P_K(d_L)$ becomes independent of d_L and N_b.

Source: Adapted from Abarbanel and Kennel (1993) p. 3063.

For the Lorenz II model we find $P_K(d_L)$ plateaus independently of N_b at $d_L = 3$, hence the procedure recovers the correct $d_A = 3$ (see Figure 3.64).

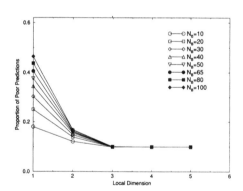

Figure 3.64 $P_k(d_L)$ for 20,000 points of $x(t)$ from the Lorenz II model for various $10 \leq N_b \leq 100$; $d_E = 6$. $d_L = 3$ is clearly the dimension where $P_K(d_L)$ becomes independent of d_L and N_b.

Source: Adapted from Abarbanel and Kennel (1993) p. 3063.

Examples 3.20, 3.21 and 3.22 using noisy data

The procedure's robustness to moderate noise[93] levels is illustrated in the plots in Figures 3.65–3.67, where artificial noise was added to clean data. For noisy data from the Lorenz II model, the procedure is still able to recover the correct $d_A = 3$.

---

[93] 1% noise means $0.01 = $ Variance(noise)/Variance(signal).

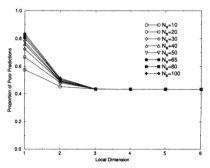

Figure 3.65 $P_k(d_L)$ for 20,000 points of $x(t)$ from the Lorenz II model for various $10 \leq N_b \leq 100$; $d_E = 6$. These data are contaminated with 1% noise relative to the global size of the attractor. $d_L = 3$ is clearly the dimension where $P_K(d_L)$ becomes independent of d_L and N_b.

Source: Adapted from Abarbanel and Kennel (1993) p. 3065.

For noisy data from the Ikeda map, the procedure is still able to recover the correct $d_A = 2$.

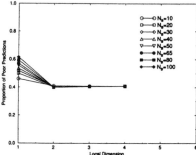

Figure 3.66 $P_k(d_L)$ for 20,000 points of the real part of $z(n)$ from the Ikeda map for various $10 \leq N_b \leq 100$; $d_E = 5$. These data are contaminated with noise of size 2% relative to the global size of the attractor. $d_L = 2$ is clearly the dimension where $P_K(d_L)$ becomes independent of d_L and N_b.

Source: Adapted from Abarbanel and Kennel (1993) p. 3066.

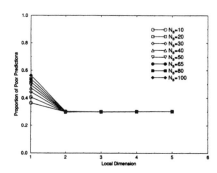

Figure 3.67 $P_k(d_L)$ for 20,000 points of the real part of $z(n)$ from the Ikeda map for various $10 \leq N_b \leq 100$; $d_E = 5$. These data are contaminated with noise of size 1% relative to the global size of the attractor. $d_L = 2$ is clearly the dimension where $P_K(d_L)$ becomes independent of d_L and N_b.

Source: Adapted from Abarbanel and Kennel (1993) p. 3066.

Finally, sufficiently high noise levels can cripple the procedure as illustrated in Figures 3.68 and 3.69.

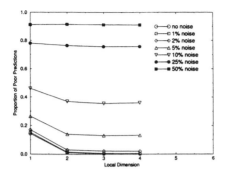

Figure 3.68 $P_K(d_L)$ for 20,000 points of the real part of $z(n)$ from the Ikeda map with $d_E = 4$ and varying amounts of noise added to the data. The percentage of noise is given relative to the global size of the attractor and ranges from 0% to 50%. When the noise level is 10% or less, $d_L = 2$ is clearly chosen. After that the method of local false nearest neighbors is overwhelmed by the contamination. In this figure $N_b = 100$ and $\beta = 0.5$.
Source: Adapted from Abarbanel and Kennel (1993) p. 3066.

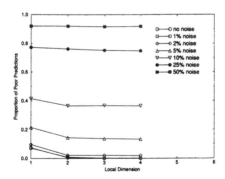

Figure 3.69 $P_K(d_L)$ for 20,000 points of the real part of $z(n)$ from the Ikeda map with $d_E = 4$, and varying amounts of noise added to the data. The percentage of noise is given relative to the global size of the attractor and ranges from 0% to 50%. When the noise level is 10% or less, $d_L = 2$ is clearly chosen. After that the method of local false nearest neighbors is overwhelmed by the contamination. In this figure $N_b = 40$ and $\beta = 0.5$.
Source: Adapted from Abarbanel and Kennel (1993) p. 3066.

Problems

- For the Lorenz I model, it appears from Figures 3.70–3.73 that $d_A = 2$ when the data is read forward in time while $d_A = 3$ when the data is read backwards. In principle, the procedure should yield the same value for d_A whether the data is read forward or backward in time. But, if there is a large negative Lyapunov exponent

143

for the system, then even points from different parts of the attractor may be getting squeezed together giving a false reading for $P_K(d_L)$ when $d_L < d_A$.

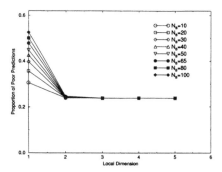

Figure 3.70 $P_K(d_L)$ for 20,000 points of $x(t)$ from the Lorenz I model for various $10 \le N_b \le 100$; $d_E = 5$. From this figure, which uses data read forward in time, one might choose $d_L = 2$.

Source: Adapted from Abarbanel and Kennel (1993) p. 3064.

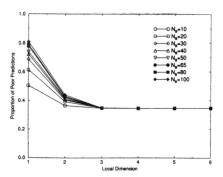

Figure 3.71 $P_K(d_L)$ for 20,000 points of $x(t)$ from the Lorenz I model for various $10 \le N_b \le 100$; $d_E = 5$. From this figure, which uses data read backward in time, one clearly would select $d_L = 3$.

Source: Adapted from Abarbanel and Kennel (1993) p. 3064.

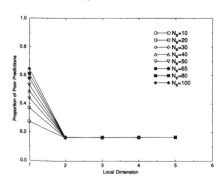

Figure 3.72 $P_K(d_L)$ for 20,000 points of $x(t)$ from the Lorenz I model, for various $10 \le N_b \le 100$; $d_E = 5$. These data are contaminated with 1% noise relative to the global size of the attractor. From this figure, which uses data read forward in time, one might choose $d_L = 2$.

Source: Adapted from Abarbanel and Kennel (1993) p. 3065.

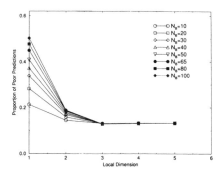

Figure 3.73 $P_K(d_L)$ for 20,000 points of $x(t)$ from the Lorenz I model, for various $10 \leq N_b \leq 100$; $d_E = 5$. These data are contaminated with 1% noise relative to the global size of the attractor. From this figure, which uses data read backward in time, one clearly would select $d_L = 3$.

Source: Adapted from Abarbanel and Kennel (1993) p. 3065.

The same problem emerges in laboratory data (see Figures 3.74 and 3.75).

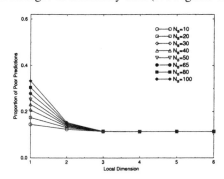

Figure 3.74 $P_K(d_L)$ for 20,000 points of a voltage from the hysteretic circuit of Pecora and Carroll for various $10 \leq N_b \leq 100$; $d_E = 6$. $d_L = 3$ is clearly the dimension where independence of d_L and N_b occurs for $P_K(d_L)$. This data is read backward in time.

Source: Adapted from Arbanel and Kennel (1993) p. 3064.

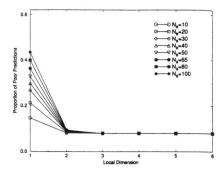

Figure 3.75 $P_K(d_L)$ for 20,000 points of a voltage from the hysteretic circuit of Pecora and Carroll for various $10 \leq N_b \leq 100$; $d_E = 6$. The dimension where independence of d_L and N_b occurs for $P_K(d_L)$ is $d_L \leq 3$. This data is set to read forward in time.

Source: Adapted from Arbanel and Kennel (1993) p. 3064.

145

However, when the procedure was applied to the Ikeda map, there was no difference in the value of d_A obtained from the data read forward and backward in time. To guarantee the correct value of d_A, the algorithm must be applied to the data read both forward and backward, and the larger of the two estimates taken for the value of d_A.

- If the calculation of $K(n; d_L)$ causes an overrun of the data set at $\underline{y}(n)$, then the offending neighborhood is eliminated from consideration, and the summation limit and divisor in the definition of $P_K(d_L)$ is adjusted downwards accordingly.
- $P_K(d_L)$ does not work as a sharp tool in identifying the minimum embedding dimension d_e.

Remarks

- Note the scale of $P_K(d_L)$ in the forward and backward runs. It is not the level of $P_K(d_L)$ or its plateau that determines d_A, but when $P_K(d_L)$ becomes level and independent of both d_L and N_b.
- Having said that, we note from Figure 3.68 and Figure 3.69 that, at noise levels greater than 10%, $P_K(d_L)$ remains high and flat independent of d_L and N_b. Recalling the definition of $P_K(d_L)$, and specifically that of $K(n; d_r)$, the implication is that predictions will be poor.

REFERENCES AND FURTHER READING

Abarbanel, H. and Kennel, M. (1993) 'Local false nearest neighbors and dynamical dimensions from observed chaotic data', *Physical Review E* 47.05, pp. 3057–3068.

Albano, A., Muench, J., Schwartz, C., Mees, A. and Rapp, P. (1988) 'Singular value decomposition and the Grassberger–Procaccia algorithm', *Physical Review A* 38.06, pp. 3017–3026.

Albano, A., Passamante, A., Hediger, T. and Farrell, M. (1992) 'Using neural nets to look for chaos', *Physica D* 58, pp. 1–9.

Brandstarter, A. and Swinney, H. L. (1987) 'Strange attractors in weakly turbulent Couette–Taylor flow', *Physical Review A* 35, pp. 2207–2220.

Broomhead, D. and King, G. (1986) 'Extracting qualitative dynamics from experimental data', *Physica D* 20, pp. 217–236.

Broomhead, D. A., Jones, R. and King, G. P. (1987) 'Topological dimension and local coordinates from time series data', *Journal of Physics A: Mathematics and General [Letters]* 20, pp. L563–L569.

Buzug, Th. and Pfister, G. (1992a) 'Optimal delay time and embedding dimension for delay-time coordinates by analysis of the global static and local dynamic', *Physical Review A* 45.1, pp. 7073–7084.

Buzug, Th. and Pfister, G. (1992b) 'Comparison of algorithms calculating optimal embedding parameters for delay time coordinates', *Physica D* 58, pp. 127–137.

Casdagli, M., Eubank, S., Farmer, D. and Gibson, J. (1991) 'State space reconstruction in the presence of noise', *Physica D* 51, pp. 52–98.

Eckmann, J. and Ruelle, D. (1985) 'Ergodic theory of chaos and strange attractors', *Reviews of Modern Physics* 57.3, pp. 617–656.

Eckmann, J., Ruelle, D., Ciliberto, S. and Oliffson Kamphorst, S. (1986) 'Lyapunov exponents from time series', *Physical Review A* 34.06, pp. 4971–4979.

Fraedrich, K., and Wang, R. (1993) 'Estimating the correlation dimension of an attractor from noisy and small datasets based on re-embedding', *Physica D* 65, pp. 373–398.

Fraser, A. and Swinney, H. (1986) 'Independent coordinates for strange attractors from mutual information', *Physical Review A* 33.02, pp. 1134–1140.

Gibson, J., Farmer, D., Casdagli, M. and Eubank, S. (1992) 'An analytical approach to practical state space reconstruction', *Physica D* 57, pp. 1–30.

Holmes, P. A. (1979) 'A nonlinear oscillator with a strange attractor', *Phil. Trans. Roy. Soc.* (London) 292A, pp. 419–448.

Kaplan, D. and Glass, L. (1992) 'Direct test for determinism in time series', *Physical Review Letters* 68.04, pp. 427–430.

Kaplan, D. (1994) 'Exceptional events as evidence for determinism', *Physica D* 73, pp. 38–48.

Kember, G. and Fowler, A. (1993) 'A correlation function for choosing time delays in phase portrait reconstructions', *Physics Letters A* 179, pp. 72–80.

Kennel, M., Brown, R. and Abarbanel, H. (1992) 'Determining embedding dimension for phase space reconstruction using a geometrical construction', *Physical Review A* 45.06, pp. 3403–3411.

Kennel, M. and Isabelle, S. (1992) 'Method to distinguish possible chaos from colored noise and to determine embedding parameters', *Physical Review A* 46.06, pp. 3111–3118.

Manuca, R. and Savit, R. (1996) 'Model misspecification tests, model building and predictability in complex systems', *Physica D* 93, pp. 78–100.

Mees, A., Rapp, P. and Jennings, L. (1987) 'Singular value decomposition and embedding dimension', *Physical Review A* 36.01, pp. 340–346.

Mindlin, G. and Gilmore, R. (1992) 'Topological analysis and synthesis of chaotic time series', *Physica D* 58, pp. 229–242.

Packard, N., Crutchfield, J., Farmer, J. and Shaw, R. (1980) 'Geometry from time series', *Physical Review Letters* 45.09, pp. 712–716.

Palus, M. and Dvorak, I. (1992) 'Singular value decomposition in attractor reconstruction: pitfalls and precautions', *Physica D* 55, pp. 221–234.

Parker, T. and Chua, L. (1987a) 'INSITE – A software toolkit for the analysis of nonlinear dynamical systems', *Proceedings of the IEEE* 75.08, pp. 1081–1089.

Parker, T. and Chua, L. (1987b) 'Chaos: a tutorial for engineers', *Proceedings of the IEEE* 75.08, pp. 982–1008.

Rosenstein, M., Collins, J. and De Luca, C. (1994) 'Reconstruction expansion as a geometry-based framework for choosing proper delay times', *Physica D* 73, pp. 82–98.

Roux, C. *et al.* (1983) 'Observation of a strange attractor', *Physica D* 8, pp. 257–266.

Sauer, T. (1992) 'A noise reduction method for signals from nonlinear systems', *Physica D* 58, pp. 193–201.

Sauer, T., Yorke, J. and Casdagli, M. (1991) 'Embedology', *Journal of Statistical Physics* 65.03, pp. 579–615.

Schouten, J., Takens, F. and van den Bleek, C. (1994) 'Estimation of the dimension of a noisy attractor', *Physical Review E* 50.03, pp. 1851–1861.

Takens, F. (1981) 'Detecting strange attractors in turbulence', *Warwick 1980 Lecture Notes in Mathematics* 898, pp. 366-381.

Wayland, R. *et al.* (1993) 'Recognizing determinism in time series', *Physical Review Letters* 70.05, pp. 580–582.

Whitney, H. (1934) 'Differentiable manifolds', *Annals of Mathematics* 37.03, pp. 645–680.

4

Fractal dimension

INTUITION

The objective of this chapter is to establish a link between the geometry of the experimental attractor and the number of degrees of freedom in the system, and thereby obtain an estimate of the number of independent variables needed to model the system.

Global charaterizations of the attractor's geometry will not help us here, as they are a function of the way in which the attractor is projected into whatever space it is being observed in. However, any reasonable (useful) projection will at least preserve the local geometery of the attractor, and so that is where we have to look to find out anything about something as fundamental as degrees of freedom. It would be nice if we could characterize the local geometery of the attractor with some familiar concept of dimension which fits easily with the intuitive idea of degrees of freedom. For instance, we might try to define dimension by saying that an object is m-dimensional if it has a finite non-zero volume when viewed in an m-dimensional space. Using that notion of dimension, a doughnut is three-dimensional while its surface is two-dimensional. But a chaotic attractor is a complex structure consisting of tightly packed interweaving trajectories that form an object with dense lacunarity, and the 'volume' of such an object is indeterminate. Hence, this approach will not get us very far. We get an approach that is fruitful by replacing the notion of volume with that of the continuity of (the points that make up) the object. The change of focus is not all that radical, since, for example, we think of a curve as one-dimensional because the points that make it up are continuous in one direction, and we think of a surface as two-dimensional because the points that make it up are continuous in two directions. But point continuity is only a useful idea if we can quantify (measure) it, and if we can show that it has something to do with the number of degrees of freedom in the system. We will look at the connection with degrees of freedom first.

A trajectory is a sequence of points generated by iterates of a discrete time system or a continuous flow of points generated by a continuous time system. Now, the time between iterates of a discrete time system could have any positive value in 'real' time so, for simplicity, we will just 'think' about a continuous time system as being a 'limit' of a discrete time system as the 'real' time between iterates goes to zero. The reason for doing that is not to make mathematical statements, but to define a concept that applies to both discrete and continuous time systems without having to continually refer to both. Having done that, we can say that, in a sense, a trajectory is a stream of points generated by 'iterating' a function defining the dynamical behavior of a deterministic system. We can now think about the attractor of such a system as being a stream of points on the asymptotic trajectories of the system. But it can also be useful to think of the attractor as just a cloud of points with a definite spatial relationship or distribution.

There are two fundamentally different mechanisms governing the distribution of points on the attractor. To illustrate[1], consider a small 'volume' of points on the attractor. In general, such a volume will contain a number of trajectory segments, that is, small sections of the trajectories that pass through the volume element (see Figure 4.1). Hence, the trajectory points in the volume element will be segregated into

---

[1]For simplicity and ease of visualization, the illustrations are done in two and three-dimensional spaces, but the arguments carry through to higher dimensions.

groups where the points in each group are distributed along a particular trajectory segment. Thus, the short term iterates of the map impart a one-dimensional structure on the geometry or distribution of points in the volume element[2]. That one-dimensional structure, which we shall call the *temporal-distribution*, is a trivial consequence of short term iterations of the state transition function of the system.

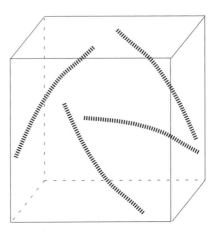

Figure 4.1 Trajectory segments passing through a small volume element on the attractor.

The second mechanism governing the distribution of points in the volume element is the recurrence property of chaotic systems discussed in Chapter 1. Namely, the property that any volume element on the attractor will be revisited by the system infinitely often. Consider the trajectory's revisits to the volume element sequentially. Each revisit creates a new trajectory segment with its own location in the volume element. Because we have assumed that the system's dynamics are continuous (in the state variable) all the trajectory segments in a small volume element will be nearly parallel. Thus, instead of the haphazard layout of trajectory segments shown in Figure 4.1, we have a bundle of parallel trajectories like that shown on the left of Figure 4.2. The temporal-distribution is just the distribution of trajectory points along the parallel trajectory segments. We obtain another, more interesting picture by passing a planar[3] surface through the volume element, transverse to the bundle of trajectory segments. As shown in Figure 4.2, the points at which the trajectory segments intercept the cutting plane *B* have their own distribution, which is unconnected with the temporal-distribution and which we call the *spatial-distribution*. It is the spatial-distribution which carries non-trivial information about the number of degrees of freedom in the system.

---

[2]The one-dimensional aspect is along each trajectory segement, unless of course the volume element only contains a single point from each trajectory segment, in which case the effect is not present.

[3]The plane is to be thought of as having a bit of thickness to avoid the distracting problems associated with its orientation in the presence of discrete points, and also to enable us to simply restrict a three-dimensional probability measure on the attractor to a surface passing through it.

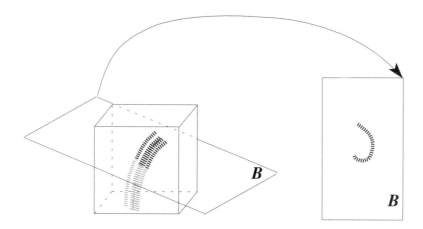

Figure 4.2 On the left, the plane B cuts the bundle of trajectory segments transversely. On the right, B is rotated so that it is easy to see the points at which the trajectory segments intercept B forming the spatial-distribution.

Keep in mind that we arrived at the diagram in Figure 4.1 by zooming in on a small volume of phase space on the attractor. Hence, the spatial-distribution we are talking about is local to that part of the attractor. Now consider Figure 4.1 again and ask what it would mean if the trajectory segments were to line themselves up in such a way that they produced a continuous one-dimensional curve on the surface of the cutting plane as shown on the right of Figure 4.2? If that were the case, implicit in the graph of the curve is a rule that relates one coordinate of a point on the curve to the other coordinate of that point. That is, there is a functional relationship between the coordinates of the point. Since the coordinates are thus dependent, there *cannot* be two degrees of freedom in the system[4]. If, on the other hand, the trajectory segments in the volume element were to pile up in such a way that they intercepted the plane B in a continuous two-dimensional fashion, then the coordinates of those points carry independent information, and there are two degrees of freedom at play. What about a distribution of points which is not as clear cut as the previous two cases; what does it imply about the degrees of freedom in the system? The answer to that question will become clear once we have developed a measure of point continuity, which we do now.

Cover the cutting plane B of Figure 4.2 with a grid of ε sized squares, and assume that the trajectory point x is near the center of one of the squares. Let ρ be the natural probability measure[5] on the attractor, and let $\rho[x; \varepsilon]$ be the probability on the square region centered at x. Now hold x fixed and consider $\rho[x; \varepsilon]$ as a function of ε. Suppose that near x the trajectory points on the attractor pile up along a smooth curve (see Figure 4.3). Since $\rho[x; \varepsilon]$ is a probability measure on the square, it must be proportional to the density of points in the square. The density of points in the square is proportional to the length of the curve along which the trajectory points lie,

---

[4]That is, in addition to the one implied by the temporal-distribution.

[5]See Chapter 1 for the definition of natural measure.

and it is easy to see that the length of a smooth curve passing through a small square is proportional to the edge size ε of the square. Hence, $\rho[x; \varepsilon]$ is proportional to ε.

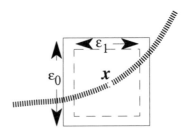

Figure 4.3 Small neighborhood of the point x on a transverse intersection with the attractor. The distribution of points is one-dimensional and the density is proportional to ε, the width of the grid square.

So if we shrink the square by decreasing ε from say ε_0 to ε_1, then

$$\frac{\rho[x; \varepsilon_0]}{\varepsilon_0} \approx \frac{\rho[x; \varepsilon_1]}{\varepsilon_1}$$

That is, $\rho[x; \varepsilon]/\varepsilon$ should remain roughly constant as $\varepsilon \to 0$. On the other hand, if the trajectory points were uniformly, and densely packed together in a neighborhood of x, then the distribution of points on the attractor near x is two-dimensional, and we conclude that $\rho[x; \varepsilon]$ is proportional to the area ε^2 of the square, and that $\rho[x; \varepsilon]/\varepsilon^2$ should remain roughly constant as $\varepsilon \to 0$ (see Figure 4.4).

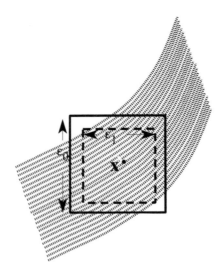

Figure 4.4 Small neighborhood of the point x on a transverse intersection with the attractor. The distribution of points is two-dimensional and the density is proportional to ε^2, the area of the grid square.

There are other possibilities of course. For example, the stacking up of trajectories in the neighborhood of x could form a Cantor-set-like structure[6] (as many chaotic attractors do), like that depicted in Figure 4.5. The result is that, while the local distribution of points is predominantly one-dimensional, it also has some two-dimensional characteristics on *all scales*.

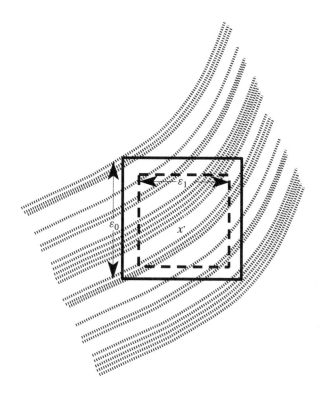

Figure 4.5 Small neighborhood of the point x on a transverse intersection with the attractor. The distribution of points is one-dimensional in one direction, and Cantor-set-like in the other, so the density is proportional to a non-integer power of ε the width of the grid square.

In that case, $\rho[x; \varepsilon]$ will not be proportional to either ε or ε^2. If the spatial-distribution is Cantor-set-like, then

$$\rho[x; \varepsilon] \underset{\varepsilon \to 0}{\sim} \varepsilon^{D(x)} \qquad \text{Eq. 4.1}$$

where $1 < D(x) < 2$. Taking logarithms and rearranging we get

$$\frac{\text{Log} \, \rho[x; \varepsilon]}{\text{Log} \, \varepsilon} \underset{\varepsilon \to 0}{\approx} D(x) \qquad \text{Eq. 4.2}$$

For Eq. 4.2 to make sense, $D(x)$ must be (eventually) independent of ε. $D(x)$

---

[6]Sets with dense lacunarity having a self-similar structure on all scales.

will be independent of ε whenever the distribution of trajectory points is in some sense uniform at any sufficiently fine level of resolution. Cantor-set-like distributions have that property because they are self-similar on all scales. Most known chaotic attractors are smooth in some directions and Cantor-set-like in other directions. We show an example of one that is not in the theoretical section below. If we take away the artificial device of the cutting plane in Figure 4.2, so that the neighborhood of a point on the attractor is now a cube, then we see that $D(x)$ measures the local continuity of the full three-dimensional distribution of trajectory points, encompassing both the spatial and temporal distributions.

Returning to the unanswered question posed earlier: what does a fractional $D(x) = 2.09$ mean? It means that all three coodinates of the recurrent trajectory points in a small neighborhood of x are carrying some independent information, and so there must be more than two, that is, at least three, degrees of freedom in the system[7]. Hence, when it exists, the asymptotic quantity $D(x)$ locally characterizes the degrees of freedom in the system.

$D(x)$, sometimes called the **pointwise dimension**, is a *fractal* dimension of the *neighborhood* of x. If we take its expected value[8] with respect to ρ, then we get a quantity D_1 which is the average pointwise dimension over the system's attractor:

$$\frac{\sum_x \rho[x; \varepsilon] \operatorname{Log} \rho[x; \varepsilon]}{\operatorname{Log} \varepsilon} \underset{\text{for small } \varepsilon}{\approx} D_1 \qquad\qquad \textbf{Eq. 4.3}$$

We have seen the expression in the numerator of the left hand side of the above relationship before. It is the Shannon entropy for an ε-grid partition of the phase space. In Chapter 2, Entropy, we arrived at the expression for D_1, the information dimension, from an information theoretic viewpoint. In this chapter we arrive at the same expression through geometric considerations.

THEORY

A spectrum of dimensions

Let $Q \subset \mathfrak{R}^n$ and let ρ be a probability measure on Q. Let $\mathfrak{R}^n$ be partitioned by a grid of n-dimensional hypercubes with edge size ε. Then every point in Q is in some hypercube $Q_i(\varepsilon)$, $i = 1, \ldots, N(\varepsilon, Q)$, that is, $\{Q_i(\varepsilon)\}$ is a covering of Q, and $N(\varepsilon, Q)$ is the number of hypercubes in the covering of the set Q.

---

[7]There could be more than three degrees of freedom if $D(x)$ is reflecting the geometry of an attractor which is not an embedding.

[8]See Chapter 1 for the definition of expected value.

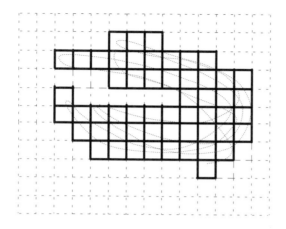

Figure 4.6 A grid partition of a two-dimensional phase space containing an attractor Q. The covering of Q is the collection of grid elements with solid edges.

The *fractal dimension spectrum* of the measure ρ (or the set Q) is defined by

$$D_q(Q, \rho) \equiv \frac{1}{q-1} \lim_{\varepsilon \to 0} \frac{\text{Log} \sum_{i=1}^{N(\varepsilon; Q)} \rho^q[Q_i(\varepsilon)]}{\text{Log}\, \varepsilon} \qquad \text{Eq. 4.4}$$

We will often use $D_q(Q)$, and sometimes $D_q(\rho)$, as shorthand for $D_q(Q, \rho)$, but it should be borne in mind that the fractal dimension of a set Q is computed relative to a measure ρ on Q.

Evaluation of the fractal dimension D_q requires passing to a limit. In practice, we use Eq. 4.4 to obtain the *scaling exponent* of the Renyi[9] entropies for the given measure ρ on Q. The scaling exponent is the slope of the graph of the Renyi entropy versus Log ε for small ε. In an abuse of language, the scaling exponent is also called the fractal dimension. In general $\cdots \geq D_q \geq D_{q+1} \geq \cdots$ so that $D_{q+n}(n > 0)$ is a lower bound on D_q. Higher positive values of q probe the high density characteristics of ρ, while high negative values of q probe the low density characteristics. A set (measure) with $D_q \neq D_p$ ($q \neq p$) is called *multi-fractal*. Since an estimate of any of the fractal dimensions will suffice for the purpose of getting a lower bound on the number of degrees of freedom in the system, the choice is made on the basis of which is more tractable and robust in the experimental situation. It seems intuitive that it will be computationally easier to deal with Eq. 4.4 for small positive values of q, so we will examine D_0, D_1, and D_2.

Capacity

When q is set equal to zero we obtain a quantity D_0 called the *capacity* or *box-counting dimension*:

$$D_0(Q) = \lim_{\varepsilon \to 0} -\frac{\text{Log}\, N(\varepsilon, Q)}{\text{Log}\, \varepsilon} = \lim_{\varepsilon \to 0} \frac{\text{Log}\, N(\varepsilon, Q)}{\text{Log}\, 1/\varepsilon} \qquad \text{Eq. 4.5}$$

[9]Note that numerator in Eq. 4.4 is just the Renyi entropy defined in Chapter 2, Entropy.

The term box-counting refers to the fact that D_0 measures the power law that governs the increase in the number of hypercubes required to cover the set Q as the size of the hypercubes is reduced, that is

$$N(\varepsilon, Q) \sim \varepsilon^{-D_0(Q)}$$

If Q is two-dimensional, then $N(\varepsilon, Q) \sim \varepsilon^{-2}$, and so on. The first problem with D_0 is that, with the limited data sets encountered in practice, some areas of the attractor may not get a true distribution of trajectory points. These areas will make unrepresentative contributions to the calculation of D_0. In particular, sparse areas of the true attractor will be visited rarely by an experimental trajectory. These areas will be under-represented in terms of the number of boxes needed to cover the attractor, so that D_0 will be underestimated. The second problem with D_0 is that the numerical procedures for estimating it have been inefficient, requiring large amounts of computer memory and processing time.

Information
The quantity

$$D_1(Q) = \lim_{q \to 1} D_q(Q)$$

is called the **information dimension**. Using L'Hôpital's rule it can be shown that

$$D_1(Q) = \lim_{\varepsilon \to 0} \frac{\sum_{i=1}^{N(\varepsilon, Q)} \rho[Q_i(\varepsilon)] \, \mathrm{Log} \, \rho[Q_i(\varepsilon)]}{\mathrm{Log} \, \varepsilon} \qquad \textbf{Eq. 4.6}$$

from which we obtain the scaling properties of the Shannon entropy in the numerator. From Eq. 4.3, $D_1(Q)$ is also the expected value of the pointwise dimension $D(x)$ where

$$D(x) = \lim_{\varepsilon \to 0} \frac{\mathrm{Log} \, \rho[x; \varepsilon]}{\mathrm{Log} \, \varepsilon}, \qquad x \in Q$$

The existence of an ergodic natural measure ρ implies that $D(x) = D_1(Q)$, except possibly on a set of ρ measure zero. So it would seem that we could easily get a good estimate of the attractor's information dimension by computing $D(x)$ for x, a typical point[10] on the attractor. Unfortunately, in practice $D(x)$ has to be estimated with a finite data set which is likely to result in $D(x) \neq D(x')$ for $x \neq x'$ even if we could determine that x and x' were typical. The set of untypical points has natural measure zero, so the pointwise dimensions of those points make a nil contribution to Eq. 4.6. For these reasons the information dimension of the attractor is typically estimated via $D_1(Q)$. Using the first order correlation integral[11] of a reconstructed trajectory of the system to approximate Eq. 4.6 indeed yields a tractable computation for obtaining the scaling exponent $D_1(Q)$.

---

[10] A typical point is a point on a typical trajectory. A typical trajectory is one which is ergodic, that is, infinitely recurrent in all neighborhoods on the attractor.

[11] See Appendix 3, Correlation integral, and the algorithm section of Chapter 2, Entropy.

Correlation

When q is set equal to 2 we obtain a quantity called the ***correlation dimension***:

$$D_2(Q) = \lim_{\varepsilon \to 0} \frac{\text{Log} \sum_{i=1}^{N(\varepsilon, Q)} \rho^2[Q_i(\varepsilon)]}{\text{Log } \varepsilon} \qquad \text{Eq. 4.7}$$

Like D_1, D_2 can be approximated by replacing the summation in Eq. 4.7 with a correlation integral of an experimental trajectory[12] $y(j) \in \mathfrak{R}^{d_r}$. In the case of the correlation dimension, however, we use the second order correlation integral $C_2(d_r; \varepsilon)$ [13], which is particularly well suited to numerical methods. The substitution yields for small ε

$$D_2(Q) \approx \frac{\text{Log } C_2(d_r; \varepsilon)}{\text{Log } \varepsilon} \qquad \text{Eq. 4.8}$$

D_2 is invariant

We obtain Eq. 4.8 from Eq. 4.7 by using the ergodicity of ρ to get a time average, and then replacing the trajectory of the true system with a trajectory reconstructed from observational data. It is not obvious that the two expressions give the same value as $N_r \to \infty$. For that to be true, the experimental attractor would have to have the same detailed geometric structure as the attractor of the true system and that needs to be proved.

Theorem 1 (Ding *et al.*, 1993)
Consider a map $G : \mathfrak{R}^n \to \mathfrak{R}^m$. Let $Q \subset \mathfrak{R}^n$, and ρ be a probability measure on Q.
The induced measure of a set $E \subset \mathfrak{R}^m$ is defined by

$$\rho_G(E) \equiv \rho[G^{-1}(E)]$$

where $G^{-1}(E)$ is the set of points mapped by G into E.
If $m \geq D_2(\rho)$, then for almost every smooth map $G : \mathfrak{R}^n \to \mathfrak{R}^m$, the correlation dimension is preserved under G, that is, $D_2(\rho_G) = D_2(\rho)$.

Theorem 2 (Ding *et al.*, 1993)
Consider a dynamical system defined by the map $f : \mathfrak{R}^n \to \mathfrak{R}^n$. Let $Q \subset \mathfrak{R}^n$ be invariant under f; that is, $Q \equiv f(Q)$. Let ρ, a measure on Q, be invariant[14] under f for all measurable sets $E \subset \mathfrak{R}^n$; that is, $\rho(E) \equiv \rho[f^{-1}(E)]$. Let $h : \mathfrak{R}^n \to \mathfrak{R}$. Finally, let Φ be the delay coordinate map[15] $\Phi(s) : \mathfrak{R}^n \to \mathfrak{R}^m$ defined by

$$\Phi(s) = [h(s), h(f^{-1}[s]), \dots, h(f^{-(m-1)}[s])]$$

If f has only finitely many periodic orbits in Q of period less than or equal to m, and if $m \geq D_2(\rho)$, then for almost every smooth h, $D_2(\rho_\Phi) = D_2(\rho)$.

---

[12] See Chapter 3, Phase space reconstruction.
[13] The dependence on the number of trajectory points N_r is suppressed.
[14] If ρ is the natural measure, then ρ is invariant.
[15] See Chapter 3, Phase space reconstruction.

Theorem 2 says that we can estimate the correlation dimension of the true system from the correlation dimension of the system reconstructed from experimental data with a delay coordinate map, provided the reconstruction dimension d_r is sufficiently large. By implication, the estimates of D_2 from the reconstructed system should be stable if d_r is sufficiently large whatever its value.

The correlation exponent

Dynamical system	Parameters	Estimate of D_2
Logistic	$a = 3.56994$	0.500 ± 0.005
Hénon	$a = 1.4, b = 0.03$	1.21 ± 0.01
Kaplan–York	$a = 0.2$	1.42 ± 0.01
Lorenz	$\alpha = 10, b = 8/3, R = 28$	2.05 ± 0.01
Rabinovich–Fabrikant	$\alpha = 1.1, \gamma = 0.87$	2.19 ± 0.01
Zaslavskii	$\Gamma = 3.0, v = 400/3, \varepsilon = 0.3,$ $\mu = (1 - e^{-\Gamma})/\Gamma$	≈ 1.53
Mackay–Glass	$a = 0.2, b = 0.1, c = 10,$ $\tau = 100$	≈ 7.1

Table 4.1 Estimates of the correlation dimension for some chaotic systems. (See text below for the equations of motion for these systems.)

Source: Mackay–Glass estimate obtained from Ding *et al.* (1993) p. 411. Other estimates obtained from Grassberger and Procaccia (1983) p. 193.

Consider the correlation integral $C_2(d_r; \varepsilon)$ computed from a trajectory reconstructed with a d_r-dimensional delay coordinate map. From the above discussion, for most systems, that is, for systems where D_2 exists, Log $C_2(\varepsilon)$[16] should scale like D_2 Log ε. In particular, it follows from Theorem 2 above that the graph of Log $C_2(d_r; \varepsilon)$ versus Log ε should be roughly linear with slope $\Delta(d_r) \approx D_2$ for small values of ε, provided d_r is sufficiently large. The estimate $\Delta(d_r)$ is the correlation exponent. The following six examples demonstrate the validity of the linear relationship for the *true* attractor of several chaotic systems. As the equations of motion are known for these systems, we can generate very long trajectories that approximate the attractor closely and yield accurate estimates of the correlation integral. In all but the last example the plots show a region where Log $C_2(\varepsilon)$ is linear in Log ε[17]. The reasons for the non-linearity at extreme values of ε in the plots are discussed below under the scaling region rubric.

Example 4.1

The Logistic map

$$x_{n+1} = ax_n(1 - x_n)$$

μ: $a = 3.56994$

---

[16] $C_2(\varepsilon)$, the theoretical order-2 correlation integral, is the probability that two randomly chosen points on the attractor will fall within ε of one another. $C_2(d_E, \varepsilon) \to C_2(\varepsilon)$ as $N_r \to \infty$

[17] Without loss of generality, we can choose a measurement scale such that the entire attractor is contained in a hypersphere of radius $\varepsilon = 1$. A consequence of this is that $C_2(d_r, \varepsilon) \to 1$ as $\varepsilon \to 1$.

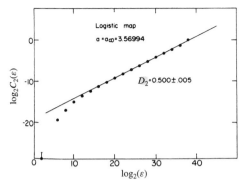

Figure 4.7 Correlation integral C_2 and exponent D_2 for the Logistic map. Initial value $x_0 = 0.5$ and $N_r = 30,000$.

Source: Adapted from Grassberger and Procaccia (1983) p. 193.

Example 4.2

The Hénon map

$$x_{n+1} = Y_n + 1 - ax_n^2$$
$$Y_{n+1} = bx_n$$

μ: $\quad a = 1.4, b = 0.03$

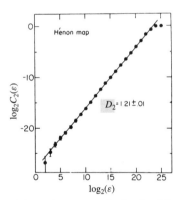

Figure 4.8 Correlation integral C_2 and exponent D_2 for the Hénon map. $N_r = 15,000$.

Source: Adapted from Grassberger and Procaccia (1983) p. 194.

Example 4.3[18]

The Kaplan–York map

$$x_{n+1} = 2x_n \, (mod \, 1)$$
$$y_{n+1} = ax_n + \cos 4\pi x_n$$

μ: $\quad a = 0.2$

---

[18]Kaplan and York (1979).

161

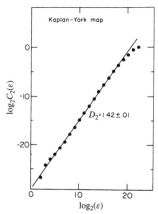

Figure 4.9 Correlation integral C_2 and exponent D_2 for the Kaplan–York map. $N_r = 15,000$.
Source: Adapted from Grassberger and Procaccia (1983) p. 194.

Examples 4.4 and 4.5[19]

The Lorenz equations	*The Rabinovich–Fabrikant equations*

$$\frac{dx}{dt} = \sigma(y - x)$$

$$\frac{dy}{dt} = -y - xz + Rx$$

$$\frac{dz}{dt} = xy - bz$$

$$\mu: \quad \sigma = 10, b = 8/3, R = 28$$

$$\frac{dx}{dt} = y(z - 1 + x^2) + \gamma x$$

$$\frac{dy}{dt} = x(3z + 1 - x^2) + \gamma y$$

$$\frac{dz}{dt} = -2z(\alpha + xy)$$

$$\mu: \quad \gamma = 0.87, \alpha = 1.1$$

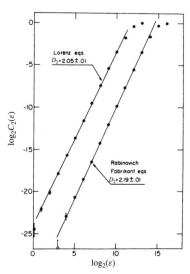

Figure 4.10 Correlation integrals C_2 and exponents D_2 for the Lorenz equations, and for the Rabinovich–Fabrikant equations. In both cases $N_r = 15,000$.
Source: Adapted from Grassberger and Procaccia (1983) p. 196.

---

[19]Rabinovich and Fabrikant (1979).

Example 4.6 [20]

But the pattern is not inevitable as demonstrated by this example (see Figures 4.11 and 4.12) where the self-similar structure of the attractor breaks down at small scales.

The Zaslavskii map

$$x_{n+1} = [x_n + \nu(1 + \beta y_n) + \varepsilon\nu\beta\cos(2\pi x_n)] \ (\text{mod } 1)$$
$$y_{n+1} = e^{-\Gamma}[y_n + \varepsilon\cos(2\pi x_n)]$$

$$\mu: \quad \Gamma = 3.0, \ \nu = 400/3, \ \varepsilon = 0.3, \ \beta = (1 - e^{-\Gamma})/\Gamma$$

Zaslavski map, 15000 pts.

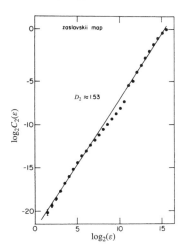

Figure 4.11 Zaslavskii attractor. $N_r = 25,000$. The y-coordinate is blown up by a factor of 25 for faster scaling. With the blow up, the attractor is square-like at low resolution. Without the blow up factor, the attractor looks almost one-dimensional at low resolution (see small insert in figure).

Source: Adapted from Grassberger and Procaccia (1983) p. 195.

Figure 4.12 Correlation integral C_2 and exponent D_2 for the Zaslavskii map. The nonlinear behavior indicates a fundamental change in the structure of the attractor even on small scales.

Source: Adapted from Grassberger and Procaccia (1983) p. 195.

[20]Zaslavskii (1978).

Convergence

In the experimental situation, d_r has to be chosen sufficiently large to stabilize $\Delta(d_r)$. In practice we do not know what 'sufficiently large' is, but based on the discussion so far, we might reasonably expect that $\Delta(d_r)$ would saturate (converge) for increasing values of d_r. From Theorem 2 above, the onset of the saturation of $\Delta(d_r)$ should be somewhere above the minimum embedding dimension d_e. To illustrate the point we consider the Mackay–Glass equation

$$\frac{dx}{dt} = \frac{ax(t - \tau)}{1 + [x(t - \tau)]^c} - bx(t)$$

μ: $\quad a = 0.2; b = 0.1; c = 10; \tau = 100$

Because it is a delay differential equation it is infinite dimensional, but its dynamics contract phase space to produce an attractor with a correlation dimension D_2 of about 7.1 . The plots in Figure 4.13 show a slope $\Delta(d_r) \approx d_r$ for $d_r \leq 7$ and a saturated $\Delta(d_r) \approx 7.1$ for $d_r \geq 8$. For this system $\Delta(d_r)$ saturates at $\lceil D_2 [21]$, the lower bound on the number of degrees of freedom in the system. Since any embedding dimension must be greater than or equal to $\lceil D_2$, the dimension required for an embedding is as small as it could be. That is fortunate, but unusual.

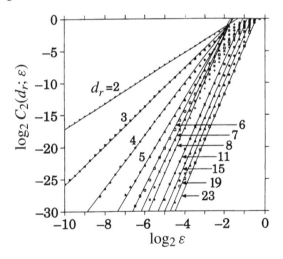

Figure 4.13 Log-log plots of the correlation integrals generated by $N_r = 50,000$ points of the Mackay–Glass equation shown at various values of the reconstruction dimension d_r.

Source: Adapted from Ding *et al.* (1993) p. 412.

Scaling region

It is clear from Figures 4.7–4.12 that the graphs are not everywhere linear, that is, they are only linear over a limited range of ε values. If it exists, the range of ε values corresponding to the linear region is called the ***scaling region***. In practice the high end of the scaling region can be significantly less than the extent of the attractor, and

[21]The expression $\lceil D_2$ denotes the smallest integer greater than or equal to D_2.

low end significantly greater than zero. The existence and size of the linear scaling region determine how successful we will be at extracting a reliable estimate of D_2, so it is important to develop some feeling for the nature, extent, and causes of any nonlinear behavior we can expect to encounter in practice.

We begin with the factors that limit the linear scaling region for small values of ε near zero. Continue to assume that ε is normalized so that it measures in units of $d(A)$, the maximum extent of the attractor. The smallest value of ε for which $C_2(d_r; \varepsilon)$ has any meaning is when ε is on the order of the smallest interpoint distance[22], which is given by

$$\varepsilon_{\min} \approx [N_r]^{-\frac{2}{D_2}}$$

At that distance

$$C_2(d_r; \varepsilon_{\min}, N_r) \approx \frac{2}{[N_r]^2}$$

The above number is the lower cutoff on the C_2 axis of the $\text{Log } C_2 - \text{Log } \varepsilon$ plots, so that all else being equal, a shorter data set will produce a smaller scaling region. This effect can be seen quite clearly in Figure 4.15. Figure 4.14 illustrates another consequence of small data sets, namely the poor covergence of D_2 and breakdown in the relationship between the minimum embedding dimension and the convergence of D_2. Note that increasing N_r by over sampling does not improve the situation as in that case we are just filling out the temporal-distribution, which will saturate at some point.

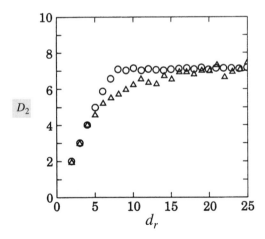

Figure 4.14 D_2 versus d_r, plotted as open circles, for the data set of $N_r = 50,000$ points used to generate Figure 4.13. D_2 versus d_r, plotted as triangles, for a smaller data set of $N_r = 2,000$ points. The convergence of D_2 for the smaller data set is slow and unstable.

Source: Adapted from Ding et al. (1993) p. 412.

To understand the curves for values of a normalized ε near 1 we have to do a bit more work. Figure 4.15 was computed from data generated by the Mackay–Glass equation with somewhat different parameters than were used to generate

---

[22]See Appendix 1, Data requirements.

Figure 4.13. This parameterization produces an attractor with a correlation dimension D_2 of roughly 5.7.

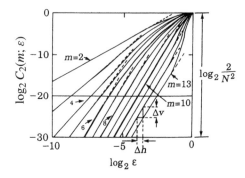

Figure 4.15 Schematic diagram of correlation integrals, where m is the reconstruction dimension d_r.

Source: Adapted from Ding *et al.* (1993) p. 412.

From Figure 4.15 we see that the graphs of the correlation integral have several notable features for larger values of ε. First, in the linear scaling region, $\Delta(d_r) \approx d_r$ until $\Delta(d_r)$ saturates. Second, the nonlinear aspects of the plots, when ε is near 1, are, for small values of d_r, quite different than they are for large values of d_r. Finally, in the linear scaling region the log-log plots have constant vertical and horizontal spacing for increasing values of d_r, after $\Delta(d_r)$ saturates. The first two features are a manifestation of the distribution of the experimental trajectory points for a given reconstruction dimension d_r. If the reconstruction dimension is not sufficiently large, the experimental attractor is projected into a space where there are insufficient degrees of freedom to accommodate its true geometry. The result is that the spatial correlations of the trajectory points are destroyed and their distribution becomes more or less random (as Figure 4.16 illustrates).

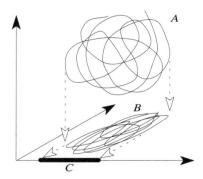

Figure 4.16 B is the projection of the three-dimension attractor A into a two-dimensional space. C is the projection of B into a one-dimensional space. As the attractor is projected into lower dimensions the distribution of trajectory points becomes increasingly random.

As a consequence, for smaller values of d_r, the distribution of trajectory points of the reconstructed system will have scaling properties similar to a random distribution of points in $\Re^{d_r}$. Trajectories for random systems tend to be isotropic, that is, they uniformly fill out whatever space they live in. As the distribution is isotropic, it looks continuous in all directions as the number of trajectory points gets large. Recall from the Intuition section that a distribution of trajectory points that is continuous in d_r dimensions will scale like ε^{d_r}. Hence, $C_2(d_r; \varepsilon)$ will scale like ε^{d_r} and the log-log plot will generally have slope $\Delta(d_r) = d_r$. However, assuming the trajectory is on an attractor, it is not free to fill up $\Re^{d_r}$ and will be confined to some bounded region. The nonlinearity in $C_2(d_r; \varepsilon)$, as $\varepsilon \to 1$, for smaller values of d_r is due to the edge effects at the boundary of the attractor. To see this, start with $d_r = 1$ and assume the trajectory points are independent and uniformly distributed on $[0,1]$. Then $C_2(1; \varepsilon)$ approximates the probability of two points, chosen at random with respect to a uniform measure, being within ε of each other. That probability, for two points y' and y'', is equal to the area of the shaded region in Figure 4.17.

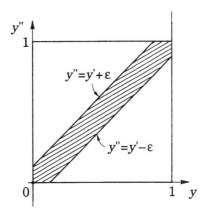

Figure 4.17 Illustration for the calculation of the correlation integral $C_2(1; \varepsilon)$ from a random data set.

Source: Adapted from Ding et al. (1993) p. 413.

So

$$C_2(1, \varepsilon) = \sqrt{2} \cdot \sqrt{2} \cdot \varepsilon - \varepsilon^2$$
$$= 2\varepsilon - \varepsilon^2$$

It follows that

$$\text{Log } C_2(1, \varepsilon) = \text{Log } \varepsilon + \text{Log}(2 - \varepsilon)$$

Thus

$$\Delta(1) = \frac{d \text{ Log } C_2(1, \varepsilon)}{d \text{ Log } \varepsilon} = 1 - \frac{\varepsilon}{2 - \varepsilon}$$

where $\Delta(1)$ is the slope of the log-log plot of $C_2(1; \varepsilon)$ versus ε. For ε small $\Delta(1) \approx 1$, while $\Delta(1) \to 0$ as $\varepsilon \to 1$.

In the general case, it can be argued that, with the max-norm used as a distance measure, the assumption of uniformly distributed trajectory points implies

$$C_2(d_r, \varepsilon) = [C_2(1, \varepsilon)]^{d_r} = (2\varepsilon - \varepsilon^2)^{d_r}$$

so

$$\Delta(d_r) = \frac{d \, \text{Log} \, C_2(d_r, \varepsilon)}{d \, \text{Log} \, \varepsilon} = d_r \frac{d \, \text{Log} \, C_2(1, \varepsilon)}{d \, \text{Log} \, \varepsilon} = d_r \left(1 - \frac{\varepsilon}{2 - \varepsilon} \right)$$

Hence, for a fixed value of d_r, $\Delta(d_r) \approx d_r$ for ε small, while $\Delta(d_r) \to 0$ as $\varepsilon \to 1$ as before. But for a fixed ε, $\Delta(d_r)$ is increasing and C_2 is decreasing with d_r, so we always get a picture like that in Figure 4.18. As long as the reconstruction dimension d_r is not large enough to completely unfold the experimental trajectory we can expect to see $\Delta(d_r)$ exhibit the characteristics of randomly distributed points in a bounded region as discussed above.

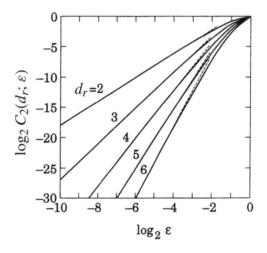

Figure 4.18 Log-log plots of the correlation integrals for a random data set.
Source: Adapted from Ding *et al.* (1993) p. 414.

When $d_r \geq d_e$ the slope $\Delta(d_r)$ remains stable at D_2, but the log-log plots will continue to exhibit nonlinear behavior as $\varepsilon \to 1$. The nonlinearity in this case arises from a sharp increase in C_2 before it converges to 1 (see Figure 4.15). Although the effect is quite different, the nonlinearities are again a function of the distribution of trajectory points over the attractor. Consider the reconstructed trajectory y. When the reconstruction is a time delay mapping[23] and $d_r \gg d_e$, the first and last vector components of an experimental trajectory point on y tend to come from points on the true trajectory that are separated by relatively large intervals of time. In this situation, the information contained in the first vector components of the reconstructed trajectory point are coming from one location on the attractor of the true system, while

---

[23]See Chapter 3, Phase space reconstruction.

the information contained in the last vector components is coming from a different location on the true attractor. Points on a chaotic attractor separated by relatively large distances are practically uncorrelated, so the head and the tail components of the reconstructed trajectory point vectors are reflecting different dynamics. The result is that the reconstructed attractor can acquire increasingly complicated folds as d_r is increased. As the folding becomes severe, the distribution of trajectory points acquires higher dimensional features on scales similar to those of the folds (see Figure 4.19). On these scales $C_2(d_r; \varepsilon)$ balloons, causing $\Delta(d_r)$ to increase rapidly at first and then drop to zero as the boundary effects take hold.

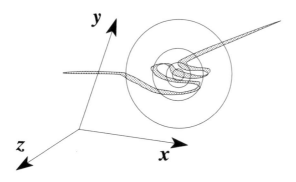

Figure 4.19 Distortion of the distribution of trajectory points caused by attractor folds induced by reconstructions in excessively high dimensions.

The last feature of Figure 4.15 that we need to explain is why, after saturation in d_r, the Log C_2 curves, for increasing d_r, maintain a constant vertical distance Δv, and horizontal distance Δh, between each other throughout the linear scaling region. This behavior is not surprising when we recall from Chapter 2: Entropy that when N_r is large, and ε is small, and $d_r \geq d_e$

$$\Delta v \equiv \text{Log} \, \frac{C_2(d_r, \varepsilon)}{C_2(d_r + 1, \varepsilon)} \approx \tau K_2$$

where τ is the time delay of the reconstruction. The fact that Δv and $\Delta(d_r)$ are constant for $d_r \geq d_e$, implies that Δh is also constant. We complete this section with a note on how noisy data will affect the scaling region.

The previous log-log plots were all constructed using noise free data. If the data is contaminated by noise with standard deviation σ_{noise}, then C_2 behaves as if the data were random noise when $\varepsilon < \sigma_{\text{noise}}$. The effect is to produce a 'knee' in the log-log plots and reduce the scaling region even more, since necessarily the smallest useful value of ε is larger than either the smallest interpoint distance ε_{min} or σ_{noise}. The plots in Figure 4.20 clearly show evidence of the knee effect. But, they also show the approximate scale at which the effect takes hold, which, for a reasonably sized data set, is σ_{noise}. Hence, as long as the noise level is not too high, we can also estimate the amount of noise in the system from the same plots we use to estimate the correlation dimension D_2. If there is a requirement for a large number of correlation dimension

169

estimates, then the estimate of the noise can be automated so that both are computed simultaneously[24].

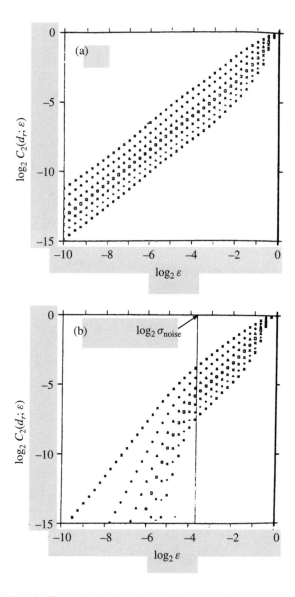

Figure 4.20 The 'knee' effect in the log-log plots of the correlation integrals of the Hénon map for $d_r = 2, 3 \ldots 8$ using (a) clean data and (b) noise contaminated data.
Source: Adapted from Ding *et al.* (1993) p. 418.

---

[24]Schouten, Takens and van den Beek (1994).

Pitfalls

It would be difficult to overstate the following point. If the system posesses a chaotic attractor, then the correlation exponent is a computationally efficient way to estimate its fractal dimension, but a *small noninteger correlation exponent does not imply low dimensional chaotic dynamics.*

The reason is that the correlation integral only reflects the spatial distribution of the experimental trajectory points, and not their temporal relationship. In other words, the correlation integral does not reflect the dynamics responsible for the spatial distribution of the trajectory points. Hence, *the correlation integral sees no difference between a fractal distribution induced by a fractal trajectory of a stochastic system, and a fractal distribution induced by the recurrence properties of a smooth trajectory of a deterministic system. Therefore, any algorithm that employs the correlation integral to estimate fractal dimension cannot distinguish between a fractal attractor and a fractal trajectory, if the two have the same dimension.*

To illustrate that we are not speaking of pathological cases, consider a time series whose Fourier power spectrum $P(\omega)$ obeys the power-law

$$P(\omega) = C\omega^{-\alpha} \qquad \alpha \geq 0$$

A variety of physical systems, both chaotic and stochastic, are known to have power spectra of the above type. A stochastic process that has a power-law power spectrum with $1 < \alpha < 3$ is called 'colored noise.' If a time series from a colored noise process is used as the basis for a d_r-dimensional state space reconstruction, the resulting experimental trajectory will be self-similar on all scales, and have a fractal dimension D given by[25]

$$D = \min\left\{\frac{2}{\alpha - 1}, d_r\right\}$$

The correlation exponent $\Delta(d_r)$ approximates D, so, as the reconstruction dimension d_r is increased, $\Delta(d_r) \approx d_r$ until $d_r = D$. When $d_r > D$, the correlation exponent saturates at $2/(\alpha - 1)$. Since increasing d_r until $\Delta(d_r)$ converges is a common method for estimating the fractal dimension[26], it is clear from the above colored noise example that a sequence of correlation exponent estimates that converges to a finite value does not imply chaotic dynamics.

We note that the convergent correlation exponent for the above colored noise processes runs contrary to the widely held view that noise always 'fills up' whatever space it is embedded in, and therefore, should not have a finite fractal dimension. In fact, the expectation is fulfilled, at least asymptotically as $N_d \to \infty$, for 'white noise' processes, that is, stochastic processes that have a power-law spectrum with $0 \leq \alpha \leq 1$. However, since the 'filling out' of the reconstructed state space is an asymptotic phenomenon, it may manifest itself only very slowly, so that short or even moderate length time series from white noise processes can also yield convergent correlation exponents.

---

[25] D is a fractal dimension called the Hausdorff dimension. It is close in value to D_1 and D_2. See Osborne and Provenzale (1989).

[26] See the GPA algorithm in the Algorithms section of this chapter.

ALGORITHMS

GPA algorithm[27]

It should be clear by now that any numerical procedure for estimating the correlation dimension D_2 will involve two distinct steps. The first is to compute the correlation integral C_2 for a given phase space reconstruction, and the second is to estimate the slope $\Delta(dr)$ in the linear scaling region of the graph of Log C_2 versus Log ε. The procedure is then repeated in increasingly higher dimensional reconstructed phase spaces until $\Delta(dr)$ converges to some value Δ^* which we take as the estimate of the correlation dimension D_2. That procedure is known as the **Grassberger–Procaccia algorithm.**

The computation of C_2 requires a reconstruction of the phase space to obtain an experimental trajectory followed by a nearest neighbor search to obtain the distribution of trajectory points on the reconstructed attractor, neither of which is a trivial task. An extensive discussion of the issues and algorithms is contained in Chapters 3, Phase space reconstruction and 6, Noise reduction, and in Appendices B, Nearest neighbor searches and C, Correlation integral. Calculating the slope of the Log C_2-Log ε plot is comparatively simple. We obtain by visual inspection the end points of the linear scaling region, and then use a linear regression of Log C_2 on Log ε to obtain the slope $\Delta(d_r)$.

As the reliability of Δ^* depends on the size of the linear scaling region, it is desirable to repeat all the calculations for several values of each of the parameters used in the phase space reconstruction and noise reduction to find a stable Δ^* with a maximum scaling region. The extended procedure is summarized in Figure 4.21.

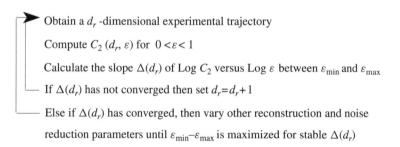

Obtain a d_r-dimensional experimental trajectory

Compute $C_2(d_r, \varepsilon)$ for $0 < \varepsilon < 1$

Calculate the slope $\Delta(d_r)$ of Log C_2 versus Log ε between ε_{min} and ε_{max}

If $\Delta(d_r)$ has not converged then set $d_r = d_r + 1$

Else if $\Delta(d_r)$ has converged, then vary other reconstruction and noise reduction parameters until $\varepsilon_{min} - \varepsilon_{max}$ is maximized for stable $\Delta(d_r)$

Figure 4.21 Grassberger–Procaccia algorithm.

Remarks

- As discussed in the Theory section of this chapter, while the GPA algorithm is an efficient way to estimate the fractal dimension of a chaotic attractor when there is one present, a convergent sequence of correlation exponents does not imply chaotic dynamics.

[27]Grassberger and Procaccia (1983).

- Since the time series from a white noise process can be slow to fill up a reconstructed state space, even they can yield finite correlation exponents. However, since the filling-up process will eventually occur, a correlation exponent that increases with the length N_d of the time series, must be taken as a warning that the system under study may be stochastic.

- We also stress how important the techniques used in noise reduction and phase space reconstruction can be in determining the accuracy of the correlation dimension estimates. For a dramatization of the point, see the examples for the re-embedding algorithm in Chapter 6, Noise reduction.

REFERENCES AND FURTHER READING

Albano, A., Muench, J., Schwartz, C., Mees, A. and Rapp, P. (1988) 'Singular value decomposition and the Grassberger–Procaccia algorithm', *Physical Review* A 38.06, pp. 3017–3026.

Auerbach, D., Cvitanovic, P., Eckmann, J., Gunaratne, G. and Procaccia, I. (1987) 'Exploring chaotic motion through periodic orbits', *Physical Review Letters* 58.23, pp. 2387–2389.

Bingham, S. and Kot, M. (1989) 'Multidimensional trees, range searching and a correlation dimension algorithm of reduced complexity', *Physics Letters* A 140.06, pp. 327–330.

Buzug, Th., Pawelzik, K., von Stamm, J. and Pfister, G. (1994) 'Mutual information and global strange attractors in Taylor–Couette flow', *Physica* D 72, pp. 343–350.

Ding, M., Grebogi, C., Ott, E., Sauer, T. and York, J. (1993) 'Estimating correlation dimension from a chaotic time series: when does plateau onset occur', *Physica* D 69, pp. 404–424.

Eckmann, J. and Ruelle, D. (1992) 'Fundamental limitations for estimating dimensions and Lyapunov exponents in dynamical systems', *Physica* D 56, pp. 185–187.

Fraedrich, K. and Wang, R. (1993) 'Estimating the correlation dimension of an attractor from noisy and small datasets based on re-embedding', *Physica* D 65, pp. 373–398.

Grassberger, P. and Procaccia, I. (1983) 'Measuring the strangeness of strange attractors', *Physica* D 9, pp. 189–208.

Grassberger, P. (1990) 'An optimized box assisted algorithm for fractal dimensions', *Physics Letters* A 148.01, pp. 63–68.

Kaplan, J. L. and York, J. A. (1979) 'Chaotic behavior of multidimensional difference equations', in H. Peitgen and H. Walther (eds) *Lecture Notes in Math.* 730, Springer, Berlin pp. 204–207.

Osborne, A. and Provenzale, A. (1989) 'Finite correlation dimension for stochastic systems with power-law spectra', *Physica* 35D, pp. 357–381.

Packard, N., Crutchfield, J., Farmer, J. and Shaw, R. (1980) 'Geometry from time series', *Physical Review Letters* 45.09, pp. 712–716.

Prichard, D. and Theiler, J. (1995) 'Generalized redundancies for time series analysis', *Physica* D 84, pp. 476–493.

Provenzale, A., Smith, L., Vio, R. and Murante, G. (1992) 'Distinguishing between low dimensional dynamics and randomness in measured time series', *Physica* D 58, pp. 31–49.

Rabinovich, M. I. and Fabrikant, A. L. (1979) 'Stochastic self-modulation of waves in nonequilibrium media', *Sov. Phys. JETP* 50, pp. 311–317.

Schouten, J., Takens, F. and van den Bleek, C. (1994) 'Estimation of the dimension of a noisy attractor', *Physical Review* E 50.03, pp. 1851–1861.

Takens, F. (1981) 'Detecting strange attractors in turbulence', *Warwick 1980 Lecture Notes in Mathematics* 898, pp. 366–381.

Theiler, J. (1987) 'Efficient algorithm for estimating the correlation dimension from a set of discrete points', *Physical Review* A 36.09, pp. 4456–4462.

Zaslavskii, G. M. (1978) 'The simplest case of a strange attractor', *Physics Letters* 69A, pp. 145–147.

5

Lyapunov exponents

INTUITION

Forecasting, even for deterministic systems, has its limitations. As discussed in Chapter 2, Entropy, the accuracy with which we can predict the behavior of a dynamical system is inversely proportional to the rate at which the system is producing information. The **Lyapunov exponents** are measures of phase space deformation as manifested in the relative movement of nearby trajectories. In the case of a one-dimensional system the Lyapunov exponent is equivalent to the rate at which the system produces information. Hence, for one-dimensional systems the Lyapunov exponent quantifies precisely the accuracy of a prediction for a given future point in time. In higher dimensions, we get a spectrum of Lyapunov exponents that characterize the deformation of the multidimensional phase space, but these are not related to the predictability of the system in a simple way. Nevertheless, even in the multidimensional case the information production rate, and hence the predictability of the system, is dominated by the largest Lyapunov exponent.

To gain a bit of intuition, we imagine a hapless homunculus wandering into a region of the attractor of a dynamical system[1] that has been temporarily shut down for repairs. He decides to build himself an enclosure by fixing the structure to a few state space points in the vicinity. He is a bit worried that this area of the universe may actually belong to someone but concludes that since he is a homunculus, the finished enclosure will only occupy a tiny (infinitesimal) volume of the state space and will probably go unnoticed (see Figure 5.1).

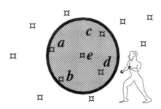

Figure 5.1 Points a, b, c, and d are on the surface of the spherical enclosure. Point e is at the center of the spherical enclosure.

On completion he climbs into the enclosure to inspect his work, when someone switches the dynamical system on. Suddenly the enclosure becomes a capsule traveling in the flow of the dynamical system like a small submarine in a river, except in this case the river consists of the flow's trajectories some of which are anchored to the walls of the capsule (see Figure 5.2). What sort of 'physical' forces will the homunculus experience in the first few moments of this unintended journey?

---

[1]For the purposes of this illustration we assume a three-dimensional continuous time dynamical system. Similar ideas apply to discrete systems and other dimensions.

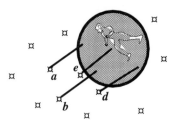

Figure 5.2 Enclosure starts to move with the flow.

Because the capsule is moving in the flow, he feels no force in the direction of the flow. But if the flow is distorting phase space volumes, then he will feel expanding and contracting pressures in other directions as the walls of the capsule try to follow the different trajectories of the initial state space points the capsule was attached to. Since the dynamics are also continuous in the state variable[2], over time the separation of nearby points is governed by a linear relationship, and an initial infinitesimal sphere is distorted into an infinitesimal ellipsoid whose principal axes are oriented in the expanding and contracting directions as well as along the direction of zero deformation. Hence, the homunculus perceives a natural reference frame whose origin is anchored to the trajectory of the initial center point of the enclosure, and whose coordinate axes coincide with the principal axes of the developing ellipsoid (see Figure 5.3).

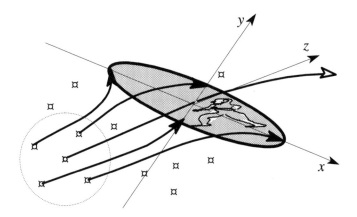

Figure 5.3 Capsule (phase space) is squeezed along the y-axis and stretched along the x-axis. There is no distortion along the z-axis which is parallel to the tangent of the trajectory of the center point.

Since there is no distortion in the direction of the flow, one of the ellipsoid's axes must coincide with the tangent line of the trajectory of the center point. This axis

[2]A standing assumption throughout the book is that the dynamical systems under consideration are continuous in the state variable.

changes direction over time, but its magnitude sustains no net long term distortion. In general, the remaining axes change in both direction and magnitude. The Lyapunov exponents are determined by the long term rate of change in the *magnitude* of the principal axes of the evolving ellipsoid. They are called exponents because the rate is measured in exponential terms. To give this some formal structure, let $p_i(t)$ represent the magnitude of the ith axis at time t. Then its exponential rate of change $\lambda_i(t)$ is given by

$$b^{t\lambda_i(t)} = \frac{p_i(t)}{p_i(0)}$$

where b is some arbitrary but fixed constant. Taking logarithms of both sides to the base b we get

$$\lambda_i(t) = \frac{1}{t} \text{Log}_b \frac{p_i(t)}{p_i(0)}$$

The limiting value of this expression as t grows large is the Lyapunov exponent corresponding to the ith axis. By convention this exponent is denoted by λ_i and is often expressed in units of binary digits per unit time by choosing $b = 2$[3]. It is also customary to label the Lyapunov exponents in decreasing order such that $\lambda_1 \geq \lambda_2 \geq$ Collectively, the Lyapunov exponents are called the **Lyapunov spectrum**.

From this illustration we can make several important observations.

1. The λ_i are *scalars* which measure the time average growth of the *magnitude* of the principal axes. The principal axes are generally changing orientation as time goes by, so the λ_i are not associated with any particular direction relative to a fixed phase space reference frame. They only measure the amount of stretching and squeezing experienced by a small rotating volume element moving in the flow.
2. An initial linear extent of magnitude $q_1(0)$ in the direction of the first principal axis grows to $b^{\lambda_1 t}q_1(0)$ after t units of time, and we say $q_1(0)$ grows like $b^{\lambda_1 t}$. The area in the plane of the first two axes grows like $b^{(\lambda_1+\lambda_2)t}$, and the volume defined by the three largest axes grows like $b^{(\lambda_1+\lambda_2+\lambda_3)t}$. In general, an initial hypervolume[4] $\Pi q_i(0)$ will grow like $b^{(\Sigma\lambda_i)t}$ to a volume of size $\Pi q_i(t) = \Pi q_i(0)b^{\lambda_i t} = b^{(\Sigma\lambda_i)t}\Pi q_i(0)$.

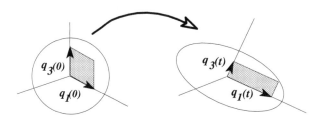

Figure 5.4 An initial area of size $q_1(0) \times q_3(0)$ gets transformed by the dynamics into an area of size $q_1(t) \times q_3(t)$ after a time interval t.

---

[3]Another base popular in the liturature in the natural number e, in which case the units of measure are called nats. Most of the results reported in the examples in this chapter are given in nats.

[4]Multidimensional 'volume' element defined by $\Pi q_i(0) = q_1(0) \times q_2(0) \times \cdots$

Hence, $\Sigma\lambda_i$ measures the time average divergence of the flow, or the long term exponential rate of change in phase space volume. A system with an attractor is dissipative, that is, net-net the system contracts phase space volume.

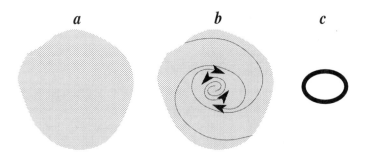

| *a* | *b* | *c* |

Figure 5.5 Phase space is contracted as (a) points in the basin of attraction (b) evolve under the dynamics (c) to the attractor.

For there to be a net long term contraction, $\Sigma\lambda_i$ must be negative, which implies that at least one $\lambda_i < 0$. Moreover, if the system is chaotic, then by definition it is producing information – expanding some parts of phase space – which implies that at least one $\lambda_i > 0$. Hence, a chaotic attractor possesses at least one positive exponent and at least one negative exponent. Any continuous time dynamical system that does not have a fixed point[5] will have at least one $\lambda_i = 0$[6]. As in the above illustration, the principal axis tangent to the flow will have a zero Lyapunov exponent. The reason is that there is no net long term increase or decrease in the separation of any two points on the same trajectory. Clearly, we can learn a great deal about the long term behavior of a dynamical system just from the signs of the Lyapunov exponents (see Table 5.1).

Sign of Lyapunov spectra $(\lambda_1, \lambda_2, \lambda_3)$	Type of attractor
$(+, 0, -)$	chaotic (strange) attractor
$(0, 0, -)$	2-torus*
$(0, -, -)$	limit cycle
$(-, -, -)$	equilibrium point

Table 5.1 Some typical Lyapunov spectra and attractors for three-dimensional continuous time dissipative dynamical systems. *Like the surface of a donut or inner tube.

3. If the attractor is bounded then there must be a folding process, along each principal axis where $\lambda_i > 0$, that brings points spread apart by stretching back together again.

[5]If there is a fixed point attractor the dynamics are degenerate and there is no motion.
[6]Discrete systems may have all non-zero Lyapunov exponents.

The result is that often the attractor will have a folded ribbon-like, or Cantor-set-like[7], structure. The Lyapunov exponents give no information on this folding process.

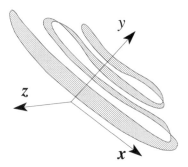

Figure 5.6 Folding of the evolving ellipsoid. The predominant contracting direction is along the y-axis and the predominant expanding direction is along the x-axis.

4. An n-dimensional attractor has n true Lyapunov exponents, but the number of degrees of freedom in the state space defines the number of principal axes of the expanding ellipsoid that is used to define the Lyapunov exponents. If an n-dimensional attractor is placed (embedded) in a higher m-dimensional space, then any numerical procedure for extracting the Lyapunov spectrum will produce m exponents. Since only n of these are the true exponents, the remaining $m-n$ exponents, sometimes referred to as **phantom exponents**, are meaningless artifacts of the estimation process[8].

As we shall see, in addition to the usual problems associated with finite accuracy and finite data, both the folding process and the existence of phantom exponents give rise to difficulties when trying to estimate Lyapunov spectra from real data.

THEORY

We saw in the previous section that given a continuous time dynamical system[9] defined by a set of autonomous ordinary differential equations $dx/dt = F(x)$, $x \in \Re^m$, an initial infinitesimal sphere on the system's attractor is distorted in finite time to an infinitesimal ellipsoid. Letting $p_i(t)$ denote the magnitude of the ith principal axis of the evolving ellipsoid, the ith Lyapunov exponent is operationally defined by the limiting value

$$\lambda_i = \lim_{t \to \infty} \frac{1}{t} \log_b \frac{p_i(t)}{p_i(0)}$$

Eq. 5.1

[7] Sets with dense lacunarity that are self-similar on all scales.

[8] In practice there is no foolproof method for identifying which are the true exponents and which are not.

[9] This chapter uses many definitions and concepts from Chapter 1, Dynamical systems.

By convention, the exponents are labeled such that $\lambda_1 \geq \lambda_2 \geq \cdots \geq \lambda_m$. While the limit may not exist, time is finite for any real data set, so in practice the average can still be defined. How do we measure $p_i(t)$? The way forward is to start with a discrete one-dimensional map $f : s_n \to s_{n+1}$ where there is only one dimension and one axis to deal with. Here, the idea is that an initial extent Δs_0 is transformed, under the n-fold iteration of the map, to the extent $\Delta s_n \equiv f^n(s_0 + \Delta s_0) - f^n(s_0)$. Thus, the time-$n$ exponential growth rate $\lambda(n)$ is the solution to

$$|\Delta s_n| = b^{n\lambda(n)}|\Delta s_0| \qquad \text{Eq. 5.2}$$

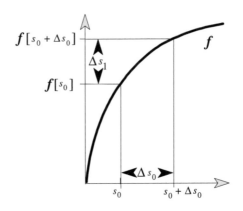

Figure 5.7 The map f takes the extent Δs_0 to the set Δs_1.

Rearranging Eq. 5.2 and expanding we get

$$b^{n\lambda(n)} = \left|\frac{\Delta s_n}{\Delta s_0}\right| = \left|\frac{\Delta s_n}{\Delta s_{n-1}} \frac{\Delta s_{n-1}}{\Delta s_{n-2}} \cdots \frac{\Delta s_1}{\Delta s_0}\right| \approx \prod_{i=0}^{n-1}\left|\frac{df(s_i)}{ds}\right| \qquad \text{Eq. 5.3}$$

where the last approximation in Eq. 5.3, using the derivative of f at s_i, is justified by the assumption that Δs_0 is small so that the following linearization holds

$$
\begin{aligned}
\Delta s_n &= f^n[s_0 + \Delta s_0] - f^n[s_0] \\
&= f(f^{n-1}[s_0 + \Delta s_0]) - f(f^{n-1}[s_0]) \\
&= f(s_{n-1} + \Delta s_{n-1}) - f(s_{n-1}) \\
&\approx \left.\frac{df(s)}{ds}\right|_{s_{n-1}} \times \Delta s_{n-1}
\end{aligned}
\qquad \text{Eq. 5.4}
$$

From Eq. 5.3 we see that the time-n exponential growth rate $\lambda(n)$ is equivalent to the growth rate of the product of the derivatives of the dynamics f, evaluated along an n-length segment of a trajectory of the system, starting at s_0[10]. After taking logarithms in Eq. 5.3, we define the Lyapunov exponent of f as the limit

$$\lambda = \lim_{n \to \infty} \frac{1}{n} \sum_{i=0}^{n-1} \mathrm{Log}_b \left| \frac{df(s_i)}{ds} \right| \qquad \text{Eq. 5.5}$$

In practice we approximate λ via

$$\lambda \underset{\text{for large } n}{\approx} \frac{1}{n} \sum_{i=0}^{n-1} \mathrm{Log}_b \left| \frac{df(s_i)}{ds} \right| \qquad \text{Eq. 5.6}$$

which is a straightforward calculation provided f is known [11], and accurate provided the set of observations $\{s_i\}$ are on, or close to, the attractor of the system. Based on a similar discussion in the theoretical section of Chapter 2, Entropy, the Lyapunov exponent defined by Eq. 5.5 is also the information production rate of the system.

Analogously, in the general m-dimensional situation, we replace the extent Δs_0, in the one-dimensional example above, by a radial vector $\delta s(0)$ of an infinitesimal m-dimensional sphere $B(s_0)$ whose initial center s_0 evolves along a reference or 'fiducial' trajectory defined by the action of the dynamics f on the initial state s_0. After a time t, the initial radial vector $\delta s(0)$ becomes a radial vector $\delta s(t)$ of the evolving ellipsoid $B(s_t)$ with center $s_t = f^t(s_0)$.

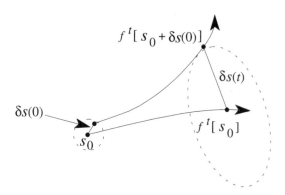

Figure 5.8 After a time t, the map f takes an initial radial vector $\delta s(0)$ to the radial vector $\delta s(t)$.

Since the points on the surface of the initial sphere $B(s_0)$ are infinitesimally separated from the center reference point s_0, their trajectories are defined by the linearized equations

$$\delta s(t) = f^t[s_0 + \delta s(0)] - f^t[s_0] = D_{s_0} f^t \cdot \delta s(0) \qquad \text{Eq. 5.7}$$

---

[10]The fact that $\lambda(n)$ may depend on the initial condition s_0 has been suppressed here and will continue to be suppressed until we tackle the issue head on below with the Multiplicative Ergodic Theorem of Oseledec.

[11]The standing assumption throughout the book is that the dynamical equations are *not* known, and have to be estimated. See Appendix 5, Linearized dynamics for an estimate based on techniques developed in this chapter.

where

$$
D_{s_0} f^t = \begin{bmatrix} \dfrac{\partial f^t_{(1)}}{\delta s_{(1)}} & \cdots & \dfrac{\partial f^t_{(1)}}{\partial s_{(m)}} \\ \vdots & \cdots & \vdots \\ \dfrac{\partial f^t_{(m)}}{\delta s_{(1)}} & \cdots & \dfrac{\partial f^t_{(m)}}{\partial s_{(m)}} \end{bmatrix}
$$

Eq. 5.8

is the Jacobian[12] matrix of f^t evaluated at s_0. Hence, the distortion of an initial infinitesimal radial vector $\delta s(0)$ is determined by the action of Df. Now recall that the Lyapunov exponents are defined by the distortion of the principal axes of an evolving infinitesimal ellipsoid. These principal axes are radial vectors about the point s_0. It follows that the Lyapunov exponents are determined by the Jacobian Df.

The domain of Df is in fact the whole infinitesimal vector space around s_0 where the linear property, Eq. 5.7, holds, and is referred to as the ***tangent space*** because infinitesimal vectors $\delta s(0)$ about s_0 are tangent to the attractor in that region. In general, the line joining any two attractor points in the linear region is a vector in the tangent space (see Figure 5.9).

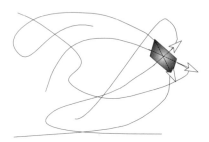

Figure 5.9 Arrows show vectors tangent to the trajectories in a small region of the attractor where the dynamics are linear. These vectors define the tangent space.

The discussion leading to Eq. 5.1 through Eq. 5.8 loosely shows that there is a connection between the Lyapunov exponents, the expansion rates of tangent vectors under the action of the dynamics, and the Jacobian of the dynamical equations. To make this relationship explicit, we begin with some simple derivations because, although we are tantalizingly close to a generalization of Eq. 5.5, we cannot make the transition from Eq. 5.7 without establishing the limiting behavior of Df, except in some special situations which we look at now.

Suppose the dynamical system $ds/dt = F(s)$ has an equilibrium point s^*. Consider the evolution of a small initial ***perturbation vector*** $s^* + \delta s(0)$ in the tangent space near s^* (Figure 5.10).

---

[12]The notation $v_{(j)}$ means the jth component of the vector v.

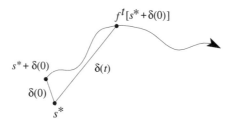

Figure 5.10 Deformation of an initial tangent vector $\delta s(0)$ near the equilibrium point s^* under the action of the flow f after time t.

Linearizing the vector field F, and using the fact that for an equilibrium point $F[s^*] = 0$, we have

$$F[s^* + \delta s] \approx F[s^*] + (D_{s^*}F) \cdot \delta s$$
$$= (D_{s^*}F) \cdot \delta s$$

By definition

$$F[s^* + \delta s] = \frac{d}{dt}(s^* + \delta s) = \frac{d\delta s}{dt}$$

Hence,

$$\frac{d\delta s}{dt} = (D_{s^*}F) \cdot \delta s$$

and this suggests that we look for solutions of the form

$$\delta s(t) = \delta s(0)b^{\lambda^* t}$$

For an m-dimensional state space, there will be m solutions $\delta s_i(0)b^{\lambda_i^* t}$, where $b^{\lambda_i^* t}$ (possibly complex) and $\delta s_i(0)$ are respectively the eigenvalues and eigenvectors of the Jacobian $D_{s^*}F$. The eigenvectors $\delta s_i(0)$ define the principal axes of the expanding ellipsoid, so the magnitude of $\delta s_i(0)$ is the same as the number $p_i(0)$ defined in Eq. 5.1. The Lyapunov exponents λ_i and the eigenvalues $b^{\lambda_i^* t}$ of $D_{s^*}F$ are related as follows. Recall, from Eq. 5.1, that the Lyapunov exponents are defined to be the average rate of exponential growth in the direction of the principal axes (eigenvectors) at s^*. So

$$\text{average growth rate} = \lim_{t \to \infty} \frac{1}{t} \text{Log}_b \frac{\|\delta s_i(t)\|}{\|\delta s_i(0)\|}$$
$$= \lim_{t \to \infty} \frac{1}{t} \text{Log}_b \frac{|b^{\lambda_i^* t}| \|\delta s_i(0)\|}{\|\delta s_i(0)\|}$$
$$= \lim_{t \to \infty} \frac{1}{t} \text{Log}_b |b^{\lambda_i^* t}|$$

Hence[13],

$$\lambda_i = \lim_{t \to \infty} \frac{1}{t} \log_b \left| b^{\lambda_i^* t} \right|$$

$$= \lim_{t \to \infty} \frac{1}{t} \operatorname{Re} \lambda_i^* t$$

$$= \operatorname{Re} \lambda_i^*$$

Suppose now, that the asymptotic set, that is, the attractor of the dynamical system, is a limit cycle rather than an equilibrium point. Recall that a surface of section induces a Poincaré first return map[14] which we denote here by P. The limit cycle will intersect the surface of section at a fixed point s^* of P (see Figure 5.11).

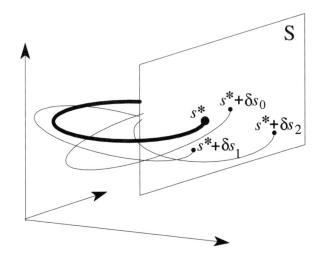

Figure 5.11 Surface of section S induces Poincaré return map P. The bold line is the limit cycle which intersects the surface of section at s^*, a fixed point of the return map P. The fine line is the trajectory whose initial point is at an infinitesimal distance δs_0 from point s^* on the limit cycle.

Now, consider a point $s^* + \delta s_0$ in the linear neighborhood of P at s^*. The small perturbation δs_0 will evolve under the action of P to δs_n according to

$$\delta s_n \equiv P^n[s^* + \delta s_0] - P^n[s^*]$$
$$= P^n[s^* + \delta s_0] - s^* \qquad \text{**Eq. 5.9**}$$

where we use the fact that for a fixed point s^* of P, $s^* = P[s^*]$. Linearizing Eq. 5.9 we get

$$\delta s_n = (D_{s^*} P^n) \cdot \delta s_0$$

---

[13] Re c denotes the real part of the complex number c.

[14] See Chapter 1, Dynamical systems.

Taking δs_0 in the eigendirection p_i, that is, in the direction of the ith eigenvector of $D_{s^*} P^n$ gives

$$\delta s_n = |m_i| \delta s_0$$

where m_i is the ith eigenvalue of $D_{s^*} P^n$. If Γ is the period of the limit cycle, then

$$m_i = b^{\lambda_i^* n \Gamma}$$

and

$$\lambda_i = \frac{1}{n\Gamma} \text{Log}_b |m_i|$$

To generalize the concept of the Lyapunov exponent beyond the special situations described above we will need the following theorem.

Theorem 1 (Multiplicative Ergodic Theorem of Oseledec[15])
Let ρ be a measure preserving ergodic probability measure[16] with respect to a map $f : M \rightarrow M$ defined on an m-dimensional space M. Let $T : M \rightarrow \{m \times m$ matrices$\}$ be a measurable map such that

$$\int \rho(ds) \, \text{Log}^+ \|T(s)\| < \infty$$

where $\text{Log}^+ u = \max [0, \text{Log}\, u]$.
Define

$$T_s^{(n)} = T[f^{n-1}s] \cdots T[fs] \cdot T[s].$$

Then, for ρ almost all s, the following limit exists[17]

$$\lim_{n \to \infty} ([T_s^{(n)}]^T T_s^{(n)})^{\frac{1}{2n}} = \Lambda_s$$

The logarithms of the magnitudes of the eigenvalues of the limit Λ_s are called the **characteristic exponents** or **Lyapunov exponents**. Note the dependence on the initial point s!

Theorem 2 (Oseledec)
If the exponents are ordered $\lambda_1 \geq \lambda_2 \geq \cdots$, and E_s^i is the subspace in $\mathfrak{R}^m$ spanned by the eigenvectors whose Lyapunov exponents are less than or equal to λ_i, then

$$\mathfrak{R}^m = E_S^1 \supset E_S^2 \supset \cdots$$

and for ρ almost all s

$$\lim_{n \to \infty} \frac{1}{n} \text{Log} \|T_s^{(n)} \cdot u\| = \lambda_i$$

when $u \in E_s^i \setminus E_s^{i+1}$. In particular, when $u \notin E_s^2$ the limit is λ_1.

---

[15]Oseledec (1968a), (1968b).
[16]See Chapter 1, Dynamical systems.
[17]The notation $[A]^T$ means the transpose of the matrix A.

The exponent's dependence on s is taken up by the sub-eigenspaces E_s^i; it does not matter which perturbation is chosen relative to s, given that it lies in E_s^i, the exponential growth is always λ_i for ρ almost any s. Any point in the basin of attraction of an attractor will have almost surely[18] the same Lyapunov exponents.

Here are some additional results which, among other things, characterize the attractors of intrinsic low dimension systems[19].

- If f^t is defined for $t < 0$, then the Lyapunov exponents for the time-reversed system are the negative of those for the original system.

- If all the Lyapunov exponents of a discrete dynamical system are negative, then there is an attracting periodic orbit or attracting fixed point for the system.

- If all the Lyapunov exponents of a continuous time system are non-zero (negative), then there is an (attracting) equilibrium point for the system, and $\rho(s) = \delta_{s^*} = \delta(s - s^*)$, where s^* is the equilibrium point[20].

- If the Lyapunov exponents of a continuous time system satisfy $\lambda_1 \geq 0 > \lambda_2 > \lambda_3 > \cdots$, then either

    (i) $\lambda_1 > 0$, and there is an equilibrium point for the system, that is, $\rho = \delta_{s^*}$, where s^* is the equilibrium point, or

    (ii) $\lambda_1 = 0$, and there is an attracting periodic orbit for the system.

    Since a system with an attractor must be contracting, at least one Lyapunov exponent is negative. In order that the system be chaotic, at least one exponent must be positive. This means that for a two-dimensional system to have a chaotic attractor we must have $\lambda_1 > 0 > \lambda_2$, but the above result implies that, in this case, the limit set of a continuous time system is an equilibrium point. Hence, continuous time systems cannot support chaotic dynamics in less than three dimensions. In contrast, discrete time systems can support chaotic dynamics in two or less dimensions.

- Since the volume element enclosed by the eigenvectors is the determinant of the Jacobian, the growth rate of an infinitesimal m-dimensional volume element is the rate of growth of the Jacobian determinant

$$\frac{1}{t} \mathrm{Log}\, |D_s f^t| = \sum_{i=1}^{m} \lambda_i$$

where

$$|D_s f^t| = \left| \det\left(\frac{\partial f_{(i)}^t(s)}{\partial s_{(j)}} \right) \right|$$

---

[18] That is, except possibly on a set with probability measure zero.

[19] Eckmann and Ruelle (1985) p. 617.

[20] Here, the delta function $\delta(u)$ is defined by $\delta(0) = 1$ and $\delta(u \neq 0) = 0$.

General formulation

We are now in a position to get a fully general formulation of the Lyapunov spectrum of a dynamical system. Consider the discrete dynamical system[21] defined by $s_n = f(s_{n-1})$ where $f : \Re^m \to \Re^m$ is a differentiable vector point function. Define[22]

$$\delta s_n \equiv f^n(s_0 + \delta s_0) - f^n(s_0)$$

Expanding the right hand side we get

$$f^n(s_0 + \delta s_0) - f^n(s_0) = D_{s_0} f^n \cdot \delta s_0 + O([\delta s_0]^2)$$

where $D_{s_0} f^n$ is the Jacobian matrix $\left[\frac{\partial f^n_{(i)}}{\partial s_{(j)}} \right]$ at s_0. So, to first order

$$\delta s_n = D_{s_0} f^n \cdot \delta s_0$$

Dividing by $\| \delta s_0 \|$ and taking norms we have

$$\frac{\| \delta s_n \|}{\| \delta s_0 \|} = \| D_{s_0} f^n \cdot u \| \qquad\qquad \textbf{Eq. 5.10}$$

where

$$u = \frac{\delta s_0}{\| \delta s_0 \|}$$

Now, from the chain rule for derivatives

$$
\begin{aligned}
D_{s_0} f^n \cdot u &= \frac{df^n[s_0]}{du} \\
&= \frac{df(f^{n-1}[s_0])}{du} \\
&= D_{s_{n-1}} f \cdot \frac{df^{n-1}[s_0]}{du} \\
&= D_{s_{n-1}} f \cdot D_{s_{n-2}} f \cdots D_{s_0} f \cdot u
\end{aligned}
$$

So, in what by now should be a familiar idea, the n-step tangent map, from the tangent space at s_0 to the tangent space at $f^n(s_0)$, is the product of the one-step tangent maps evaluated along the trajectory with initial point s_0. Finally, letting $T_s^{(1)} = D_s f$, so that $T_s^{(n)} = D_s f^n$ in Oseledec's theorem, we have that if f induces a natural ergodic measure ρ, then for ρ almost all s_0 the following limit exists

$$\lambda_u = \lim_{n \to \infty} \frac{1}{n} \mathrm{Log}_b \| D_{s_0} f^n \cdot u \| \qquad\qquad \textbf{Eq. 5.11}$$

and we obtain λ_i by choosing $u \in E^i_{s_0} \setminus E^{i+1}_{s_0}$. Eq. 5.11 is the generalization of Eq. 5.5. From Eq. 5.10 and Eq. 5.11 we have

$$\| \delta s_n \| \approx b^{n\lambda_u} \| \delta s_0 \| \qquad\qquad \textbf{Eq. 5.12}$$

---

[21]We lose no generality by considering a discrete time system since the arguments are valid if we replace the discrete map f with the time-1 map of a flow f of a continuous time system.

[22]See Figure 5.8.

Note that $E_{s_0}^2$ is small relative to $E_{s_0}^1$, so in practice most $\delta s_0 \notin E_{s_0}^2$, and

$$\delta s_n \approx e^{n\lambda_1} \delta s_0$$

As noted earlier, if f is a flow, the same analysis goes through for the flow's time-1 map f^1 in Oseledec's theorem. As it is an arbitrary constant, we will often suppress the base of the logarithm, b, in the notation, and just note that it is customary to choose b equal to 2 or the natural number e.

Lyapunov dimension

The *Lyapunov* or *Kaplan-York dimension* is defined by

$$D_{KY} \equiv k + \frac{\sum_1^k \lambda_i}{|\lambda_{k+1}|}$$ **Eq. 5.13**

where k is determined by the relation

$$\sum_1^k \lambda_i > 0 \text{ and } \sum_1^{k+1} \lambda_i < 0$$

Kaplan and York[23] conjectured that D_{KY} was equal to the information dimension D_1, but there are counter examples to this. However, in many cases D_{KY} does seem to be very close to D_1 which suggests that the fractal dimension of a dynamical system's attractor is determined by the $k+1$ largest Lyapunov exponents as defined above. That means that the first $k + 1$ exponents are fundamental. If any of the first k exponents are negative, then they are insufficient to overcome phase space volume expansion. The $k + 1$st exponent finally makes the system dissipative, causing it to collapse into a fractal structure[24].

Kolmogorov entropy

Theorem[25]

If ρ, the natural measure induced by the dynamics f, is absolutely continuous on the unstable manifold (of dimension m') then the information creation rate, that is the *Kolmogorov–Sinai entropy*[26], is the same as the mean rate of expansion of infinitesimal m'-dimensional phase space volume elements. Thus,

$$K_1 = \sum_i \lambda_i^+; \quad \lambda_i^+ = \max[\lambda_i, 0]$$

If ρ is singular along the unstable manifold then

$$K_1 < \sum_i \lambda_i^+$$

---

[23] Kaplan and York (1979) and (1983).
[24] See Chapter 4, Fractal dimension.
[25] See Eckmann and Ruelle (1985).
[26] See Chapter 2, Entropy.

Local Lyapunov exponent

It cannot be emphasized enough that the Lyapunov exponents are not local quantities in either the temporal or the spatial sense. They are rather, an average of local deformations of a small volume element, with respect to a dynamic reference frame that is 'tumbling' in the flow of the system's trajectories. Nevertheless, a local measure of the phase space deformation can be a very useful adjunct to a prediction model.

Recall that phase space expansion rates are closely related to information production rates of a system. A local measure of the information production rate at some point on the experimental attractor gives a rough guide to the accuracy that we can expect from a prediction made at that point. Algorithms for computing estimates of a local version of the largest Lyapunov exponent are given in the next section. It is sufficient to consider the largest exponent since it will dominate the information production rate of the system.

ALGORITHMS

We make (restate) certain basic assumptions about the raw material we have available to extract estimates of the Lyapunov spectrum. The most important assumption is that we do not know the dynamical equations. Instead we are given only a time series of observations on some measurable aspect of the system. We assume that the methods of Chapter 3, Phase space reconstruction, have been used to reconstruct a trajectory for the system from the time series data, and that possibly some of the methods from Chapter 6, Noise reduction, have been used to clean the data[27]. We call the reconstructed trajectory the experimental trajectory and denote it by $\underline{y}(n) \equiv \underline{y}_n \in \Re^{dr}, n = 1, \ldots, N_r$.

Since the Lyapunov exponents are basically the exponential growth rates of tangent vectors, working in the tangent space of the attractor is unavoidable. The fate of the tangent vectors is governed by the Jacobian Df of the dynamical equations defining the system, which we have assumed we do not know. We could proceed by constructing an approximation to the dynamical equations from the reconstructed trajectory and from these compute the Jacobian, but this approach has a fatal flaw. Even if we were to succeed in modeling the dynamics accurately in the sense that the model faithfully reproduces the out-of-sample[28] experimental trajectory, there is no guarantee that the Jacobian computed from such a model will accurately model the Jacobian of the true system, and for our purposes here, it is the Jacobian we need and not a predictive model.

Having eliminated the approach of modeling the dynamics, we are left with little choice but to deal directly with what is happening to the tangent vectors in the tangent space. We obtain the necessary tangent vectors along the experimental trajectory by computing vectors $\delta_i(t) \in \Re^{dr}, i = 1, \ldots, N_b$, joining $\underline{y}(t)$ to the trajectory points

---

[27]In addition to the material in Chapters 3, Phase space reconstrucion and 6, Noise reduction, readers are advised to familiarize themselves also with the material covered in Appendix 1, Data requirements and Appendix 2, Nearest neighbor searches.

[28]That portion of the data that was not used to build the model, but is used to test the model.

contained in a small neighborhood[29] $B_\varepsilon[\underline{y}(t)]$ of $\underline{y}(t)$. To avoid confusion, we denote the image of $\delta_i(t)$ under the n-fold mapping Df^n as $\delta_i^*(t+n)$, which is in general different from a tangent vector $\delta_i(t+n)$ computed at $\underline{y}(t+n)$. The technique for identifying neighborhood points is used widely in the analysis of chaotic dynamics so the details are developed separately in Appendix 2, Nearest neighbor searches. We assume here that the neighborhood points have been identified and that the tangent vectors have been computed (see Figure 5.12).

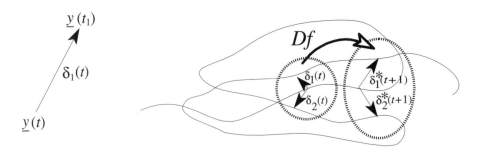

Figure 5.12 Left: Tangent vector $\delta_1(t)$ constructed from $\underline{y}(t)$ and a near neighbor $\underline{y}(t_1)$. Right: Tangent vectors $\delta_i(t)$ at $\underline{y}(t)$ mapped by Df to tangent vectors $\delta_i^*(t+1)$ at $\underline{y}(t+1)$.

In the sequel we will speak of near neighbors of an experimental trajectory point as if they were arbitrary points on the attractor which could lie on different nearby trajectories. In fact, all the trajectory points reconstructed from a single time series will constitute a single experimental trajectory, so all near neighbors of a reference point[30] will necessarily be on the same trajectory as the reference point. In practice this will not cause severe difficulties provided there is ample data, since most near neighbors will come from parts of the experimental trajectory that are seperated by relatively long periods of time. Nearby points from the same trajectory, separated by relatively long periods of time, will be practically independent and behave much like nearby points on different trajectories.

With the tangent vectors in hand we can proceed in two ways. The first is to use the tangent vectors to estimate the one-step Jacobian Df at each trajectory point, and then compute the eigenvalues of the product Jacobian Df^n. The one-step Jacobians are obtained as the solution to the familiar inverse problem

$$\delta_i^*(t+1) = Df \cdot \delta_i(t)$$

Various techniques for structuring and solving the problem are discussed below under the rubric Indirect Methods.

Before attempting an indirect approach via Df we will explore two direct methods. We can avoid working with the actual matrix Df by noting that its effects are already contained in the experimental trajectory. That is, we can calculate the evolution

---

[29]For example, a hypersphere of radius ε centered at $\underline{y}(t)$ or a hypercube of edge size ε centered at $\underline{y}(t)$.

[30] A reference point (orbit), also known as a fiducial point (orbit), is an experimental trajectory point (orbit).

$\delta_i^*(t + n)$ of an initial tangent vector $\delta_i(t)$ directly from the data without knowing Df explicitly. In principle, if $\delta_i(t)$ were infinitesimal and lay in the eigenspace E^j corresponding to λ_j, we could estimate λ_j from Eq. 5.12.

$$b^{n\lambda_j} \approx \frac{\|\delta_i^*(t + n)\|}{\|\delta_i(t)\|}$$

A naive application of this idea would soon reveal the problems associated with having to operate with finite objects. The first problem is that, even if we could determine E^j, $j \neq 1$, it is relatively small and we have no hope of finding an experimentally determined tangent vector $\delta_i(t)$ that lies precisely in E^j. By the same token, since almost every $\delta_i(t)$ will lie in E^1/E^2, under iteration it will rapidly align itself with the eigendirection corresponding to the largest exponent λ_1, so that at least the idea could be used to get λ_1 from

$$b^{n\lambda_1} \approx \frac{\|\delta_i^*(t + n)\|}{\|\delta_i(t)\|}$$

The second problem with this approach is that experimental tangent vectors are not infinitesimal and, unlike in theory where infinitesimal quantities remain infinitesimal over finite time, will in practice diverge and find the global folding of the attractor. Nevertheless, for n large enough to let the tangent vector align itself with the eigendirection of λ_1, and small enough that $\delta_i^*(t + n)$ does not find the global folding of the attractor, we might hope to get a reasonable local estimate of λ_1.

1 Direct method I[31]

Formally, let

$$\|\delta_0(i)\| = \min_{y(k):\ |k-i|>\tau_c} \|y(k) - y(i)\|$$

be the initial distance[32] from the point $y(i)$ on the reference orbit to its nearest neighbor. From the definition of λ_1 we would expect, after a suitable evolution time $T_e = N_e \tau_s$[33], that

$$\|\delta_0^*(i + T_e/\tau_s)\| \approx \|\delta_0(i)\| b^{\lambda_1 T_e}$$

Taking the logarithm of both sides gives

$$\text{Log}_b \|\delta_0^*(i + T_e/\tau_s)\| \approx \text{Log}_b \|\delta_0(i)\| + \lambda_1 T_e$$

which represents a set of parallel lines with slope λ_1 for each i. Averaging over i, that is, over the experimental trajectory, we get one line

$$\langle\text{Log}_b \|\delta_0^*(i + T_e/\tau_s)\|\rangle_i \approx \langle\text{Log}_b \|\delta_0(i)\|\rangle_i + \lambda_1 T_e \qquad \text{Eq. 5.14}$$

---

[31]Rosenstein, Collins and De Luca (1993).

[32]τ_c is the mean cycle time on the attractor. The temporal separation means that neighbors are a function of the recurrence property of the system only.

[33]τ_s is the sampling time of the measurement time series.

which yields the following estimate for λ_1

$$\lambda_1 = \frac{1}{T_e} \frac{1}{(N_r - T_e/\tau_s)} \sum_{i=1}^{N_r - T_e/\tau_s} \mathrm{Log}_b \frac{\|\delta_0^*(i + T_e/\tau_s)\|}{\|\delta_0(i)\|}$$ **Eq. 5.15**

The averaging tends to smooth out errors in small noisy data sets. It is clear from Figure 5.13 that Eq. 5.15 saturates for large T_e due to folding of the attractor.

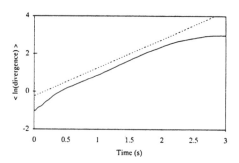

Figure 5.13 Typical plot of the average of the natural logarithm of the divergence $\|\delta_0^*(i + T_e/\tau_s)\|$ versus T_e, Eq. 5.14, for the Lorenz attractor. The solid curve is the experimentally determined result. The dashed curve has the (theoretical) slope we would expect to see.
Source: Adapted from Rosenstein, Collins and De Luca (1993) p. 122.

Parameter values
Use Eq. 5.14 over a range of values for T_e.

Remark
Although the method only yields an estimate of the largest Lyapunov exponent, it is effective and efficient.

System	Equations	Parameters	τ_s	Expected λ_1
Logistic	$x_{\mu_1} = \mu x_1(1 - x_1)$	$\mu = 4.0$	1	0.693
Hénon	$x_{\mu_1} = 1 - \mu x_1^2 + y_1$	$\mu = 1.4$	1	0.418
	$y_{\mu_1} = b x_1$	$b = 0.3$		
Lorenz	$\dot{x} = \sigma(y - x)$	$\sigma = 16.0$	0.01	1.50
	$\dot{y} = x(R - z) - y$	$R = 45.92$		
	$\dot{z} = xy - bz$	$b = 4.0$		
Rössler	$\dot{x} = -y - z$	$a = 0.15$	0.10	0.090
	$\dot{y} = x + ay$	$b = 0.20$		
	$\dot{z} = b + z(x - c)$	$c = 10.0$		

Table 5.2 Analytically determined values of λ_1 for comparison with the experimentally determined values in Examples 5.1–5.6.
Source: Rosenstein, Collins and De Luca (1993) p. 122. Values for the logistic equation come from Eckmann and Ruelle (1985), and the remainder of the values come from Wolf et al. (1985).

Example 5.1

Effects of varying d_r, the dimension of the reconstruction.

System	N_r	$J = \tau/\tau_s$	d_r	Calculted λ_1	% Error
Logistic	500	1	1	0.675	−2.6
			2	0.681	−1.7
			3	0.680	−1.9
			4	0.680	−1.9
			5	0.651	−6.1
Hénon	500	1	1	0.195	−53.3
			2	0.409	−2.2
			3	0.406	−2.9
			4	0.399	−4.5
			5	0.392	−6.2
Lorenz	5000	11	1	−	−
			3	1.531	2.1
			5	1.498	−0.1
			7	1.562	4.1
			9	1.560	4.0
Rössler	2000	8	1	−	−
			3	0.0879	−2.3
			5	0.0864	−4.0
			7	0.0853	−5.2
			9	0.0835	−7.2

Table 5.3 Experimental results for several embedding dimensions. The number of data points, the reconstruction time delay, and the dimension of the reconstruction are denoted by N_r, τ and d_r, respectively.

Source: Adapted from Rosenstein, Collins and De Luca (1993) p. 124.

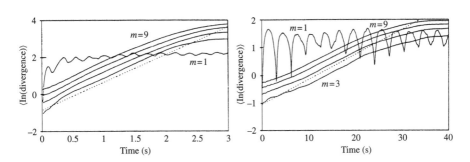

Figure 5.14 Effects of reconstruction dimension $d_r = m$. For each plot the solid curves are the calculated results, and the slope of the dashed curve is the theoretical λ_1. Left: Lorenz attractor. Right: Rössler attractor. See Table 5.3 for details.

Source: Rosenstein, Collins and De Luca (1993) p. 123.

195

Example 5.2
Effects of varying N_r, the size of the data set.

System	N_r	$J = \tau/\tau_s$	d_r	Calculted λ_1	% Error
Logistic	100	1	2	0.659	−4.9
	200			0.705	1.7
	300			0.695	0.3
	400			0.692	−0.1
	500			0.686	−1.0
Hénon	100	1	2	0.426	1.9
	200			0.416	−0.5
	300			0.421	0.7
	400			0.409	2.2
	500			0.412	−1.4
Lorenz	1000	11	3	1.751	16.7
	2000			1.345	−10.3
	3000			1.372	−8.5
	4000			1.392	−7.2
	5000			1.523	1.5
Rössler	400	8	3	0.0351	−61.0
	800			0.0655	−27.2
	1200			0.0918	2.0
	1600			0.0984	9.3
	2000			0.0879	−2.3

Table 5.4 Experimental results for various sizes of the data set. The number of data points, the reconstruction time delay, and the dimension of the reconstruction are denoted by N_r, τ and d_r, respectively.

Source: Adapted from Rosenstein, Collins and De Luca (1993) p. 124.

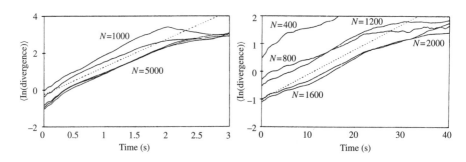

Figure 5.15 Effects of data set size $N_r = N$. For each plot the solid curves are the calculated results, and the slope of the dashed curve is the theoretical λ_1. Left: Lorenz attractor. Right: Rössler attractor. See Table 5.4 for details.

Source: Adapted from Rosenstein, Collins and De Luca (1993) p. 125.

Example 5.3

Effect of varying the time delay of the phase space reconstruction.

System	N_r	$J = \tau/\tau_s$	d_r	Calculted λ_1	% Error
Logistic	500	1*	2	0.681	−1.7
		2		0.678	−2.2
		3		0.672	−3.0
		4		0.563	−18.8
		5		0.622	−10.2
Hénon	500	1*	2	0.409	−2.2
		2		0.406	−2.9
		3		0.391	−6.5
		4		0.338	−19.1
		5		0.330	−21.1
Lorenz	5000	1	3	1.640	9.3
		11*		1.561	4.1
		21		1.436	−4.3
		31		1.423	−5.1
		41		1.321	−11.9
Rössler	2000	2	3	0.0699	−22.3
		8*		0.0873	−3.0
		14		0.0864	−4.0
		20		0.0837	−7.0
		26		0.0812	−9.8

Table 5.5 Experimental results for several reconstruction delay times $\tau = J\tau_s$. The number of data points and the dimension of the reconstruction are denoted by N_r and d_r, respectively. The asterisks denote a value of the delay time equal to the autocorrelation time, that is, the time taken for the autocorrelation function to fall to $1 - 1/e$ of its initial value.

Source: Rosenstein, Collins and De Luca (1993) p. 126.

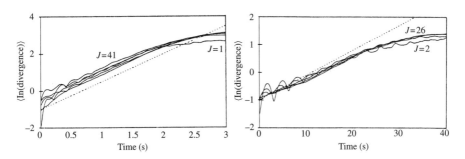

Figure 5.16 Effects of reconstruction delay time $\tau = J\tau_s$. For each plot the solid curves are the calculated results, and the slope of the dashed curve is the theoretical λ_1. Left: Lorenz attractor. Right: Rössler attractor. See Table 5.5 for details.

Source: Rosenstein, Collins and De Luca (1993) p. 127.

197

Example 5.4

Noise tends to flatten the curves and shorten the estimation region.

System	N_r	$J = \tau/\tau_s$	d_r	SNR	Calculted λ_1	% Error
Logistic	500	1	2	1	0.704	1.6
				10	0.779	12.4
				100	0.856	23.5
				1000	0.621	−10.4
				10000	0.628	−9.4
Hénon	500	1	2	1	0.643	53.8
				10	0.631	51.0
				100	0.522	24.9
				1000	0.334	−20.1
				10000	0.385	−7.9
Lorenz	5000	11	3	1	0.645	−57.0
				10	1.184	−21.1
				100	1.110	−26.0
				1000	1.273	−15.1
				10000	1.470	−2.0
Rössler	2000	8	3	1	0.0106	−88.2
				10	0.0394	−56.2
				100	0.0401	−55.4
				1000	0.0659	−26.8
				10000	0.0836	−7.1

Table 5.6 Experimental results for several noise levels. The number of data points, the reconstruction time delay, and the dimension of the reconstruction are denoted by N_r, τ and d_r, respectively. SNR (Signal to noise ratio $= -10 \, \mathrm{Log}$ (signal variance/noise variance)): 1–10 = High; 100–1000 = Medium; 10,000 = Low.

Source: Rosenstein, Collins and De Luca (1993) p. 129.

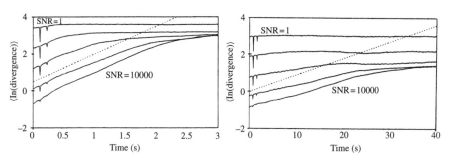

Figure 5.17 Effects of noise. For each plot the solid curves are the calculated results, and the slope of the dashed curve is the theoretical λ_1. Left: Lorenz attractor. Right: Rössler attractor. See Table 5.6 for details.

Source: Rosenstein, Collins and De Luca (1993) p. 128.

The predominance of negative bias in estimation errors above is probably due to near neighbors, and hence the tangent vectors, exploring Lyapunov directions other than that of the largest exponent λ_1 over short periods of time.

Example 5.5

System	Equations	Parameters	τ_s	Expected λ_1, λ_2
Rössler-hyperchaos	$\dot{x} = -y - z$	$a = 0.25$	0.1	$\lambda_1 = 0.111$
	$\dot{y} = x + ay + w$	$b = 3.0$		$\lambda_2 = 0.021$
	$\dot{z} = b + xz$	$c = 0.05$		
	$\dot{w} = cw - dz$	$d = 0.5$		
Mackay–Glass	$\dot{x} = \dfrac{ax(1+s)}{1 + [x(t+s)]} - bx(t)$	$a = 0.2$	0.75	$\lambda_1 = 4.37\text{E} - 3$
		$b = 0.1$		$\lambda_2 = 1.82\text{E} - 3$
		$c = 10.0$		
		$s = 31.8$		

Table 5.7 Analytically determined values of λ_1 and λ_2 from systems with two positive exponents for comparison with the experimentally determined values in Table 5.8 and Figure 5.18 below.

Source: Rosenstein, Collins and De Luca (1993) p. 129. Values are from Wolf *et al.* (1985).

System	N_r	$J = \tau/\tau_s$	d_r	Calculated λ_1	% error
Rössler-hyperchaos	8000	9	3	0.048	−56.8
			6	0.112	0.9
			9	0.112	0.9
			12	0.107	−3.6
			15	0.102	−8.1
Mackay–Glass	8000	12	3	4.15E−3	−5.0
			6	4.87E−3	11.4
			9	4.74E−3	8.5
			12	4.80E−3	9.7
			15	4.85E−3	11.0

Table 5.8 Experimental results for various reconstruction dimensions for systems with two positive exponents. The number of data points, the reconstruction time delay, and the dimension of the reconstruction are denoted by N_r, τ and d_r, respectively.

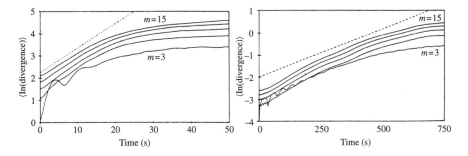

Figure 5.18 Effects of reconstruction dimensions $d_r = m$. For each plot the solid curves are the calculated results, and the slope of the dashed curve is the theoretical λ_1. Left: Rössler-hyperchaos attractor. Right: Mackay–Glass attractor. See Table 5.8 for details.

Source: Rosenstein, Collins and De Luca (1993) p. 128.

Example 5.6

For quasi-periodic and stochastic systems (or high dimensional chaos which is stochastic in any practical sense) we would expect flat plots of $\langle \text{Log}_b \, \| \delta_0^*(i + T_e/\tau_s) \| \rangle$ versus T_e. That is, we anticipate $\lambda_1 \approx 0$, since for these systems near neighbors should neither diverge nor converge on average. In addition, for a stochastic system expect an initial jump from a small separation at $T_e \approx 0$, since the small distance, implied by the linear relationship, is atypical of these systems. For example, in the plots in Figure 5.19, the anomalous scaling region is *not* linear in T_e and for the stochastic systems, the average initial distance grows rapidly and then saturates with increasing d_r.

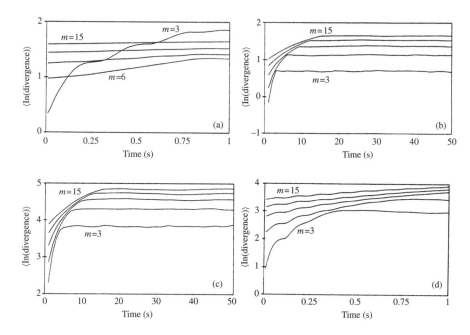

Figure 5.19 Effects of reconstruction dimension $d_r = m$ for non-chaotic systems. (a) Two-torus. (b) White noise. (c) Bandlimited noise. (d) 'Scrambled' Lorenz.
Source: Rosenstein, Collins and De Luca (1993) p. 130.

Local version

An obvious local estimate of λ_1 at $\underline{y}(n)$ is given by

$$\overset{\oplus}{\lambda_1}(n) = \frac{1}{T_e} \text{Log}_b \frac{\| \delta_0^*(n + T_e/\tau_s) \|}{\| \delta_0(n) \|}$$

2 Direct method II[34]

The second direct method we will investigate also exploits the operational definition of the Lyapunov spectrum, but uses a different algorithm to extract the exponents. The

---

[34] Wolf *et al.* (1985).

initial tangent vector is chosen to be the vector between the initial trajectory point of the data and its closest neighbor. The image of this tangent vector is then tracked until it is larger than some predefined level, at which time the growth rate in the magnitude of the initial tangent vector is computed, and a replacement tangent vector is chosen. The replacement vector has an orientation approximately the same as the final tangent vector from the previous step and has the smallest magnitude from among all such tangent vectors. If the predefined replacement level is chosen correctly, this process allows the initial tangent vector to find the eigendirection of λ_1 without finding the global folding of the attractor. When the data set is exhausted λ_1 is calculated from the geometric average of the growth rates computed at each replacement stage (see Figure 5.20).

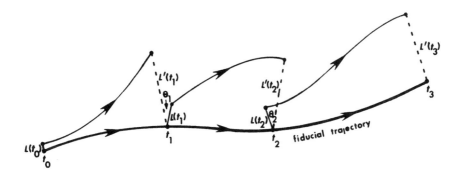

Figure 5.20 A schematic representaion of the evolution and replacement used to estimate Lyapunov exponents from experimental data. The largest Lyapunov exponent is computed from the growth of tangent vectors δ_i. When the length $L(t_{i-1}) = \delta_i(t_{i-1})$ of the initial tangent vector between two points becomes too large, say $L'(t_i) = \delta_i^*(t_i)$, a new point is chosen near the reference trajectory, minimizing both the length of $L(t_i) = \delta_{i+1}(t_i)$, and the change in phase space orientation θ_i between the original and replacement vectors.

Source: Adapted from Wolf *et al.* (1985) p. 296.

Formally, we take the first point on the experimental trajectory $y(1)$, find its closest neighbor $y(u)$ such that $|u - 1| > \tau_c$, and form the initial tangent vector

$$\delta_1(1) = y(u) - y(1)$$

To keep things on an 'infinitesimal' scale as we move forward in time, we need to do something like a renormalization when $\delta_1^*(t)$ becomes 'large'. So we monitor

$$\delta_1^*(t) = y(t + u - 1) - y(t)$$

until for some $t_1 > 1$

$$\|\delta_1^*(t_1)\| \geq \varepsilon_{\max}$$

and then replace the non-reference point $y(t_1 + u - 1)$ with a new point $y(v)$ that is close to $y(t_1)$, but temporally separated from it by at least τ_c, and lies approximately in the same direction as $\delta_1^*(t_1)$. This step replaces $\delta_1^*(t_1)$ with the new tangent vector

$$\delta_2(t_1) = y(v) - y(t_1)$$

which has a small angle with $\delta_1^*(t_1)$ and magnitude

$$\|\delta_2(t_1)\| < \varepsilon_{max}$$

Continuing this process until the last experimental trajectory point is reached, we get an estimate for the largest Lyapunov exponent λ_1 from the average growth rate in the $\delta_i(t)$

$$\lambda_1 \approx \frac{1}{t_m - 1} \sum_{i=1}^{m} \text{Log}_b \frac{\|\delta_i^*(t_i)\|}{\|\delta_i(t_{i-1})\|} \qquad \text{Eq. 5.16}$$

where $t_0 = 1$; $t_i, i = 1, \ldots, m - 1$, are the replacement times; and t_m is the index of the last trajectory point.

We can get an estimate of λ_2 by simultaneously tracking two tangent vectors and noting that the area defined by them should grow like $b^{(\lambda_1+\lambda_2)'}$. Since we have an estimate for λ_1 in Eq. 5.16, it is straightforward to extract an estimate for λ_2 from

$$\lambda_1 + \lambda_2 \approx \frac{1}{t_m - 1} \sum_{i=1}^{m} \text{Log}_b \frac{\alpha_i^*(t_i)}{\alpha_i(t_{i-1})}$$

where $\alpha_i(t)$ is the area between the fiducial trajectory point and two near neighbors.

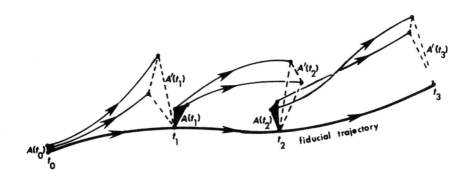

Figure 5.21 The sum of the two largest Lyapunov exponents is computed from the growth of area elements. When the area $A'(t_i) = \alpha_i^*(t_i)$ becomes too large or too skewed, two new points are chosen near the reference trajectory, minimizing both the replacement area $A(t_i) = \alpha_{i+1}(t_i)$ and the change in phase space orientation between the original and replacement area elements.
Source: Adapted from Wolf *et al.* (1985) p. 296.

The procedure can be extended to k-dimensional volume elements to estimate $\Sigma \lambda_i$ by picking k neighbors of the fiduciary point and monitoring the volume defined by the evolving $k + 1$ points. In this case the tangent vector growth is replaced by the k-volume growth

$$\sum_{1}^{k} \lambda_i \approx \frac{1}{t_m - 1} \sum_{i=1}^{m} \text{Log}_b \frac{v_i^*(t_i)}{v_i(t_{i-1})} \qquad \text{Eq. 5.17}$$

Local version

A local estimate of λ_1 at $y(n)$ is given by

$$\overset{\oplus}{\lambda_1}(n) = \frac{1}{t_n - t_{n-1}} \operatorname{Log}_b \frac{\|\delta_n^*(t_n)\|}{\|\delta_n(t_{n-1})\|}$$

Problems

The line of reasoning leading to Eq. 5.17 would suggest that we could obtain the entire Lyapunov spectrum in this fashion, but the estimates become unreliable for negative and small positive exponents. The problem with small exponents is simply the accumulation of errors on the same scale as the exponent. The difficulty with negative exponents is that the contraction in those directions is exponential, and we are already trying to work at distances as small as the data allow. Under these conditions any procedure will be intractable.

Parameter values

Implementation requires a specification of $\varepsilon_{\max}$ and the replacement times t_k. Choosing the replacement times involves a trade off. Frequent replacement is CPU costly and increases the cumulative effect of orientation errors – the angle θ between the ingoing and outgoing tangent vectors – but long propagation times allow the volume elements to find the folds of the attractor.

Reasonable ranges for the parameters seem to be

- $\varepsilon_{\max} \approx 3\% - 5\%$ of $d(A) = \max[x_i] - \min[x_i]$
- $t_k - t_{k-1} \approx 50\% - 150\%$ of τ_c

Example 5.7

The method yields estimates for λ_1 within about 5% of the theoretical values for the Hénon, Rössler, and Lorenz systems using 128, 2048 and 8192 data points respectively.

3 Indirect methods – the linear problem[35]

Let us recap some assumptions and notation. $f : Q \to Q$ is the true dynamics on the true attractor Q. f and Q are unknown and cannot be directly observed. Using the methods of Chapter 3, Phase space reconstruction, we reconstruct a trajectory of the system in $\mathfrak{R}^{d_r}$ with a $(d_r, \tau = k\tau_s)$-delay coordinate map Φ using an observational time series $x(j\tau_s)$, $j = 1, \ldots, N_d$. The reconstructed trajectory points $\underline{y}(n\tau) \in \mathfrak{R}^{d_r}$, $1, \ldots, N_r$, constitute what we know about $A = \Phi(Q)$, the representation of Q in $\mathfrak{R}^{d_r}$. Without loss of generality, we can choose a time unit such that $\tau_s = 1$, and we will as usual suppress the dependence on the delay time τ and write $\underline{y}(n\tau)$ as $\underline{y}(n)$ or $\underline{y}_n$. We postulate a dynamics $\varphi[\underline{y}(n)] = \underline{y}(n + 1)$ on the reconstructed attractor A, and use the experimental trajectory $\underline{y}(n)$ to estimate φ[36]. We denote by d_A the

[35] Darbyshire and Broomhead (1996).

[36] See Appendix 5, Linearized dynamics for an estimate based on techniques developed in this chapter.

dimension of the manifold containing the reconstructed attractor A, and note that d_A is invariant under Φ. We denote the maximum extent of the attractor A by $d(A) = \max x_i - \min x_i$. A neighborhood $B_\varepsilon[y(n)]$ of an experimental trajectory point $y(n)$ is defined to be all experimental trajectory points $y(j)$ such that $\|y(n) - y(j)\| < \varepsilon$, and denote the number of experimental trajectory points in the neighborhood $B_\varepsilon[y(n)]$ by $K_\varepsilon(n)$.

We begin the discussion of the indirect approach with a formulation and solution to the general problem. The general problem is to estimate the tangent mapping $D_s f : \mathfrak{R}^{d_A} \to \mathfrak{R}^{d_A}$ and compute the eigenvalues of the product mapping $D_s f^j$ given only the experimental trajectory y_n. That means we are stuck with trying to do the job by estimating $D_u \varphi \equiv T_u^{(1)} : \mathfrak{R}^{d_r} \to \mathfrak{R}^{d_r}$, and then extracting the Lyapunov exponents from the eigenvalues of the product mapping $T_u^{(j)}$. The first step is to approximate $T_u^{(1)}$.

Now the action of $T_u^{(1)}$ is implicit in the transformation of experimental tangent vectors, that is, the domain and range of $T_u^{(1)}$ are the tangent spaces at u and $\varphi[u]$, so we need to build vectors in these spaces from y_n. Clearly, we can only do that for u and $\varphi[u]$ among the points y_n. We get the sample tangent vectors in the domain of $T_{y_n}^{(1)}$ by identifying N_b points y_{n_k}, $1 \le k \le N_b$, in $B_\varepsilon[y(n)]$, an ε-neighborhood[37] of y_n, and forming the tangent vectors

$$\{\delta_k(n) = y_{n_k} - y_n \mid \|\delta_k(n)\| \le \varepsilon; |n_k - n| > \tau_c; 1 \le k \le N_b\}$$

We get the sample tangent vectors in the range of $T_{y_n}^{(1)}$ by computing the images of the domain vectors under the tangent mapping as follows

$$\{\delta_k^*(n+1) = y_{n_k+1} - y_{n+1} \mid 1 \le k \le N_b\}$$

We adopt the convention that $\delta_k(n)$ and $\delta_k^*(n+1)$ are column vectors in $\mathfrak{R}^{d_r \times 1}$[38]. By defining the **trajectory matrices**

$$B_{y_n} = [\delta_k(n)]^T \in \mathfrak{R}^{N_b \times d_r}$$

and

$$B_{y_n}^{(1)} = [\delta_k^*(n+1)]^T \in \mathfrak{R}^{N_b \times d_r}$$

we can state the first part of the problem as solving the following inverse problem for $T_{y_n}^{(1)}$

$$B_{y_n}^{(1)} = B_{y_n} \cdot (T_{y_n}^{(1)})^T$$

---

[37]In general, the number of points $K_\varepsilon(n)$ in an ε-neighborhood of $y(n)$ is variable. We assume for the moment that $N_b \le K_\varepsilon(n)$, $n = 1, \ldots, N_r$, so that we can actually find N_b neighborhood points.
[38]We use the notation $(\ )_{a \times b}$ or $\mathfrak{R}^{a \times b}$ to denote real valued matrices with a rows and b columns, and $(\ )^T$ to denote matrix transpose.

In practice we need $N_b \geq d_r$ tangent vectors, otherwise the problem is underdetermined. In fact, N_b is normally chosen considerably larger than d_r in the hope of averaging out some of the noise expected in real data. In this case, the inverse problem is recast as a least squares problem where $T_{\underline{y}_n}^{(1)}$ is chosen to minimize $S(n)$, the average squared error given by

$$S(n) = \frac{1}{N_b} \sum_{i=1}^{N_b} \| \delta_i^*(n+1) - T_{\underline{y}_n}^{(1)} \delta_i(n) \|^2$$

It is worth pausing here to consider the relationship between the quantities N_b, d_r, and d_A. d_r is the dimension of the reconstructed phase space and, therefore, also the dimension of the experimental tangent space vectors; d_A is the dimension of the tangent space of the true attractor of the system and, hence, also the local dimension of the (manifold containing the) true attractor; and $N_b \times d_r$ is the dimension of the trajectory matrices. The rank of $B_{\underline{y}_n}$ gives the dimension of the space spanned by the N_b experimental tangent vectors chosen at $y(n)$. Since these tangent vectors are d_r-dimensional, they can span no more than d_r dimensions, and thus rank $B_{\underline{y}_n} \leq d_r$. Since d_A is the dimension of the true tangent space, in principle rank $B_{\underline{y}_n} \leq d_A$. For reasons given in Chapter 3: Phase space reconstruction, the reconstructed phase space dimension d_r may have to be chosen such that $d_r > 2d_A$, and in any case such that $d_r \geq d_A$. In summary, we have that, at least in principle, $N_b \geq d_r \geq d_A \geq$ rank $B_{\underline{y}_n}$. The real situation may be different and we discuss that possibility below along with other problems.

Pseudo-inverse solution
One way of extracting $T_{\underline{y}_n}^{(1)}$ is by local singular value decomposition (SVD) of $B_{\underline{y}_n}$ which yields

$$B_{\underline{y}_n} = V_{\underline{y}_n} \Sigma_{\underline{y}_n} (U_{\underline{y}_n})^T$$

where

$$V_{\underline{y}_n} \in \mathfrak{R}^{N_b \times d_r}; \; \Sigma_{\underline{y}_n} \in \mathfrak{R}^{d_r \times d_r}; \; U_{\underline{y}_n} \in \mathfrak{R}^{d_r \times d_r};$$

$V_{\underline{y}_n}$ is orthogonal ; $U_{\underline{y}_n}$ is orthonormal ;

$$\Sigma_{\underline{y}_n}(i, j) = \sigma_i \delta_{ij}; \sigma_1 \geq \sigma_2 \geq \ldots \geq \sigma_{d_r} \geq 0;$$

and

$$\delta_{ij} \begin{cases} 1 & \text{if } i = j \\ 0 & \text{otherwise} \end{cases}$$

The σ_i, known as singular values, measure the extent to which the tangent vectors explore $\mathfrak{R}^{d_r}$. If $\sigma_1 \geq \sigma_2 \geq \cdots \geq \sigma_p > 0 = \sigma_{p+1} = \ldots = \sigma_{d_r}$, then the tangent vectors span a subspace of dimension $p < d_r$ and the above decomposition yields the following *pseudo-inverse* of $B_{\underline{y}_n}$

$$B_{\underline{y}_n}^+ = U_{\underline{y}_n} \Sigma_{\underline{y}_n}^+ (V_{\underline{y}_n})^T \qquad\qquad \textbf{Eq. 5.18}$$

where

$$\Sigma_{\underline{y}_n}^+ (i, j) = \begin{cases} \frac{\delta_{ij}}{\sigma_i} & \text{if } \sigma_i > 0 \\ 0 & \text{otherwise} \end{cases}$$

If rank $B_{\underline{y}_n} = d_r$, then

$$\sigma_i > 0, 1 \le i \le d_r,$$

$\Sigma_{\underline{y}_n}$ is invertible,

$$\Sigma_{\underline{y}_n}^+ = \Sigma_{\underline{y}_n}^{-1},$$

$$\begin{aligned} B_{\underline{y}_n}^+ B_{\underline{y}_n}^{(1)} &= B_{\underline{y}_n}^+ B_{\underline{y}_n} (T_{\underline{y}_n}^{(1)})^T \\ &= U_{\underline{y}_n} (\Sigma_{\underline{y}_n})^{-1} (V_{\underline{y}_n})^T V_{\underline{y}_n} \Sigma_{\underline{y}_n} (U_{\underline{y}_n})^T (T_{\underline{y}_n}^{(1)})^T \\ &= (T_{\underline{y}_n}^{(1)})^T \end{aligned}$$

Eq. 5.19

and, at least in principle, $T_u^{(j)}$ will yield $d_r = d_A$ estimates of the d_A true exponents of the system. On the other hand, if rank $B_{\underline{y}_n} < d_r$, then

$$B_{\underline{y}_n}^+ B_{\underline{y}_n}^{(1)} = U_{\underline{y}_n} P_{\underline{y}_n} (U_{\underline{y}_n})^T (T_{\underline{y}_n}^{(1)})^T = \wp_{\underline{y}_n} (T_{\underline{y}_n}^{(1)})^T$$

where

$$P_{\underline{y}_n} \in \mathfrak{R}^{d_r \times d_r} \text{ and } P_{\underline{y}_n} (i, j) = \begin{cases} \delta_{ij} & \text{if } \sigma_i > 0 \\ 0 & \text{otherwise} \end{cases}$$

In this case the solution is restricted to the local model $T_{\underline{y}_n}^{(1)} \wp_{\underline{y}_n}$. As $\wp_{\underline{y}_n}$ is the projection of a tangent vector at $\underline{y}_n$ into the linear subspace spanned by the tangent vectors $\delta_k(n)$, the above solution involves modeling the dynamics only in this subspace. Thus if rank $B_{\underline{y}_n} = m' < d_r$ there will only be m' meaningful exponents. Unfortunately, since the analysis takes place in a d_r-dimensional space, the method also generates $d_r - m'$ spurious or phantom exponents. One way to identify the spurious exponents is to run the analysis forward and backward in time. For noiseless data, the true exponents will just change sign and the spurious ones will do something else. But this approach fails when there is even a moderate level of noise in the data. On the other hand, if the data is clean to start with, then *adding* a bit of noise may reveal the phantom exponents, as they are more sensitive to the noise level.

Tangent space solution
An alternative approach, when rank $B_{\underline{y}_n} < d_r$, is to consider directly the product of local models derived above, that is, we consider

$$\check{T}_{\underline{y}_1}^{(n)} = T_{\underline{y}_n}^{(1)} \wp_{\underline{y}_n} T_{\underline{y}_{n-1}}^{(1)} \wp_{\underline{y}_{n-1}} \cdots T_{\underline{y}_1}^{(1)} \wp_{\underline{y}_1}$$

Eq. 5.20

where it can be seen that the evolving tangent vector is projected into the appropriate local tangent space before the application of the tangent map at that location. This suggests we could reformulate the problem by projecting the neighborhoods $B_{\underline{y}_n}$, and their images $B_{\underline{y}_n}^{(1)}$, into local q-*dimensional* tangent spaces[39] defined by the $d_r \times q$ matrices $\hat{U}_{\underline{y}_n}$ consisting of the first q columns[40] of $U_{\underline{y}_n}$. This produces

$$\hat{B}_{\underline{y}_n} \equiv B_{\underline{y}_n}\hat{U}_{\underline{y}_n} \in \mathfrak{R}^{N_b \times q}$$

$$\hat{B}_{\underline{y}_n}^{(1)} \equiv B_{\underline{y}_n}^{(1)}\hat{U}_{\underline{y}_{n+1}} \in \mathfrak{R}^{N_b \times q}$$

In this basis the inverse problem takes the following form

$$\hat{B}_{\underline{y}_n}^{(1)} = \hat{B}_{\underline{y}_n}(\hat{T}_{\underline{y}_n}^{(1)})_{q \times q}^T \qquad\qquad \text{Eq. 5.21}$$

Since

$$
\begin{aligned}
(\hat{B}_{\underline{y}_n})^T \hat{B}_{\underline{y}_n} &= (\hat{U}_{\underline{y}_n})^T (B_{\underline{y}_n})^T B_{\underline{y}_n} \hat{U}_{\underline{y}_n} \\
&= (\hat{U}_{\underline{y}_n})^T U_{\underline{y}_n} \Sigma_{\underline{y}_n} (V_{\underline{y}_n})^T V_{\underline{y}_n} \Sigma_{\underline{y}_n} (U_{\underline{y}_n})^T \hat{U}_{\underline{y}_n} \\
&= (\hat{U}_{\underline{y}_n})^T U_{\underline{y}_n} (\Sigma_{\underline{y}_n})^2 (U_{\underline{y}_n})^T \hat{U}_{\underline{y}_n} \\
&= (\hat{\Sigma}_{\underline{y}_n})^2
\end{aligned}
$$

where

$$\hat{\Sigma}_{\underline{y}_n} \in \mathfrak{R}^{q \times q} \text{ and } \hat{\Sigma}_{\underline{y}_n}(i, j) = \sigma_i \delta_{ij},$$

the normal equations for Eq. 5.21 reduce to

$$(\hat{B}_{\underline{y}_n})^T \hat{B}_{\underline{y}_n}^{(1)} = (\hat{\Sigma}_{\underline{y}_n})^2 (\hat{T}_{\underline{y}_n}^{(1)})^T$$

By design $\hat{\Sigma}_{\underline{y}_n}$ is invertible ($\sigma_i > 0, 1 \leq i \leq q$). This yields the following solution

$$(\hat{T}_{\underline{y}_n}^{(1)})^T = (\hat{\Sigma}_{\underline{y}_n})^{-2}(\hat{B}_{\underline{y}_n})^T \hat{B}_{\underline{y}_n}^{(1)} \qquad\qquad \text{Eq. 5.22}$$

Where the normal equations have a solution, the solution corresponds to the pseudo-inverse, so we also have

$$(\hat{T}_{\underline{y}_n}^{(1)})^T = (\hat{B}_{\underline{y}_n})^+ \hat{B}_{\underline{y}_n}^{(1)}$$

Now it can be shown that

$$(\hat{B}_{\underline{y}_n})^+ = (\hat{U}_{\underline{y}_n})^T (\hat{B}_{\underline{y}_n})^+$$

Hence, in terms of the simpler quantities

$$
\begin{aligned}
(\hat{T}_{\underline{y}_n}^{(1)})^T &= (\hat{U}_{\underline{y}_n})^T (B_{\underline{y}_n})^+ \hat{B}_{\underline{y}_n}^{(1)} \\
&= (\hat{U}_{\underline{y}_n})^T (B_{\underline{y}_n})^+ B_{\underline{y}_n}^{(1)} \hat{U}_{\underline{y}_{n+1}}
\end{aligned}
$$

[39]Identification of nearest neighbors must be done in the d_r-dimensional reconstructed phase space.
[40]The first q columns correspond to the q largest singular values.

Finally, it can also be shown that

$$\hat{T}_{\underline{y}_1}^{(n)} = (\hat{U}_{\underline{y}_{n+1}})^T \check{T}_{\underline{y}_1}^{(n)} \hat{U}_{\underline{y}_1}$$

So the least squares tangent space estimate $\hat{T}_{\underline{y}_1}^{(n)}$ is the pseudo-inverse estimate $\check{T}_{\underline{y}_1}^{(n)}$ (Eq. 5.20) projected onto the bases at $\underline{y}_1$ and $\underline{y}_{n+1}$. The point of this exercise is that the analysis here is taking place in an q-dimensional space, so if we choose q correctly, that is, $q = d_A$, then we get the d_A true exponents and only the d_A true exponents.

Total least squares solution

The tangent space and pseudo-inverse least squares estimates of the tangent map $T_{\underline{y}_n}^{(1)}$ both assume that the errors in the data only occur in the range space. Clearly, in the case of a trajectory reconstructed from a time series the assumption is not valid. An alternative method, which assumes errors occur in both domain and range, is called *total least squares (TLS)*, but is also known as **orthogonal** or **errors-in-variables regression**[41].

In TLS the inverse problem for the tangent space approach takes the form[42]

$$\left[\left(\hat{B}_{\underline{y}_n} \,;\, \hat{B}_{\underline{y}_n}^{(1)} \right)_{N_b \times 2q} \right] \left[\left(\hat{T}_{\underline{y}_n}^{(1)} \,;\, -\Im_q \right)_{q \times 2q} \right]^T = 0$$

If rank $\left[\hat{B}_{\underline{y}_n} \,;\, \hat{B}_{\underline{y}_n}^{(1)} \right] = q$, then there is a solution to the above equation. If not, TLS attempts to find a rank-q matrix that is 'closest' to $\left[\hat{B}_{\underline{y}_n} \,;\, \hat{B}_{\underline{y}_n}^{(1)} \right]$ in some norm to get an approximate solution. If the norm is the Frobenious or L^2 norm, then the rank-q matrix which is closest to $\left[\hat{B}_{\underline{y}_n} \,;\, \hat{B}_{\underline{y}_n}^{(1)} \right]$ is found by truncating the singular value decomposition of $\left[\hat{B}_{\underline{y}_n} \,;\, \hat{B}_{\underline{y}_n}^{(1)} \right]$ to include only the columns corresponding to the q largest singular values.

We can use singular value decomposition to write

$$\left[\hat{B}_{\underline{y}_n} \,;\, \hat{B}_{\underline{y}_n}^{(1)} \right] = \left[S_{\underline{y}_n} \,;\, \tilde{S}_{\underline{y}_n} \right]_{N_b \times 2q} \begin{pmatrix} \Sigma_{\underline{y}_n} & 0 \\ 0 & \tilde{\Sigma}_{\underline{y}_n} \end{pmatrix}_{2q \times 2q} \begin{bmatrix} (C_{\underline{y}_n}^{(0)})^T & (C_{\underline{y}_n}^{(1)})^T \\ (\tilde{C}_{\underline{y}_n}^{(0)})^T & (\tilde{C}_{\underline{y}_n}^{(1)})^T \end{bmatrix}_{2q \times 2q}$$

where $S_{\underline{y}_n}$ are the q largest, and $\tilde{S}_{\underline{y}_n}$ are the q smallest (left) singular vectors, that is, the vectors corresponding to the largest and smallest singular values. Similar notation applies to the remaining matrices.

In this notation replacing $\left[\hat{B}_{\underline{y}_n} \,;\, \hat{B}_{\underline{y}_n}^{(1)} \right]$ with the closest rank-q matrix yields

$$S_{\underline{y}_n} \Sigma_{\underline{y}_n} \left[(C_{\underline{y}_n}^{(0)})^T \,;\, (C_{\underline{y}_n}^{(1)})^T \right] \left[\hat{T}_{\underline{y}_n}^{(1)} \,;\, -\Im_q \right]^T = 0$$

and the error in making this adjustment is $\left\| \tilde{\Sigma}_{\underline{y}_n} \right\|$ in the appropriate norm. Whereas

---

[41]The errors in variables analysis tells us that ordinary least squares estimates of the Jacobian produce a downward bias in the derivatives which is independent of N_d.

[42]$\Im_q$ is the $q \times q$ identity matrix.

least squares minimizes the prediction error, total least squares minimizes the error that is normal to the graph of the linear predictor.

Generically, there is an exact solution to this equation provided we have chosen the rank of $\left[\hat{B}_{\underline{y}_n} ; \hat{B}_{\underline{y}_n}^{(1)} \right]$ correctly. The solution is

$$\hat{T}_{\underline{y}_n}^{(1)} = C_{\underline{y}_n}^{(1)} \left(C_{\underline{y}_n}^{(0)} \right)^{-1}$$

QR decomposition

Whatever method is used to estimate the tangent maps $T_{\underline{y}_i}^{(1)}$, we still need an efficient and numerically stable way to extract the Lyapunov exponent spectrum from the product mapping $T_{\underline{y}_1}^{(n)}$.

As noted previously, while the eigenvectors corresponding to the Lyapunov exponents at $\underline{y}_n$ form an orthogonal basis for the tangent space at $\underline{y}_n$, this basis generally changes as we move along the trajectory – expansion and contraction is measured 'in flow.' Since we do not know what the correct reference frame is at $\underline{y}_1$, and almost all tangent vector's growth is dominated by the largest exponent, almost any initial reference frame we choose will fold up along the eigendirection of λ_1. Hence, we must continually re-orthogonalize the initial reference frame after each application of the tangent map $T_{\underline{y}_i}^{(1)}$ in order to extract the whole Lyapunov spectrum.

A well known result from linear algebra is that, given a matrix A with linearly independent columns, there exists a unique factorization $A = QR$ where R is square upper right triangular with positive diagonal elements and the columns of Q are orthogonal. Now, suppose we have an orthogonal basis for the tangent space at $\underline{y}_1$ in the columns of a matrix Q_1. The columns of Q_1 under the action of $T_{\underline{y}_1}^{(1)}$ are still linearly independent, but not orthogonal. However, the unique factorization property gives

$$T_{\underline{y}_1}^{(1)} Q_1 = Q_2 R_1$$

where Q_2 is an orthogonal basis for the tangent space at $\underline{y}_2$, and R_1 relates this basis to the image of the basis at $\underline{y}_1$ under the action of $T_{\underline{y}_1}^{(1)}$. Upon repeated application of the re-orthogonalization we get the general expression

$$T_{\underline{y}_n}^{(n)} Q_1 = Q_{n+1} R_n R_{n-1} \ldots R_2 R_1$$

Finally, it can be shown that the Lyapunov spectrum is given by

$$\lambda_i = \lim_{n \to \infty} \frac{1}{n} \sum_{j=1}^{n} \mathrm{Log}(R_j)_{ii}$$

where $(R_j)_{ii}$ is the ith diagonal element of R_j, and the limit exists for almost all Q_1 with respect to the Haar measure. Of course, in practice the data is finite, so our estimate becomes

$$\lambda_i = \frac{1}{N_{QR}} \sum_{j=1}^{N_{QR}} \mathrm{Log}(R_j)_{ii}; \qquad N_{QR} \leq N_r$$

It should be kept in mind that if we allow an evolution time $T_e > 1$ between trajectory points on the reference trajectory, then this must be reflected in the estimate for λ_i as follows

$$\lambda_i = \frac{1}{T_e(N_{QR})} \sum_{j=1}^{N_{QR}} \text{Log}(R_j)_{ii}; \qquad N_{QR} \le N_r \qquad \textbf{Eq. 5.23}$$

where T_e is some multiple of the sampling time τ_s.

This method eliminates the need to calculate a large number of matrix products to get the product tangent map, avoiding the inevitable accumulation of rounding errors. The technique also allows us to compute intermediate running estimates of the λ_i as a function of trajectory length.

Several efficient algorithms exist for solving the inverse problem including singular value and QR decomposition. For some excellent sources see the references at the end of the chapter and at the end of the book.

Local version
We can obtain a local estimate of λ_1 at $\underline{y}(i)$ by forming the matrix product

$$\overset{\oplus(n_L)}{T}_{\underline{y}(i)} = T^{(1)}_{\underline{y}(i+n_L-1)} \cdots T^{(1)}_{\underline{y}(i)} \qquad 1 \le n_L \ll N_r$$

and solving for the exponent using the same techniques that were used to get the exponent from $T^{(n)}_{\underline{y}(1)}$, except that $\overset{\oplus(n_L)}{T}_{\underline{y}(i)}$ is calculated at each point $\underline{y}(i)$ on the experimental trajectory for $i < N_r - n_L$.

Parameters
- Use one of the methods in Appendix 2, Nearest neighbor searches to get neighborhood populations $K_\varepsilon(n)$ in the range $\min(2d_A, d_A + 4) < K_\varepsilon(n) \le 30\text{–}40$

  But in any case, keep ε as small as possible in the range 1–3% of $d(A)$. Having done that, the fixed neighborhood size N_b can be replaced with the actual variable neighborhood size $K_\varepsilon(n)$.

- If the reconstructed trajectory is a good representation of the true attractor, then the method is probably not all that sensitive to the exact values of either T_e or ε.

- The proper choice of 'local dimension' q for the tangent space formulation is either d_A obtained from the analysis done in reconstructing the experimental trajectory as given in Chapter 3, Phase space reconstruction, or $\lceil D_2$ obtained by methods from Chapter 4, Fractal dimension[43].

- For the QR decomposition choose $Q_1 = \Im$ (the identity matrix) and N_{QR} sufficiently large that the estimates converge, which seems to be on the order of 10^3.

---

[43]The correlation exponent D_2 is an estimate of the fractual dimension of the system's attractor. The notation $\lceil D_2$ means the smallest integer greater than equal to D_2.

Theoretical and estimated Lyapunov Spectra for some chaotic systems

System	Spectra determined from equations of motion	Spectra determined from pseudo-inverse approximation
Hénon	0.417 ± 0.006	0.408 ± 0.003
$a = 1.4, b = 0.03$	-1.58 ± 0.006	-1.58 ± 0.002
$N_r = 10{,}000, 2 \le K_\varepsilon(n) \le 20$	$K = 0.417, D_{KY} = 1.26$	$K = 0.408, D_{KY} = 1.26$
$N_{QR} = 1000$, precision=n/a		
Lorenz	1.50	1.63 ± 0.15
$\sigma = 16, b = 4.0, R = 45.92$	0.00	0.05 ± 0.25
$N_r = 5000, K_\varepsilon(n) \approx 10, \tau = 20\tau_s = 0.2$	-22.46	-3.59 ± 0.41
$N_{QR} \ge 2000$, precision $= 10^{-1}$	$K = 1.50, D_{KY} = 2.07$	$K = 1.68, D_{KY} = 2.47$
Rössler	0.09	0.096 ± 0.008
$a = 0.15, b = 0.2, c = 10$	0.00	-0.006 ± 0.004
$N_r = 5000, K\varepsilon(n) \approx 10, \tau = 12\tau_s = 1.2$	-9.8	-0.735 ± 0.057
$N_{QR} \ge 2000$, precision $= 10^{-1}$	$K = 0.090, D_{KY} = 2.01$	$K = 0.096, D_{KY} = 2.12$
Mackay–Glass	0.0071	0.0075 ± 0.0007
$a = 0.2, b = 0.1, c = 10, T = 30$	0.0027	0.0030 ± 0.0010
$N_r = 10{,}000, K\varepsilon((n) \approx 10, \tau = 12\tau_s = 13.5$	0.000	-0.0027 ± 0.00010
$N_{QR} \ge 2000$, precision $= 10^{-2}$	-0.0167	-0.0156 ± 0.0006
	-0.0245	-0.0394 ± 0.0064
	$K = 0.010, D_{KY} = 3.59$	$K = 0.011, D_{KY} = 3.50$

Table 5.9 Lyapunov spectra, Kolmogorov–Sinai entropy K, and Lyapunov dimension D_{KY} for various known chaotic systems. Table entries for the spectra come from Zeng, Eykholt and Pielke (1991) p. 3231, except for Hénon entries which come from Sano and Sawada (1985) p. 1084. Estimates for the Hénon model were obtained from (probably) high precision data with the pseudo-inverse approach and spherical neighborhoods with $\varepsilon_{min} \approx 1.0\%$ of $d(A)$. Estimates for the other models were obtained from low precision data with the pseudo-inverse approach and shell-type (see Appendix 2, nearest Neighbor searches) neighborhoods with $\varepsilon_{min} \approx 1.0\%$ of $d(A)$ and $\varepsilon_{max} \approx 5.0$–$10\%$ of $d(A)$ to obtain $K_\varepsilon(n) \approx 10$.

Example 5.8

Figures 5.22–5.26 are based on noise free data generated by the x variable of the Lorenz I system:

$$\frac{dx}{dt} = \sigma(y - x)$$

$$\frac{dy}{dt} = x(R - z) - y$$

$$\frac{dz}{dt} = xy - bz$$

$$\sigma = 16, b = 4, R = 40$$

In each case the delay time for the reconstruction $\tau = 0.05$, the trajectory length $N_r = 38,000$, the neighborhood size $N_b = 20$, and the horizontal dashed lines in the figures represent the theoretical values of the three true Lyapunov exponents. Depending on the method used, the calculated values of both the true and phantom Lyapunov exponents are marked as indicated in the legends of the plots.

Pseudo-inverse approach

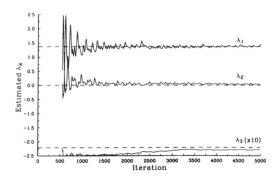

Figure 5.22 Convergence, as a function of calculation length N_{QR}, of estimates of the Lyapunov exponents of the Lorenz I model. Estimates of the tangent maps where made using the pseudo-inverse method in a ***three-dimensional projection*** of a global SVD basis constructed using a window length $\tau_w = \tau \times (d_r - 1) = 20$. Note the difference in scale for the third exponent to keep the plot in a tight range.

Source: Adapted from Darbyshire and Broomhead (1996) p. 296.

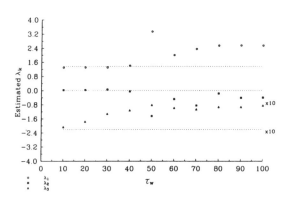

Figure 5.23 Estimates of Lyapunov exponents of the Lorenz I model. The tangent maps were approximated using the pseudo-inverse method in a ***three-dimensionsal projection*** of a global SVD basis constructed using a range of window lengths τ_w. The calculation length $N_{QR} = 3000$. Note how the estimates degenerate for large window sizes as the trailing vector components of the trajectory points decorrelate. Note also the difference in scale for the third exponent to keep the plot in a tight range.

Source: Darbyshire and Broomhead (1996) p. 297.

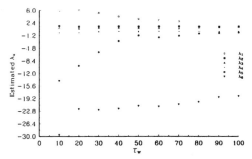

Figure 5.24 Estimates of Lyapunov exponents of the Lorenz model. The tangent maps were approximated using the pseudo-inverse method in a ***six-dimensionsal projection*** of a global SVD basis constructed using a range of window lengths τ_w. The calculation length $N_{QR} = 3000$. Note the three phantom exponents.

Source: Adapted from Darbyshire and Broomhead (1996) p. 297.

Tangent space approach

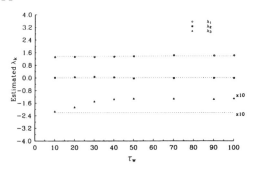

Figure 5.25 Estimates of Lyapunov exponents of the Lorenz model. The tangent maps were approximated using the least squares tangent space method with $q = 3$ ***in a six-dimensionsal projection*** of a global SVD basis constructed using a range of window lengths τ_w. The calculation length $N_{QR} = 3000$. Note the difference in scale for the third exponent to keep the plot in a tight range.

Source: Adapted from Darbyshire and Broomhead (1996) p. 298.

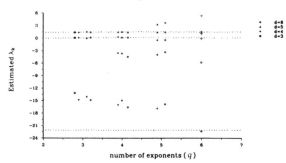

Figure 5.26 Estimates of Lyapunov exponents of the Lorenz model. The tangent maps were approximated using the least squares tangent space method with $3 \leq q \leq d$, ***in a d-dimensional projection***, $3 \leq d \leq 6$, of a global SVD basis constructed using a window length $\tau_w = 30$. The calculation length $N_{QR} = 3000$.

Source: Adapted from Darbyshire and Broomhead (1996) p. 298.

213

Problems

- As noted, the pseudo-inverse method forces us to compute d_r exponents which include the true exponents and possibly phantom exponents as well – while the true exponents are computed correctly, in practice it is difficult to pick them out. The same is true of the tangent space method unless we pick the local dimension q correctly as d_A.

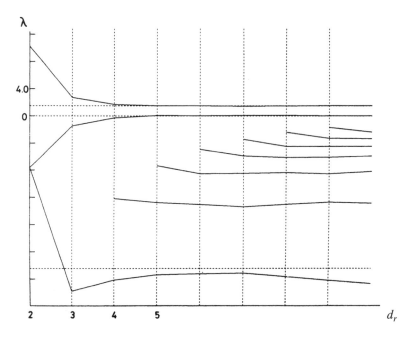

Figure 5.27 Lyapunov exponents of the Lorenz equations versus the reconstruction dimension d_r computed using the pseudo-inverse method. Exponents are connected in a 'natural' way with increasing d_r. $K_\varepsilon(n) \geq 2d_r$; $N_r = 64{,}000$; and $\tau = 0.03$.

Source: Adapted from Eckmann *et al.* (1986) p. 4973.

- In order to contract the phase space to $\mathfrak{R}^{d_A}$ the phantom exponents are usually negative. However, attractor curvature in non-infinitesimal neighborhoods, introduced by finite data and measurement accuracy, will be noticed by a linear model fit. This will cause the fitting procedure to adjust terms in the model to compensate, and can result in exponents that are even larger than the largest true exponent. Both effects can be seen in Figures 5.24 and 5.26 where phantom exponents have been produced.

- We can expect that all real data is corrupted to some extent by noise and measurement error which can have a significant effect on the estimates (see Figures 5.28 and 5.29).

214

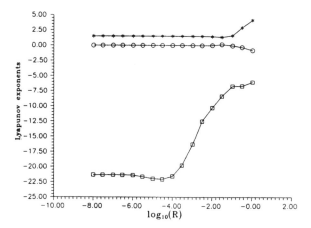

Figure 5.28 The effect of additive white noise on the determination of the Lyapunov exponents for the Lorenz system. In the figure the local dimension $q = 3$. On the horizontal axis is the $\log_{10}$ of R, the standard deviation of the noise divided by the standard deviation of the signal.
Source: Adapted from Bryant, Brown and Abarbanel (1991) p. 2803.

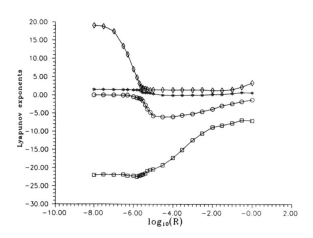

Figure 5.29 The effect of additive white noise on the determination of the Lyapunov exponents for the Lorenz system. In this figure the local dimension $q = 4$. The phantom exponent wanders from about $+19$ to -6 as the noise level is varied. Note that the exponents do not cross each other, but prefer to switch roles as they become close. In the $q = 3$ case (Figure 5.28) the correct exponents are more robust against the addition of noise. On the horizontal axis is the $\log_{10}$ of R, the standard deviation of the noise divided by the standard deviation of the signal.
Source: Adapted from Bryant, Brown and Abarbanel (1991) p. 2804.

- When the data is corrupted by noise or measurement error, it is possible that $d_A <$ rank $B_{\underline{y}_n} \leq d_r$. In this situation, even if rank $B_{\underline{y}_n} = d_r$, we wind up with more estimates than there are true exponents.

- In general, the experimental trajectory points $y(n)$ will not be uniformly distributed in the neighborhood of a given reference point $y(i)$, so that the experimental tangent vectors may only rarely explore some directions of the true tangent space. In particular, the directions corresponding to the negative Lyapunov exponents often have a fractal distribution of trajectory points that are poorly resolved by experimental data and result in an attractor that appears very thin in those directions. As a consequence, the estimated tangent mapping $T_{y_i}^{(1)}$ does not reflect the dynamics in these directions, so we can determine with confidence only the positive Lyapunov exponents.
- Getting an estimate for negative exponents depends on their magnitude and the signal-to-noise ratio of the data. Big negative exponents tend to squeeze the attractor in some directions to thicknesses that are below measurement accuracy. Note the error in the estimates for the large negative exponents in Table 5.9.
- Note that we may not be able to find $K_\varepsilon(n) \geq d_r$ neighbors at some point $y(i)$ on the fiducial trajectory, or that the tangent vectors at that point do not span the tangent space. Thus, at some points the tangent mapping may be only partially determined. The indeterminacy of the tangent maps does introduce phantom exponents, but this is not fatal to the algorithm, which should be continued even when the $T_{y_i}^{(1)}$ are singular as they will still contain useful information.
- In practice Eq. 5.23 may show significant fluctuations when iterated to get a running estimate of λ_i. The effect can be mitigated by windowing the summands with a function that smoothly dampens the first few and last few terms in the summation.

4 Indirect method with nonlinear terms[44]

For infinitesimal tangent vectors $\delta_i(t)$, the linear Jacobians Df embody the true dynamics of the system. In practice nothing is infinitesimal, and data sets are finite in accuracy and quantity. In small but non-infinitesimal neighborhoods, points on an experimental attractor will generally lie on a curved surface which may not be well represented by a linear model.

Hence, estimates of Df will be corrupted by nonlinearities as well as by lack of information in the direction of the negative exponents.

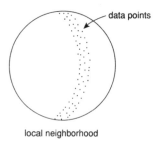

Figure 5.30 Local two-dimensional neighborhood for a data set that is nearly singular in one direction. Illustrates data 'curvature' which can be a source of severe errors in the calculation of the Jacobian matrices when a strictly linear analysis is undertaken.

Source: Adapted from Bryant, Brown and Abarbanel (1991) p. 2791.

[44]Bryant, Brown and Abarbanel (1991).

The basis of the linearization of the dynamics in the tangent space is a Taylor expansion truncated at 1st order terms. That puts, on Df, the full burden of estimating the stretching and squeezing of the map (the Lyapunov exponents), as well as the curvature introduced by finite data and measurement accuracy. The burden of estimating curvature can be lifted from Df by continuing the Taylor expansion past order-1 terms, leaving Df to measure just the linear component that defines the Lyapunov exponents. Thus, the model for the tangent space mapping becomes

$$\delta_{k,\alpha}^*(i) = \sum_{\beta} D_{\underline{y}_i} f(\alpha, \beta)\delta_{k,\beta}(i) + \sum_{\beta,\gamma} D_{\underline{y}_i}^{(2)} f(\alpha, \beta, \gamma)\delta_{k,\beta}(i)\delta_{k,\gamma}(i)$$

where $1 \le \alpha, \beta, \gamma \le d_r$; $1 \le k \le N_b \le K_\varepsilon(i)$; $\delta_{k,\alpha}(i)$ and $\delta_{k,\alpha}^*(i+1)$ are the α components of the vectors $\delta_k(i)$ and $\delta_k^*(i+1)$ respectively; and

$$D_{\underline{y}_i} f(\alpha, \beta) = \left.\frac{\partial f_{(\alpha)}}{\partial y_{(\beta)}}\right|_{\underline{y}(i)} \quad ; \quad D_{\underline{y}_i} f \in \mathfrak{R}^{d_r \times d_r}$$

$$D_{\underline{y}_i}^{(2)} f(\alpha, \beta, \gamma) = \frac{1}{2}\left(\left.\frac{\partial^2 f_{(\alpha)}}{\partial y_{(\beta)}\partial y_{(\gamma)}}\right|_{\underline{y}(i)}\right) \quad ; \quad D_{\underline{y}_i}^{(2)} f \in \mathfrak{R}^{d_r \times d_r \times d_r}$$

By defining

$$V^\alpha = \begin{bmatrix} \delta_{1,\alpha}^*(n + T_e) \\ \delta_{2,\alpha}^*(n + T_e) \\ \vdots \\ \delta_{N_b,\alpha}^*(n + T_e) \end{bmatrix} \qquad B^\alpha = \begin{bmatrix} D_{\underline{y}_n} f(\alpha, 1) \\ \vdots \\ D_{\underline{y}_n} f(\alpha, d_r) \\ D_{\underline{y}_n}^{(2)} f(\alpha, 1, 1) \\ \vdots \\ D_{\underline{y}_n}^{(2)} f(\alpha, 1, d_r) \\ D_{\underline{y}_n}^{(2)} f(\alpha, 2, 1) \\ \vdots \\ D_{\underline{y}_n}^{(2)} f(\alpha, d_r, d_r) \end{bmatrix}$$

$$\underline{X} = \begin{bmatrix} \delta_{1,1}(n) & \cdots & \delta_{1,d_r}(n) & \delta_{1,1}(n)\delta_{1,1}(n) & \delta_{1,1}(n)\delta_{1,2}(n) & \cdots & \delta_{1,d_r}(n)\delta_{1,d_r}(n) \\ \delta_{2,1}(n) & \cdots & \delta_{2,d_r}(n) & \delta_{2,1}(n)\delta_{2,1}(n) & \delta_{2,1}(n)\delta_{2,2}(n) & \cdots & \delta_{2,d_r}(n)\delta_{2,d_r}(n) \\ \vdots & \vdots & \vdots & \vdots & \vdots & \vdots & \vdots \\ \delta_{N_b,1}(n) & \cdots & \delta_{N_b,d_r}(n) & \delta_{N_b,1}(n)\delta_{N_b,1}(n) & \delta_{N_b,1}(n)\delta_{N_b,2}(n) & \cdots & \delta_{N_b,d_r}(n)\delta_{N_b,d_r}(n) \end{bmatrix}$$

we can state the inverse problem as a sequence of least squares problems of the form $V^\alpha = \underline{X}B^\alpha$, where V^α is known, and $\underline{X}$ is known and fixed. The solution B^α is obtained by the same methods as the linear problem and its first d_r components give the α row of the Jacobian. Solving for each α gives the full Jacobian $D_{\underline{y}_n} f$ at each point on the fiducial trajectory.

217

Finally, a QR decomposition can be used to obtain the λ_i, viz

$$D_{\underline{y}_n} f \cdot Q_n = Q_{n+1} \cdot R_n; \qquad Q_1 = \Im_{d_r}$$

$$\lambda_i = \frac{1}{T_e(N_{QR})} \sum_{j=1}^{N_{QR}} (R_k)_{ii}; \qquad i = 1, \ldots, d_r; \qquad N_{QR} \leq N_r$$

In the manner of the linear model, we can project the tangent vectors (after the nearest neighbor search) into a q-dimensional subspace. The formulation and solution method remains the same except that the dimension of the space where the calculations are made is q instead of d_r.

Parameters

The number of unknowns N_p, and hence the number of neighbors needed, when a Taylor expansion of order N_{Tay} is used is given by

$$N_p = \left[\prod_{k=1}^{N_{Tay}} \frac{d_r + k}{k} \right] - 1$$

The guidelines for choosing parameters follow those used for the linear model except that the lower bound on the size of the neighborhoods is initially set such that $K_\varepsilon(i) \geq 2N_p$.

Remarks

It will become apparent to the reader, on inspection of the results using the more complex nonlinear algorithm reported in Table 5.10 through Table 5.14, that the estimates do not demonstrate any obvious improvement over the results obtained with the linear approach. We conjecture that such a comparison is, unfortunately, not possible due to the use of sub-optimal phase space reconstruction parameters and tangent space projection methods employed in obtaining the tables for the nonlinear model. For example, here the trajectory matrices are projected from $\Re^{d_r}$ into $\Re^q$ by truncating the original tangent vectors to the first q components without having first used a global SVD to organize the components in decreasing order of significance.

Example 5.9

Order of polynomial	λ_1	λ_2
Linear	0.43451	−1.5849
Quadratic	0.44707	−1.5096
Cubic	0.44685	−1.5486
Quartic	0.45142	−1.4679

Table 5.10 The Lyapunov exponents for the Hénon map ($a = 1.4, b = 0.3, N_r = 11,000, N_{QR} = 1000$). These were computed in a local dimension $q = 2$ from data with **four** digits of accuracy. The global dimension used in the nearest neighbor seach is $d_r = 2$. The table shows the effects of varying the order of the local polynomial fit to the neighborhood-to-neighborhood map. The correct values of the exponents are $\lambda_1 \approx 0.41$ and $\lambda_2 \approx -1.62$.

Source: Adapted from Bryant, Brown and Abarbanel (1991) p. 2796.

Example 5.10

If the local dimension q in the estimation procedure is larger than d_A, then as in the case of the linear model $q - d_A$ phantom exponents will be produced.

d_r	λ_1	λ_2	λ_3	λ_4	λ_5	λ_6
2	0.44490	−1.6091				
3	0.44073	−0.89324	−1.6535			
4	0.44199	−0.30700	−0.80362	−1.6246		
5	0.46312	−0.049209	−0.38960	−0.76073	−1.6352	
6	0.46482	0.14395	−0.22719	−0.42457	−0.87154	−1.6449

Table 5.11 The Lyapunov exponents for the Hénon map ($a = 1.4$, $b = 0.3$, $N_r = 11,000$, $N_{QR} = 1000$). These were computed with a cubic neighborhood-to-neighborhood map on data with *six* digits of accuracy. The table shows the effects of varying the dimensions $q = d_r$. The correct values of the two real exponents are 0.41 and −1.62.
Source: Adapted from Bryant, Brown and Abarbanel (1991) p. 2796.

d_r	λ_1	λ_2	λ_3	λ_4	λ_5	λ_6
3	1.5169	−0.007994	−23.093			
4	1.5375	−0.070352	−22.147	−108.21		
5	1.5631	−0.015967	−21.600	−77.403	−114.83	
6	1.6123	0.0095171	−21.340	−60.057	−80.231	−115.50

Table 5.12 The Lyapunov exponents for the Lorenz system ($a = 16$, $b = 4$, $R = 45.92$, $N_r = 50,000$, $N_{QR} = 5000$, $\tau = 0.1$). These were computed with a cubic neighborhood-to-neighborhood map on data with *six* digits of accuracy. The table shows the effects of varying the dimensions $q = d_r$. The correct values of the exponents are $\lambda_1 \approx 1.51$, $\lambda_2 = 0.0$ and $\lambda_3 \approx -22.5$. Hence, the phantom exponents appear as λ_4, λ_5 and λ_6 in the table.
Source: Adapted from Bryant, Brown and Abarbanel (1991) p. 2800.

Example 5.11

If the data is contaminated by noise or measurement error the estimates are affected. In Tables 5.13 and 5.14 the number of retained digits of accuracy in the data was used to simulate measurement error which, like noise, results in phase space points not being properly located. Tables 5.13 and 5.14 compare results for the Ikeda map when various nonlinear models are used first on data with five digits of accuracy, and then on data with eight digits of accuracy.

Order of polynomial	λ_1	λ_2	λ_3	λ_4
Linear	0.87959	0.46944	−0.68081	−1.1596
Quadratic	0.49544	0.019259	−0.61109	−0.80987
Cubic	0.49597	−0.19415	−0.652624	−0.76963
Quartic	0.50354	−0.15418	−0.62712	−0.78807

Table 5.13 The Lyapunov exponents for the Ikeda map ($p = 1, B = 0.9, \kappa = 0.4, \alpha = 6.0, N_r = 21,000, N_{QR} = 3000$). These were computed with $q = d_r = 4$ on data with *five* digits of accuracy. The table shows how the order of the neighborhood-to-neighborhood polynomial fit affects the two legitimate and the two spurious exponents. For the fourth-order fit the spurious exponents appear as λ_2 and λ_3. Correct values for the two real exponents are 0.503 and −0.719. Notice that the spurious positive exponent decays as the order of the polynomial fit is increased.

Source: Adapted from Bryant, Brown and Abarbanel (1991) p. 2798.

Order of polynomial	λ_1	λ_2	λ_3	λ_4
Linear	0.92179	0.47199	−0.67251	−1.1499
Quadratic	1.5205	0.48759	−0.58714	−0.80199
Cubic	0.64834	0.42411	−0.57930	−0.73951
Quartic	0.61181	0.30823	−0.60121	−0.72202

Table 5.14 The Lyapunov exponents for the Ikeda map ($p = 1, B = 0.9, \kappa = 0.4, \alpha = 6.0, N_r = 21,000, N_{QR} = 3000$). These were computed with $q = d_r = 4$ on data with *eight* digits of accuracy. Correct values for the two real exponents are 0.503 and −0.719. For the fourth-order fit the spurious exponents appear as λ_2 and λ_3. The table shows how the order of the neighborhood-to-neighborhood polynomial fit affects the two legitimate and the two spurious exponents. Notice that the negative Lyapunov exponent is determined to greater accuracy than in the previous case, but the spurious positive exponent does not decay as fast as in the previous case.

Source: Adapted from Bryant, Brown and Abarbanel (1991) p. 2798.

5 Cross-correlation of local stretching[45]

The major problem with estimating Lyapunov exponents is their sensitivity to noise. This algorithm attempts to exploit that fact to discriminate between prossible sources of the data, or quantify the amount of noise present in the data.

Recall that

$$S_1(t) = \frac{1}{N_e \tau_s} \text{Log} \frac{\delta y_1(t + N_e \tau_s)}{\delta y_1(t)} \qquad \textbf{Eq. 5.24}$$

where $\delta y_1(t)$ is a small initial distance between two reconstructed trajectory points

---

[45]C. Bertram and X. Tian, *Physica D* 58, 1992.

in the neighborhood of $y(t)$, and $\delta y_1(t + N_e \tau_s)$ is the distance between them after an evolutionary time $N_e \tau_s$, is a local measure of the divergence rate of nearby trajectories. Similarly,

$$S_2(t) = \frac{1}{N_e \tau_s} \text{Log} \frac{\delta y_2(t + N_e \tau_s)}{\delta y_2(t)}$$

based on a different initial perturbation $\delta y_2(t)$ in the neighborhood of $y(t)$, also estimates trajectory divergence rates locally. If the equations of motion are continuous in the state variable, then $S_1(t)$ and $S_2(t)$ will be governed by the same local dynamics. Thus, the behavior of $S_1(t)$ and $S_2(t)$ should be correlated if the process is deterministic, and uncorrelated if the process is based on noise. The cross-correlation between $S_1(t)$ and $S_2(t)$ defined by

$$c_{12} = \frac{E[(S_1 - \langle S_1 \rangle)(S_2 - \langle S_2 \rangle)]}{\sigma_{S_1} \sigma_{S_2}}$$

measures the degree to which local stretching is organized in phase space, or equivalently, the extent to which the data are corrupted by noise or high dimensional attractor dynamics.

The divergence estimates $S_1(t)$ and $S_2(t)$ can be computed easily from Eq. 5.24 for short evolution times $N_e \tau_s$, but generally, because they will be dominated by the largest Lyapunov exponent for longer evolution times, they can be constructed by either of the direct methods described at the begining of this algorithm section[46]. The cross-correlation c_{12} is estimated with the standard approximation

$$c_{12} = \frac{\sum_{n=1}^{N_r - N_e}[S_1(n) - \langle S_1 \rangle][S_2(n) - \langle S_2 \rangle]}{\hat{\sigma}_{S_1} \hat{\sigma}_{S_2}}$$

where

$$\langle S_j \rangle = \sum_{n=1}^{N_r - N_e} S_j(n)/(N_r - N_e)$$

and

$$\hat{\sigma}_{S_j}^2 = \sum_{n=1}^{N_r - N_e} [S_j(n) - \langle S_j \rangle]^2$$

Remarks

- Note that the computation of c_{12} involves zero lag between $S_1(t)$ and $S_2(t)$.
- Maximizing c_{12} with respect to the embedding parameters, maximizes determinism, and also seems to give an independent quantitative criterion for the optimal choice of the other computational parameters involved in the estimation of the largest Lyapunov exponent – see the examples below.
- Note that there is no precise level of c_{12} that divides noise from determinism. Thus, it seems best to use c_{12} as a discriminating statistic for a hypothesis test, or derive confidence intervals for it, as described in the appendix on surrogate data and in Chapter 7, Concluding remarks.

[46]If the Wolf algorithm is used, employ the same trajectory replacement criterea for S_1 and S_2, and make the replacements simultaneously.

Parameters

Choose the number of steps N_e such that $N_e\tau_s < \tau_c$, the average cycle time of the system.

Examples

For these examples, S_1 and S_2 were computed by the Wolf algorithm (Direct Method II, above).

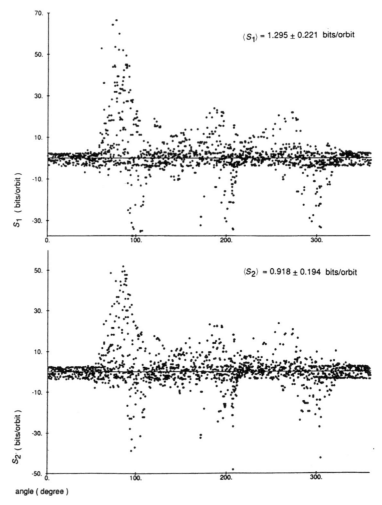

Figure 5.31 The statistics S_1 and S_2 versus angular location (θ) for different neighborhoods on the reversed time Rössler attractor – see chapter on Dynamical Systems for equation of motion. A location on the attractor can be defined in terms of an angle θ due to the single-centredness of the Rössler attractor. Thus, the angle variable θ can be used as a surrogate for the evolution time $N_e\tau_s$. The time series consisted of 2048 data points obtained by 'sampling' the x-component of the trajectory at $\tau_s = 0.3$ sec. The reconstruction was done in 3-dimensions with a delay time $\tau = \tau_s$. The evolution time $N_e\tau_s = 5\tau_s$.

Source: Adapted from Bertram and Tian (1992) p.474

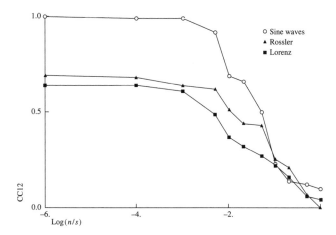

Figure 5.32 The cross-correlation c_{12} as a function of additive noise plotted as the Log (noise-to-signal) ratio versus the cross-correlation of the S_1 and S_2 time series. Note that the periodic system has a markedly higher cross-correlation than the chaotic systems, and that the breakdown in determinism for all the systems occurs when the noise exceeds 1%.

Source: Adapted from Bertram and Tian (1992) p.476

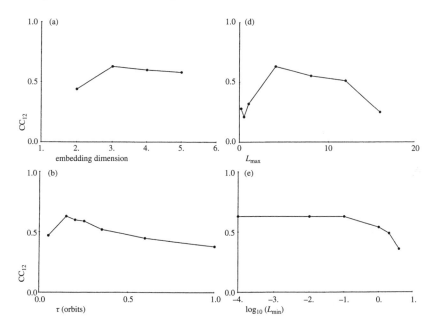

Figure 5.33 Optimal parameter settings seem to occur at maximum cross-correlation. Plotted are the cross-correlation c_{12} for the Rössler attractor versus (a) embedding dimension, (b) embedding time delay, (c) evolution time $N_e \tau_s$ in the definition of S_i, (d) maximum length of $\delta y_i(t + N_e \tau_s)$ before replacement in the Wolf algorithm (e) minimum length of $\delta y_i(t)$ after replacement in the Wolf algorithm, and (f) maximum angular deviation allowed for replacements in the Wolf algorithm.

Source: Adapted from Bertram and Tian (1992) p.479

223

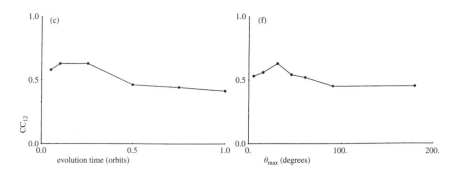

Figure 5.33 Continued.

REFERENCES AND FURTHER READING

Abarbanel, H. (1994) 'Analysing and utilizing time series observations from chaotic systems' in J. Thompson and S. Bishop (eds), *Nonlinearity and Chaos in Engineering Dynamics*, John Wiley & Sons, pp. 379–392.

Bertram, C. and Tian, X. (1992) 'Correlation of local stretchings as a way of characterising chaotic dynamics amid noise', *Physica D* 58, pp. 469–481.

Briggs, K. (1990) 'An improved method for estimating Lyapunov exponents of chaotic time series', *Physics Letters A* 151.01, pp. 27–32.

Bryant, P., Brown, R. and Abarbanel, H. (1990) 'Lyapunov exponents from observed time series', *Physical Review Letters* 65.13, pp. 1523–1530.

Brown, R., Bryant, P. and Abarbanel, H. (1991) 'Computing the Lyapunov spectrum of a dynamical system from an observed time series', *Physical Review A* 43.06, pp. 2787–2806.

Darbyshire, A. and Broomhead, D. (1996) 'Robust estimation of tangent maps and Lyapunov spectra', *Physica D* 89, pp. 287–305.

Doerner, R., Hubinger, B. and Martienssen, W. (1994) 'Advanced chaos forecasting', *Physical Review E: Rapid Communications* 50.01, pp. 12–15.

Dressler, U. and Farmer, D. (1992) 'Generalized Lyapunov exponents corresponding to higher derivatives', *Physica D* 59, pp. 365–377.

Eckhardt, B. and Yao, D. (1993) 'Local Lyapunov exponents in chaotic systems', *Physica D* 65, pp. 100–108.

Eckmann, J. and Ruelle, D. (1985) 'Ergodic theory of chaos and strange attractors', *Reviews of Modern Physics* 57.3, pp. 617–656.

Eckmann, J. and Ruelle, D. (1992) 'Fundamental limitations for estimating dimensions and Lyapunov exponents in dynamical systems', *Physica D* 56, pp. 185–187.

Eckmann, J., Ruelle, D., Ciliberto, S. and Oliffson Kamphorst, S. (1986) 'Lyapunov exponents from time series', *Physical Review A* 34.06, pp. 4971–4979.

Grassberger, P. and Procaccia, I. (1983) 'Measuring the strangeness of strange attractors', *Physica D* 9, pp. 189–208.

Kaplan, J. L. and York, J. A. (1979) 'Functional differential equations and approximation of fixed points: Numerical solution of a generalized eigenvalue problem for even mappings', *Lecture Notes in Mathematics* 730, eds. Dold, A. and Eckman, D. pp. 228–237.

Kaplan, J. L. and York, J. A. (1983) 'The Lyapunov dimension of strange attractors', *J. Differential Equations* 49, pp. 185–207.

Oseledec, V. I. (1968a) 'A multiplicative ergodic theorem. Lyapunov characteristic numbers for dynamical systems', *Trudy Mosk. Mat. Obsc.* 19, p. 179.

Oseledec, V. I. (1968b) 'A multiplicative ergodic theorem. Lyapunov characteristic numbers for dynamical systems', *Moskow Math. Soc.* 9, p. 197.

Parker, T. and Chua, L. (1987) 'Chaos: a tutorial for engineers', *Proceedings of the IEEE* 75.08, pp. 982–1008.

Rosenstein, M., Collins, J. and De Luca, C. (1993) 'A practical method for calculating largest Lyapunov exponents from small data sets', *Physica D* 65, pp. 117–134.

Sano, M. and Sawada, Y. (1985) 'Measurement of the Lyapunov spectrum from a chaotic time series', *Physical Review Letters* 55.1, pp. 1082–1085.

Takens, F. (1981) 'Detecting strange attractors in turbulence', *Warwick 1980 Lecture Notes in Mathematics* 898, pp. 366–381.

Tong, H. (1983) 'Threshold models in nonlinear time series', *Lecture Notes in Stats.* 21, Springer, Heidleberg.

Wilkonson, H. and Reinsch, C. (1971) *Linear Algebra*, Springer-Verlag, Berlin.

Wolf, A., Swift, J., Swinney, H. and Vastano, J. (1985) 'Determining Lyapunov exponents from a time series', *Physica D* 16, pp. 285–317.

Zeng, X., Eykholt, R. and Pielke, R. (1991) 'Estimating the Lyapunov exponent spectrum from short time series of low precision', *Physical Review Letters* 66.25, pp. 3229–3232.

Noise reduction

INTUITION

The material world is a complicated place made up of countless interacting processes. In that world an isolated system is only an ideal, which we can sometimes get close to, but never fully realize. There are always some residual phenomena which are physically beyond our control. Any observation we make on a system will include information on the residual phenomena that coexist with the system in space and time. In other words, residual phenomena, also known as noise, contaminate all real data. Noise is a subjective designation not an attribute. Noise may be a nuisance in the context of a given experiment, but the constituents of noise are not without interest or beyond comprehension in principle. Nevertheless, the methodology developed in the previous chapters presumes that the measurement time series we are working with contains accurate information about the evolving states of the system. Given the sensitivity of chaotic systems to both the system's initial state, and the parameters of its dynamical equations, any noise could undermine our efforts to model and predict the behavior of the system of interest.

Since we hope that the system of interest is low dimensional, what will constitute noise for us will be contamination from high dimensional systems, by which we mean either stochastic (random) processes or chaotic dynamical systems with many degrees of freedom[1]. We will assume that the signal from the system of interest dominates the information contained in the measurement time series. In particular, we assume that the signal from the system of interest is not buried in another much stronger signal, and that its amplitude is well above the level of measurement accuracy. If this assumption does not hold, and there really is an observable system, then the methods that are being used to measure the system are inadequate.

In the context of chaotic dynamics, noise leaves us somewhere between a hard place and a rock. On the one hand, as mentioned above, noise can seriously affect system modeling and prediction because of the system's sensitivity to parameters and initial states. On the other hand, naive approaches to noise reduction can alter the data in such a way that the experimental trajectory no longer reflects the chaotic dynamics of the underlying system. Unfortunately, that means that classical methods of noise reduction for linear systems are generally inappropriate, and in that case the only way to deal with the noise problem is with new techniques developed from first principles.

The effect of noise contaminated measurement time series on the modeling and prediction of nonlinear systems, especially those operating in a chaotic mode, is a complex issue, easier to demonstrate than explain. Notwithstanding, we can get an intuitive feeling for at least one important aspect of that issue by making a few simplifying assumptions about the nature of the noise contamination.

We have seen numerous examples in previous chapters of how noise can spoil our ability to estimate the invariants of a system accurately. Here, in order to illustrate the spoiling mechanism clearly, we will look at how noise affects a prediction

$$y^{(p)}(t + L)$$

made at time t for the state of the system L time units in the future. For the purpose of

[1]In any practical sense, high dimensional chaos is unpredictable and can be thought of as a random process. See the discussion on system redundancy and irrelevance times in Chapter 3, Phase space reconstruction.

229

this illustration only, we assume that the true dynamics f are known, for the simple reason that we do not wish to introduce another source of error, namely the error in estimating the dynamics, which is not central to the point we are trying to make.

Assume that we have used our measurement time series $x(t)$ as the basis of a (d_r, τ)-reconstruction[2] of the state space to obtain the experimental trajectory $\underline{y}(t)$. Denote the delay coordinate map of the reconstruction by Φ. Then $\underline{y}(t) = \Phi[s(t)]$, that is, Φ maps $s(t)$, the system's true trajectory in the original state space, to a representation $\underline{y}(t)$ in $\Re^{d_r}$. If Φ is an embedding, then it is invertible, and in the absence of noise, $s(t) = \Phi^{-1}[\underline{y}(t)]$. In other words, for a noiseless embedding we can retrieve the true state of the sytem from the experimental trajectory. Hence, we can say precisely that

$$y^{(p)}(t + L) = \Phi f^L \Phi^{-1}[\underline{y}(t)]$$

because

$$
\begin{aligned}
y^{(p)}(t + L) &= \Phi[s(t + L)] \\
s(t + L) &= f^L[s(t)]
\end{aligned}
$$

and

$$s(t) = \Phi^{-1}[\underline{y}(t)]$$

Now, suppose that a measurement taken at time t yields the observed value

$$x(t) = h[s^*(t)] = x^*(t) + \xi(t)$$

where h describes the output of the measuring device we use to observe the true state $s^*(t)$ in the original state space, $x^*(t)$ is the noise free component of the measured signal, and $\xi(t)$ is the error due to noise. When noise enters a system in this way, it is called **additive noise**. Since

$$
\begin{aligned}
\underline{y}(t) &= [x(t), x(t - \tau), \dots, x(t - (d_r - 1)\tau)] \\
&= [x^*(t), x^*(t - \tau), \dots, x^*(t - (d_r - 1)\tau)] \\
&\quad + [\xi(t), \xi(t - \tau), \dots, \xi(t - (d_r - 1)\tau)] \qquad \text{Eq. 6.1} \\
&= y^*(t) + \xi'(t)
\end{aligned}
$$

the additive errors in $x(t), x(t - \tau), \dots$ create an additive error in $\underline{y}(t)$. Suppose that we try to use the noisy experimental trajectory point $\underline{y}(t)$ as the initial condition for a prediction. How good is that prediction? If the initial condition for the prediction is wrong, then the error in the prediction $y^{(p)}(t + L)$ can grow very rapidly in L because of the characteristic exponential divergence of nearby trajectories in chaotic systems. We expect errors in the prediction $y^{(p)}(t + L)$ because we know there are errors in $\underline{y}(t)$. The temptation here is to identify the error in $\underline{y}(t)$ with the error in the initial condition, but that would be incorrect. It is easy to lose sight of the fact that even a noise free reconstructed attractor A is just a representation of the original attractor Q obtained through the delay coordinate mapping $A = \Phi[Q]$. The real system does not

---

[2]See Chapter 3, Phase space reconstruction.

live in the reconstructed state space; it lives in the original state space. Thus, the error in $y^{(p)}(t + L)$ actually depends on the error in locating the true current state $s^*(t)$ in the original state space, and as we will show below, the image $s(t) = \Phi^{-1}[y(t)]$ may not even be close to $s^*(t)$, even when the error in $y(t)$ is small as measured in the reconstructed state space. We will demonstrate that effect by building up a picture of the probability distribution of the true state $s^*(t)$ of the system, given a noisy reconstructed state $y(t)$. Ideally, that probability distribution should form a sharp peak centered on the true state $s^*(t)$, if the prediction error is to be minimal.

For simplicity, assume that both the original attractor Q, and a noise free reconstructed attractor A, are flat two-dimensional surfaces. In that case, $y^*(t) = [x^*(t), x^*(t - \tau)]$ is a point on A, and $s^*(t) = [s_1^*(t), s_2^*(t)]$ is a point on Q. Let y be a noisy experimental trajectory point, which will generally not lie on the noise free attractor A. Imagine we fix one end of a string of length r to the point y, and to the other end we attach a piece of chalk. Holding the string taut, we move the head of the piece of chalk across the attractor A tracing out a curve $C_A^{(r)}$. The curve $C_A^{(r)}$ is made up of all the points on the attractor that are at a distance r from the observation y (see Figure 6.1).

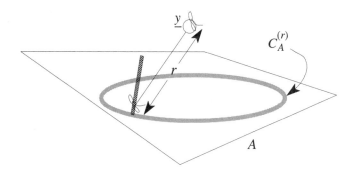

Figure 6.1 Tracing out the points on the attractor A that are consistent with the observation y and an error of magnitude r.

Dropping the time index and rearranging Eq. 6.1 gives

$$y^* = y - \xi'$$

Now, fix the magnitude of the noise vector ξ' in Eq. 6.1 at r, that is, let $\|\xi'\| = r$, and consider all the noise free points y^* on the attractor which could have given rise to the observation y. Those are just the set of points along the curve $C_A^{(r)}$. If we use Φ^{-1} to map the curve $C_A^{(r)}$ back to the original state space, then we get a curve $C_Q^{(r)}$ on the true attractor Q that contains all the points in the original state space which are consistent with the observation y, and an error of magnitude r.

To keep things simple, assume that the scalar noise variables $\xi(t)$ have zero mean, and are independent and identically distributed, so that all noise vectors $\xi'(t)$ with magnitude r have the same probability. In other words, $\xi'(t)$ has a constant probability

231

density $\rho_\xi(r)$ on the surface of a sphere of radius r centered at the origin. If we translate that sphere so that its center is at $\underline{y}$, then the intersection of its surface with the attractor A is just the curve $C_A^{(r)}$. Moreover, the conditional[3] probability density $p[C_A^{(r)}; \underline{y}]$ of the points on the attractor that make up the curve $C_A^{(r)}$ is equal to that part of the probability density surface $\rho_\xi(r)$ that coincides with the curve $C_A^{(r)}$. Furthermore, $p[C_A^{(r)}; \underline{y}]$ implies, via Φ^{-1}, a conditional probability density $p[C_Q^{(r)}; \underline{y}]$ for the curve $C_Q^{(r)}$ in the original state space. (see Figure 6.2).

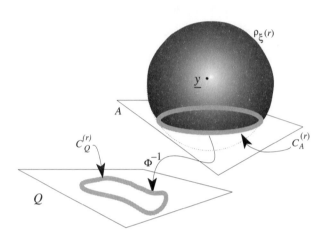

Figure 6.2 The probability density $\rho_\xi(r)$ lies on the surface of a sphere of radius r centered at $\underline{y}$. The intersection of that spherical surface with the reconstructed attractor A is the curve $C_A^{(r)}$ containing points on A that are consistent with the observation $\underline{y}$ and an error of magnitude r. When $C_A^{(r)}$ is mapped back to Q with Φ^{-1}, it produces a set $\overline{C}_Q^{(r)}$ containing points in the original state space that are consistent with the observation $\underline{y}$ and an error of magnitude r.

Now, there is nothing special about the magnitude r in the above discussion, so we can argue in the same way for all values of r, that is, for a solid probability density sphere $S_\xi[\underline{y}]$ centered at $\underline{y}$ instead of the surface $\rho_\xi(r)$. The end result is that the intersection of A with $\rho_\xi(\overline{r})$, the curve $C_A^{(r)}$, is replaced by the intersection of A with $S_\xi[\underline{y}]$, which is a surface C_A. When C_A is mapped back to the true attractor with Φ^{-1}, we get a surface C_Q which contains all the points in the original state space that are compatible with the observation $\underline{y}$. The probability density sphere $S_\xi[\underline{y}]$ imparts a conditional probability density $p(\underline{y}^*; \underline{y})$ for points in C_A, which in turn implies a conditional probability density $p(s^*; \underline{y})$ for points in C_Q. Graphically, the picture would look something like the diagram in Figure 6.3 below.

---

[3]Since we are considering a particular obervation $\underline{y}$, all probability densities are conditioned on the value of $\underline{y}$.

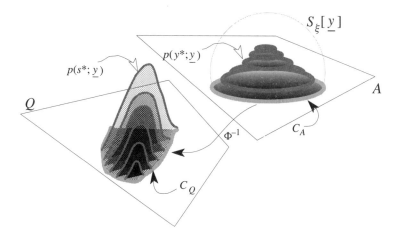

Figure 6.3 In the figure, the attractors A and Q are flat two-dimensional surfaces. The sets C_A and C_Q lie in the surfaces A and Q respectively. The conditional probability densities $p(y^*; \underline{y})$ and $p(s^*; \underline{y})$ are drawn out of the plane above C_A and C_Q, respectively.

We have kept the illustrations simple so far by drawing the attractors A and Q as flat two-dimensional surfaces. To get a more realistic picture, we will assume that while the true attractor Q is a flat two-dimensional surface as before, the reconstruction of the attractor A is done in three dimensions. In that situation, the reconstructed attractor may not have a simple geometry. When the reconstructed attractor A becomes complicated it can have a dramatic effect on the set C_Q, and therefore, the localization of $s^*(t)$, even though the error probability density sphere $S_\xi[\underline{y}]$ in the reconstructed state space remains unchanged. Thus, instead of the simple well localized probability density $p(s^*; \underline{y})$ we had before, we could get a probability density for $s^*(t)$ that looks something like the one depicted in Figure 6.4.

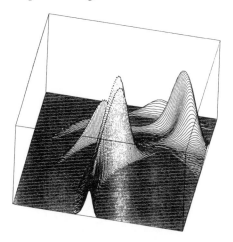

Figure 6.4 The conditional probability density $p(s^*; \underline{y})$ for a bad reconstruction. The position of the true state $s^*(t)$ is not well localized.

Source: Adapted from Casdagli *et al.* (1991) p. 63.

To see how this situation can arise, consider Figure 6.5.

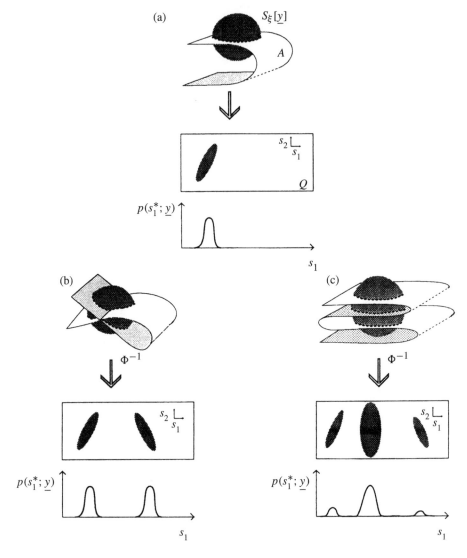

Figure 6.5 Good and bad reconstructions. The quality of the reconstruction depends on the shape of the surface $A = \Phi(Q)$. In (a) the surface $\Phi(Q)$ is well behaved within the 'noise ball' $S_\xi[\underline{y}]$ about $\underline{y}$ and the resulting conditional probability $p(s^*; \underline{y})$ is well localized. In (b) $\underline{y}$ is near a self-intersection and $p(s^*; \underline{y})$ is bimodal. Even when Φ is a global embedding, problems can occur if $\Phi(Q)$ is tightly folded, as illustrated in (c). In each of the three panels a, b, and c, the top figure shows C_A, the intersection of A with the sphere $S_\xi[\underline{y}]$; vertically below that, the middle figure shows $C_Q = \Phi^{-1}[C_A]$ on the original two-dimensional attractor Q; and the final figure shows a projection of the conditional density $p[s^*; \underline{y}]$ onto the first coordinate of the original state space.

Source: Adapted from Casdagli *et al.* (1991) p. 62.

From Figure 6.5 we can see two distinct causes for the delocalization of $s^*(t)$. The first is the geometry of the reconstructed attractor A. Specifically, at locations where the reconstructed attractor is tightly folded, or intersects itself, C_A is a collection of disjoint sets which produce a multi-modal error distribution for $s^*(t)$ as shown in Figure 6.4. Those situations can be eliminated or mitigated by choosing the reconstruction parameters along the lines discussed in chapter three: Phase Space Reconstruction. The second cause for the delocalization of $s^*(t)$ is the distribution of errors around $y(t)$ which is depicted in Figure 6.5 as the sphere $S_\xi[y(t)]$. The effective radius of the sphere varies with the variance of the errors in $y(t)$. Clearly, reducing the variance of the errors in $y(t)$ will result in a better localization of $s^*(t)$, and although we have made a number of assumptions to arrive at that simple picture, in general any error reduction scheme must ultimately result in a reduction of the mean square errors in $y(t)$.

The example for additive noise that we have been considering illustrates the importance of the error made in the original state space over the error made in the reconstructed state space and, in particular, how the geometry of the reconstructed attractor can transform and amplify the latter. Fortunately, we have some control over the geometry of the attractor in the choice of reconstruction parameters, so the example is helpful in pointing to practical steps we can take to minimize some of the effects of noise. On the other hand, the example has nothing to say on how we might go about reducing the mean square error in $y(t)$. The usual way to approach that problem, when the noise is additive, is to replicate the measurements and take an average of the resulting experimental trajectories. With fairly mild restrictions on the noise process, the averaging will tend to cancel the error, precipitating the deterministic noise free trajectory $y^*(t)$. But duplicating the measurements requires a restart of the system using the same initial condition. In practice it is unlikely that we would have the luxury of being able to do that. While we may be able to control the length of the time series of observations, reinitialization is simply not an option in the experimental situation. A time series of observations on a system constitutes a *single* observation on a segment of a particular trajectory in the original state space. Hence, there is nothing obvious to average, and we have to adopt a different approach to reduce the noise.

THEORY

The theoretical section of this chapter will be relatively short for two reasons. The natural way to approach noise reduction is to characterize the noise free signal of the system of interest so that we can distinguish it from noise. Unfortunately, it has proven difficult to establish theoretical properties of trajectories from systems that do not have certain 'nice' attributes[4], and most systems encountered in practice do not fall into that category. Additionally, some of the theory is best developed within the presentation of particular noise reduction algorithms in the next main section. The two theoretical topics singled out for discussion here motivate most of the noise reduction methods. The first concerns the existence of 'shadowing orbits', and the

---

[4] Here 'nice' means being hyperbolic, which is defined later in this section.

second concerns the relationship between periodic orbits and the structure of chaotic attractors. We begin with a description of two forms of noise often encountered in the literature, followed by the definition of a shadowing orbit and some measures of noise, and end with the theoretical results promised above.

Origins of noise contamination

Consider a discrete time dynamical system $s(t + 1) = f_\mu[s(t)]$ parameterized by the vector μ, and let $\xi(t)$ be a noise vector[5] independent of $s(t)$[6].

- *Additive noise* occurs when either the environment in which the system is operating obscures the true state of the system from measurement, or the device with which we observe the system is inaccurate. In either case, the apparent trajectory of the system will be $f_\mu[s(t)] + \xi(t + 1)$ even though the true trajectory is generated by $f_\mu[s(t)]$.
- *Dynamical noise* is fundamentally different from additive noise, and enters into a system in two ways. The first is by altering the system's dynamics from f_μ to $f_{\mu+\xi(t)}$[7]. The second way that dynamical noise arises is through a perturbation of state from $s(t)$ to $s(t) + \xi(t)$. Either phenomonen could occur, for example, if the system was subject to exogenous forces. While the distinction between the two forms of dynamical noise is somewhat artificial, the fact is that, in the presence of dynamical noise, a 'dynamical system' cannot be said to have a true trajectory, since in the first case, the trajectory followed by the system is $f_{\mu+\xi(t)}[s(t)]$, while in the second case, the trajectory followed by the system is $f_\mu[s(t) + \xi(t)]$, and neither of those is a deterministic function of the initial state $s(t)$.

We will assume, unless otherwise stated, that both dynamical and additive noise are always present in the measurement time series, with the proviso that instabilities in the dynamics of the system consist only of small perturbations of the parameter μ, which do not induce any bifurcations[8].

Noise measurement and shadowing orbits

If the reconstruction is an embedding, then $\underline{y}(t)$ is deterministic, that is, $\underline{y}(t) = \varphi[\underline{y}(t - 1)]$, only in the absence of noise. Hence, we can measure the departure from the noise free situation by computing the quantity

$$\sigma^2_{\text{dyn}} = \frac{1}{N_r} \sum_{k=2}^{N_r} \|\underline{y}_k - \varphi[\underline{y}_{k-1}]\|^2$$

[5]We use $\xi(t)$ to denote a general noise variable whose dimension is implied by the context in which it is used.

[6]Dynamical and additive noise can be defined similarly in terms of the vector fields and flows of continuous time systems.

[7]In Chapter 1, Dynamical systems we already assumed that μ is not a deterministic function of time $\mu(t)$.

[8]This is the same assumption made at the end of the discussion on bifurcations in Chapter 1, Dynamical systems.

where φ is the dynamics in the reconstructed state space[9]. σ_{dyn}^2 is the mean square dynamical error or just the **dynamical error**. In principle, if φ is known then, in the absence of noise, $\sigma_{\mathrm{dyn}} = 0$. In practice, the trajectory $\underline{y}(t)$ is noisy so, even if φ is known, $\sigma_{\mathrm{dyn}} > 0$.

A **shadowing orbit** u_k is a deterministic trajectory that closely follows the noisy experimental trajectory $\underline{y}_k$. Formally, a deterministic orbit u_k is said to ε-**shadow** $\underline{y}_k$ if

$$\|\underline{y}_k - \varphi^{k-1}[u_1]\| = \|\underline{y}_k - \underline{u}_k\| < \varepsilon; \quad 1 \le k \le N_r$$

Clearly, we can reduce the dynamical error to zero by replacing the noisy trajectory $\underline{y}_k$ with a shadowing orbit. But noisy trajectories are not deterministic so, even if a shadowing orbit does exist, we do not know if the behavior of the shadowing orbit truly reflects the behavior of the underlying purely deterministic system. Intuitively it should, provided the shadowing orbit is typical[10] and the noise level is not too high, even though the shadowing orbit and the noisy orbit may have originated from different initial conditions (see Figure 6.6).

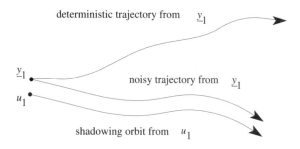

Figure 6.6 Diagram of the shadowing phenomenon. The deterministic trajectory veers away from the noisy trajectory starting from the same initial condition $\underline{y}_1$, but there may be a deterministic orbit, starting from a different initial condition u_1, that shadows the noisy trajectory.

Note that, even if the system does have a true noise free trajectory, when the noise level is high, it is possible that a shadowing orbit of $\underline{y}_k$ is not close to the true trajectory. But, if the noise level is moderate, and if there is a deterministic trajectory u_k which is typical and shadows the noisy orbit $\underline{y}_k$, then we can characterize the noise in the system as follows:

- In the additive noise case, u_k should be close to the noise free trajectory y_k^*. In fact, in principle, there is a shadowing orbit which is identical to y_k^* so that $\underline{y}_k = y_k^* + \xi_k \equiv \varphi[u_{k-1}] + \xi_k$.
- In the dynamical noise case, a true noise free trajectory does not exist. However, since u_k is typical by assumption, we are free to imagine that u_k is the true trajectory and, hence, the deterministic orbit of an additive noise process that is identical to $\underline{y}_k = \varphi[u_{k-1}] + \xi_k$.

[9]For example, we can take it to be the estimate derived in Appendix 5, Linearized dynamics.
[10]A typical trajectory is one which evolves onto the attractor. See discussion below on hyperbolic and non-hyperbolic systems.

It follows that we have some justification for simply defining noise as the pointwise fitting error $\xi_k = \underline{y}_k - \varphi[u_{k-1}] = \underline{y}_k - u_k$. Thus, it makes sense to define the total *fitting error*

$$\sigma_{\text{fit}}^2 = \frac{1}{N_r} \sum_{k=2}^{N_r} \|\xi_k\|^2$$

Now note that, if the ξ_k are identically distributed with zero mean, and if the shadowing orbit coincides with the true noiseless trajectory of the system, then the total fitting error is also the noise variance of the system.

Hyperbolic systems

A system is *everywhere hyperbolic* if, at each point on the attractor, the dynamics can be factored into directions[11] where the motion is *either* exponentially expanding *or* exponentially contracting, and the exponential rate is bounded away from zero[12]. That implies, among other things, that at each point of a hyperbolic attractor, the stable and unstable manifolds are distinct, so, while *homoclinic intersections* are admissible, *homoclinic tangencies* are not (see Figure 6.7).

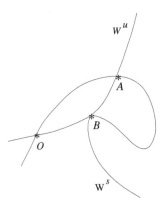

Figure 6.7 The stable manifold W^s and the unstable manifold W^u of the trajectory point O in a homoclinic intersection at A, and a homoclinic tangency at B.

The reason that hyperbolic systems are interesting is that they can be proved to possess a number of important properties. First, the existence of shadowing orbits for experimental trajectories can be established for such systems.

Theorem (Anosov (1967) and Bowen (1975))
For everywhere hyperbolic systems with bounded noise, that is $\|\xi_t\| < \delta, \delta > 0$, shadowing orbits which are typical always exist. In particular, if $\underline{y}_t$ is a noisy trajectory, then for every $\varepsilon > 0$ there exists a $\delta > 0$ such that $\|u_t - \underline{y}_t\| < \varepsilon$ for all

---

[11]Stable and unstable eigenspaces/manifolds. See Chapter 1, Dynamical systems.
[12]In the case of continuous time systems, the statement is ammended to allow for the zero Lyapunov exponent in the direction of the flow.

t, for some deterministic orbit u_t. Moreover, the shadowing orbit u_t is unique in the limit as $t \to \infty$.

Second, the unstable (saddle) periodic orbits of hyperbolic chaotic systems are densely embedded in the attractor. Thus, the closure of the set of unstable periodic orbits is the set of chaotic orbits, that is, the chaotic attractor. The short saddle periodic orbits give a rough outline of the attractor, while the long unstable n-periodic orbits fill it out as $n \to \infty$. Knowledge of the number, distribution, and properties of the unstable periodic orbits gives a detailed characterization of the chaotic orbits and system invariants[13].

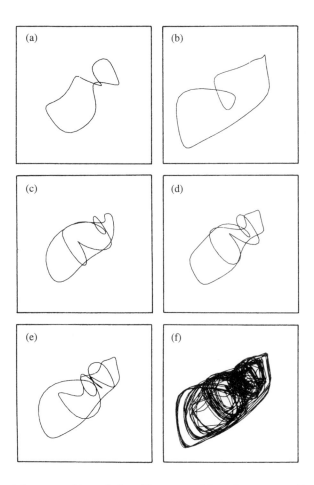

Figure 6.8 (a–e) Some unstable periodic orbits extracted from data generated by the Mackay–Glass equation in the chaotic regime. (f) Projection of all the unstable periodic orbits up to order-5 yields an object that is a good representation of the true chaotic attractor.
Source: Adapted from Pawelzic and Schuster (1991) p. 1810.

[13] Auerbach *et al.*, (1987); Cvitanovic (1988); Grebogi *et al.* (1988).

Non-hyperbolic systems

Unfortunately, everywhere hyperbolic systems are *not typical* in the real world. However, knowledge of the periodic orbits and their organization should severely constrain the recurrence properties and dynamics of chaotic attractors even when they are not hyperbolic. Thus, a relatively small number of unstable periodic orbits should be sufficient to characterize the typical chaotic attractor.

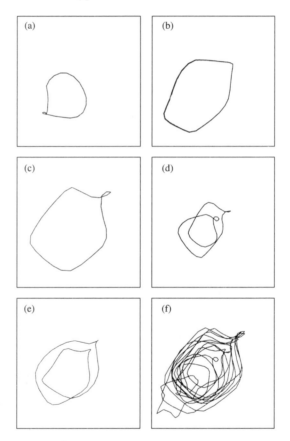

Figure 6.9 (a–e) Some unstable periodic orbits extracted from data from a driven pendulum experiment. (f) Projection of all the unstable periodic orbits up to order-5 yields an object that is likely to be a good representation of the true chaotic attractor.

Source: Adapted from Pawelzic and Schuster (1991) p. 1810.

As for shadowing orbits, it has been shown[14] that, for non-hyperbolic two-dimensional maps at low noise levels, there exist deterministic shadowing orbits for finite segments of noisy trajectories. The length of those segments is limited by the presence of homoclinic tangencies of a given severity, the probability of running into which increases as the noisy trajectory segment length increases. Currently, we can only speculate that a similar result holds in higher dimensions.

---

[14]Hammel *et al.* (1987).

ALGORITHMS

Some words of warning before proceeding.

- Using singular value decomposition or band width filters to isolate the deterministic signal in the experimental trajectory can produce a trajectory whose coordinates no longer have the same nonlinear coupling as the original. Thus, these techniques are capable of altering the dynamics of the system. Some of the methods described in this section use singular value decomposition and low pass filters to reduce dimension as part of the algorithm, and each of these inherits the above hazard.
- Another type of problem arises when the experimental trajectories are subjected to any kind of transformation. While the transformed trajectory may still be diffeomorphic to the true trajectory in the original state space, it is unlikely to correspond to a delay coordinate map of a time series. That is, there is no guarantee that the first component of the transformed trajectory point y_i' is the same as the second component of y_{i+1}', and so on. Depending on our objectives, that may not matter. For example, when estimating invariants it does not matter; all we need is the transformed trajectory to be diffeomorphic to the original. But if we need to maintain a relationship with the original time series, which we would if, for example, our objective was forecasting, then care must be taken to ensure that the adjusted vectors y_i' correspond to the delay vectors of a time series x_i'. That is, if the adjustment produces

$$y_i' = [x_i', \ldots, x_{i-d_E+1}'],$$

then $x_i', \ldots, x_{i-d_E+2}'$ must be the last $d_E - 1$ components of y_{i+1}'.
- In most practical situations the data is noisy and the dynamical mapping φ is unknown as we have assumed. In that case, noise reduction and function estimation are perforce interdependent, and the amount of noise reduction will depend on how well we can estimate the dynamics. If the noise levels are not too high an iterative scheme will probably work: namely, approximate dynamics φ on the reconstructed state space, apply a noise reduction algorithm, then approximate φ again on the noise reduced data, and so on, until things hopefully converge.

Some of the algorithms which follow just assume that an approximation for φ is in hand. Hence, a prerequisite for implementing those algorithms is a model of the dynamics. For those situations, we will adopt the local linear estimate of φ developed from concepts discussed in chapters 1, Dynamical Systems and 5, Lyapunov Exponents[15].

1 Statistical noise reduction[16]

A simplistic approach to noise reduction is to fit a model to the noisy data, such as the above local linear map, then pick an arbitrary initial condition and iterate the map to create a new trajectory. In many cases the resulting trajectory will have statistical properties closer to the true trajectory than those of the original noisy data.

---

[15]See Appendix 5, Linearized dynamics.

[16]Farmer and Sidorowich (1991) p. 373.

If the function approximation is good, then the new trajectory can give information at smaller length scales than the original experimental data, making it ideal for measuring invariants.

Example 6.1

The noisy data for this example (see Figure 6.10a) was obtained by adding uniformly distributed noise of magnitude 0.02 to each component of a clean trajectory generated by the Ikeda map[17].

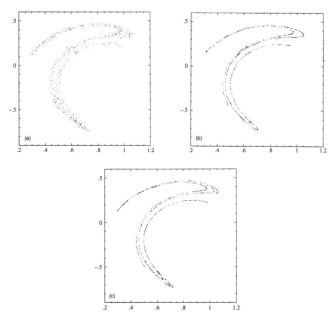

Figure 6.10 (a) A phase space plot showing 1500 points of a noisy time series, obtained by adding uniformly distributed noise of magnitude 0.02 to each component of the data shown in (c). (b) The result of applying the simple statistical noise reduction technique to the data in (a); the overall features are quite similar to (c), *but there is no detailed correspondence between the points in (b) and those in (c)*. (c) A plot of 1500 successive iterates of the Ikeda map with parameter $\mu = 0.7$. These are the 'true' data for (a) before adding noise.

Source: Adapted from Farmer and Sidorowich (1991) p. 384–385.

Remarks

The technique is similar to the boot strapping approach to extending a time series.

Problems

- When the initial condition is chosen in the basin of attraction and close to the attractor, it should not take long for the new trajectory to trace out a good representation of the attractor. However, for some initial conditions in the basin of attraction it may take a long time for the new trajectory to approach the attractor. If the initial condition is not in the basin of attraction then, clearly, the new trajectory will not resemble the attractor.

---

[17]See Chapter 1, Dynamical systems for the equations of motion.

- Statistical noise reduction can be very misleading when the dynamics are acutely sensitive to the parameters of the map. In some cases the parameter values corresponding to periodic orbits are dense in the parameter space corresponding to chaotic attractors, so it is not difficult to produce the wrong limit set.

2 Manifold decomposition[18]

A logical approach to finding a shadowing orbit for the experimental trajectory is to locate a trajectory u_t which minimizes the fitting error σ_{fit}^2 subject to a constraint requiring the trajectory to be deterministic, namely that $u_{t+1} = \varphi[u_t]$, so that $\sigma_{\text{dyn}}^2 = 0$. That problem can be solved using the method of Lagrange multipliers which yields the equivalent ordinary minimization of

$$Q = \sum_1^{N_r} \| \underline{y}_t - u_t \|^2 + 2 \sum_1^{N_r-1} [\varphi[u_t] - u_{t+1}]^T \cdot \lambda_t$$

where the 2 is just for simplifying the following normal equations

$$\frac{\partial Q}{\partial u_t} \;=\; 0 = -(\underline{y}_t - u_t) + (D_u\varphi[u_t])^T \lambda_t - \lambda_{t-1}$$

$$t = 1, \ldots, N_r; \lambda_0 = \lambda_{N_r} = 0$$

$$\frac{\partial Q}{\partial \lambda_t} \;=\; 0 = \varphi[u_t] - u_{t+1}$$

$$t = 1, \ldots, N_r - 1$$

Let $u_t^{(0)}$ be a candidate shadow orbit. Now define

$$\Delta_t \;=\; u_t^{(0)} - u_t = \text{deviation from the true shadowing orbit}$$

$$\gamma_t \;=\; \underline{y}_t - u_t^{(0)} = \text{deviation from the observed noisy orbit}$$

$$\varepsilon_t \;=\; u_{t+1}^{(0)} - \varphi[u_t^{(0)}] = \text{deviation from determinism}$$

$$J_t \;=\; D_u\varphi[u_t^{(0)}] = \text{Jacobian of } \varphi \text{ at } u_t^{(0)}$$

Since $u_t^{(0)}$ is assumed to be close to u_t, from the first order (linear) Taylor expansion of φ,

$$D_u\varphi[u_t^{(0)}] \approx D_u\varphi[u_t]$$

and

$$\varphi[u_t^{(0)}] - \varphi[u_t] \approx J_t \cdot \Delta_t.$$

Hence, the *linearized* normal equations are

$$\gamma_t \;=\; J_t^T \cdot \lambda_t - \lambda_{t-1} - \Delta_t \qquad t = 1, \ldots, N_r; \lambda_0 = \lambda_{N_r} = 0 \qquad \textbf{Eq. 6.2}$$

$$\Delta_{t+1} \;=\; J_t \cdot \Delta_t + \varepsilon_t \qquad t = 1, \ldots, N_r - 1 \qquad \textbf{Eq. 6.3}$$

[18]Farmer and Sidorowich, (1991) p. 373.

Equivalently, the linearized normal equations can be written in matrix form as follows

$$
\begin{bmatrix}
-\Im & J_1^T & & & & & & \\
J_1 & 0 & -\Im & & & & & \\
& -\Im & -\Im & J_2^T & & & & \\
& & J_2 & 0 & -\Im & & & \\
& & & & \cdots & & & \\
& & & & \cdots & & & \\
& & & & & -\Im & -\Im & J_{N_r-1}^T & \\
& & & & & & J_{N_r-1} & 0 & -\Im \\
& & & & & & & -\Im & -\Im
\end{bmatrix}
\begin{bmatrix}
\Delta_1 \\
\lambda_1 \\
\Delta_2 \\
\lambda_2 \\
\Delta_3 \\
\vdots \\
\vdots \\
\lambda_{N_r-1} \\
\Delta_{N_r}
\end{bmatrix}
=
\begin{bmatrix}
\gamma_1 \\
-\varepsilon_1 \\
\gamma_2 \\
-\varepsilon_2 \\
\gamma_3 \\
\vdots \\
\vdots \\
-\varepsilon_{N_r-1} \\
\gamma_{N_r}
\end{bmatrix}
$$

Abbreviating, the above becomes $Mv = \omega$, where M is $(2N_r - 1) \times (2N_r - 1)$ and the elements of M are themselves $d_E \times d_E$ matrices[19]. If we could invert M to solve for Δ_t, then a new and hopefully better estimate of u_t would be $u_t^{(1)} = u_t^{(0)} - \Delta_t$. Replacing the initial guess $u_t^{(0)}$ with $u_t^{(1)}$ and recomputing, and so on, we get the recursion

$$G[u_t^{(m)}] = u_t^{(m+1)}$$

The shadowing orbit is a fixed point of the above equation and we can use Newton's method[20] to locate it, if the initial guess $u_t^{(0)}$ is close enough to u_t. Unfortunately, M can be difficult to invert. When the underlying dynamics are chaotic, M becomes ill-conditioned[21], and the presence of homoclinic tangencies also causes it to become rank deficient as the Jacobians become so. The ill-conditioning arises because the eigenvalues of the product Jacobians ΠJ_i, involved in some inversion methods, diverge exponentially for a chaotic system. The rank deficiency arises because, at homoclinic tangencies, the expanding and contracting eigenspaces of the Jacobians coincide. These difficulties can be overcome with singular value decomposition, but SVD is a computer resource pig requiring on the order of n^3 calculations and n^2 memory locations to decompose an $n \times n$ matrix. In contrast, the following method, called **manifold decomposition**, requires on the order of n calculations and n memory locations.

Observe that what we really want are the corrections Δ_t. Note that Δ_1, the direction and length from u_1 to $u_1^{(0)}$, is the only unknown in the recursion

$$\Delta_t = \left(\prod_{i=1}^{t-1} J_i\right) \Delta_1 + \varepsilon_{t-1} + \sum_{i=1}^{t-2} \left(\prod_{k=i+1}^{t-1} J_k\right) \varepsilon_i \qquad \textbf{Eq. 6.4}$$

obtained by iterating Eq. 6.3. The recursive relationship defining the Δ_t, Eq. 6.4, becomes simple provided we require additional constraints on the end points of the trajectory $u_t^{(0)}$. To wit, suppose, fortuitously, that the perturbation Δ_1 lies precisely

---

[19] $\Im$ is the identity matrix.

[20] See Appendix 6, Equilibrium points and cycles, or any good numerical analysis text.

[21] Ill-conditioning occurs when numerical operations involve numbers with vastly different magnitudes.

on the *stable* manifold of $u_1^{(0)}$. That makes Δ_1 approximately a tangent vector in $E^s[u_1^{(0)}]$, the stable eigenspace[22] at $u_1^{(0)}$ (see Figure 6.11). Thus, for large t[23]

$$\left\| \left(\prod_{i=1}^{t} J_i \right) \Delta_1 \right\| \approx [\Lambda_s]^t \|\Delta_1\| \approx 0 \qquad \text{Eq. 6.5}$$

where Λ_s is the largest Lyapunov *number*[24] less than 1. Hence, under these assumptions, the forward recursion, Eq. 6.4, becomes independent of $\|\Delta_1\|$ for large t.

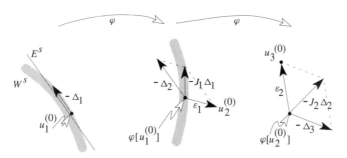

Figure 6.11 Schematic showing the progression of the trajectory adjustments under iteration of the dynamics φ.

Similarly, we can get a backward recursion from Eq. 6.3 by utilizing the invertibility of φ to get

$$\Delta_t = \left(\prod_{k=t}^{N_r-1} J_k^{-1} \right) \Delta_{N_r} - \sum_{i=t}^{N_r-1} \left(\prod_{k=t}^{i} J_k^{-1} \right) \varepsilon_i \qquad \text{Eq. 6.6}$$

If this time we pick the perturbation Δ_{N_r} on the *unstable* manifold of $u_{N_r}^{(0)}$, then for small t

$$\left\| \left(\prod_{k=t}^{N_r-1} J_k^{-1} \right) \Delta_{N_r} \right\| \approx [\Lambda_u]^{t-N_r} \|\Delta_{N_r}\| \approx 0 \qquad \text{Eq. 6.7}$$

where Λ_u is the smallest Lyapunov *number* greater than 1. As before, under these conditions Δ_t becomes independent of $\|\Delta_{N_r}\|$ for $N_r \gg t$.

Now, assume that the stable and unstable manifolds are transverse everywhere and consider the decomposition

$$\Delta_t = [\Delta_t]_{\text{stable}} + [\Delta_t]_{\text{unstable}}$$

where $[\ ]_{\text{(un)stable}}$ means projection onto the (un)stable eigenspace at $u_t^{(0)}$. Taking the stable direction first, Eq. 6.4 and Eq. 6.5, imply

$$[\Delta_t]_{\text{stable}} = \left[\varepsilon_{t-1} + \sum_{i=1}^{t-2} \left(\prod_{k=i+1}^{t-1} J_k \right) \varepsilon_i \right]_{\text{stable}} \qquad t \gg 1 \qquad \text{Eq. 6.8}$$

---

[22] See Chapter 1, Dynamical systems.

[23] See Oseledec's theorem in Chapter 5, Lyapunov exponents.

[24] The Lyapunov number is e^{λ_s}, the exponential of the Lyapunov exponent.

Next, in the unstable direction Eq. 6.6 and Eq. 6.7 imply

$$[\Delta_t]_{\text{unstable}} = \left[-\sum_{i=t}^{N_r-1} \left(\prod_{k=t}^{i} J_k^{-1} \right) \varepsilon_i \right]_{\text{unstable}} \qquad t \ll N_r \qquad \text{Eq. 6.9}$$

Hence, as long as the stable and unstable manifolds are transverse, for $1 \ll t \ll N_r$, Eq. 6.8 and Eq. 6.9 yield estimates for the components of the correction vector Δ_t along the stable and unstable manifolds respectively. These stable and unstable correction vectors are then added together to form the complete correction Δ_t which is accurate to the extent that the quantities on the right hand side of Eq. 6.5 and Eq. 6.7 are negligible. Finally, having obtained the full corrections Δ_t, we calculate the new trial trajectory $u_t^{(1)} = u_t^{(0)} - \Delta_t$, and repeat the whole process again for $u_t^{(1)}$, and so on, until $u_t^{(m+1)}$ shows no significant improvement over $u_t^{(m)}$ as measured by σ_{dyn}.

Problems

- We note from Figure 6.11 that since $J_1 \Delta_1 + \varepsilon_1$ is generally not in the stable direction of J_2, and so on, the corrections $J_t \Delta_t$, that is, the quantity within the brackets on the right hand side of Eq. 6.8 tends to relax onto the unstable manifold. The result is that the right hand side of Eq. 6.8 tends to vanish. The problem can be corrected by projecting Δ_t onto the stable direction of J_t at each step (see Figure 6.12). That stabilizes the algorithm, but it also introduces a new source of errors which will accumulate. Similar comments apply to Eq. 6.9.

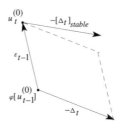

Figure 6.12 An illustration of the approximation needed to iterate Eq. 6.3 stably. At each stage Δ_t is projected onto the stable manifold at $u_t^{(0)}$, that is, in the stable direction of J_t. The geometry here coincides with the right hand side of Figure 6.11.

- In stretches that are free of homoclinic tangencies, manifold decomposition works well, but blows up when the angle between the stable and unstable manifolds gets too small. While singular value decomposition is not an efficient way to obtain corrections, that is, invert the matrix M along the entire trajectory, it can be useful around homoclinic tangencies where manifold decomposition breaks down. An effective way to combine the two methods is as follows. Compute the stable and unstable directions of the Jacobians at each trajectory point, and apply SVD to invert M for short trajectory segments centered at points where the angle between the tangent manifolds is small. Use the resulting corrections to adjust the troublesome trajectory segments. Then apply one iteration of manifold decomposition to the whole, partially corrected trajectory. Then apply SVD again to segments around points that are not deterministic, and so on, until the dynamical error ceases to improve.

Another way to proceed, when we have some measure of the noise level, is to monitor how far $u_t^{(1)}$ has wandered from $u_t^{(0)}$ after manifold decomposition. If the move, that is $\|\Delta_t\|$, is much larger than the noise level, it is probably due to a tangency. Having thus obtained the location of the problematic points, we have several options before further iteration with manifold decomposition. We can use singular value decomposition on the segment of the experimental trajectory near the offending point. Or, we can reset the offending point to its original position at $u_t^{(0)}$, or add some noise to the trajectory segment near the offending point $u_t^{(0)}$. Alternatively, we can just settle for using manifold decomposition on its own, without adjusting the experimental trajectory, but obtaining only piecewise clean orbits between the offending trajectory points.

Parameter values

Set $u_t^{(0)} = \underline{y}_t$.

Example 6.2

The noisy data for this example is the same as that used in example 6.1 for the Statistical Noise Reduction algorithm above. Note that while panels (a) and (c) are the same in Figures 6.10 and 6.13, panel (b) is not the same for the two figures.

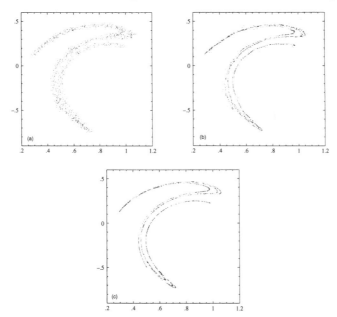

Figure 6.13 (a) A phase space plot showing 1500 points of a noisy time series, obtained by adding uniformly distributed noise of magnitude 0.02 to each component of the data shown in (c). (b) The result of applying the noise reduction technique described in the text to the data in (a); the overall features are quite similar to (c), and, in contrast to Figure 6.10, *there is a detailed correspondence between the points in (b) and those in (c)*. (c) A plot of 1500 successive iterates of the Ikeda map with parameter $m = 0.7$. These are the 'true' data for (a) before adding noise.

Source: Adapted from Farmer and Sidorowich (1991) p. 384.

247

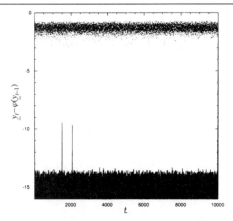

Figure 6.14 Pointwise dynamical error versus time for a trajectory of the Hénon map. The scattered points at the top are the original noisy time series (uniformly distributed noise of magnitude 0.04 was added to both components of the deterministic time series.) After two iterates of the mixed manifold decomposition and singular value decomposition technique described in the text, the noise has been reduced to the level indicated by the solid curves at the bottom. Logarithms are taken to base 10.

Source: Adapted from Farmer and Sidorowich (1991) p. 380.

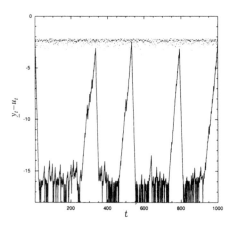

Figure 6.15 Pointwise fitting error versus time for a trajectory of 1000 points from the Hénon map. Scattered points at the top of the figure are the original noisy data (uniform noise of magnitude 0.005 was added to both components of the deterministic time series). Solid curve is the noise level after two iterates of the mixed manifold decomposition and singular value decomposition technique.

Source: Adapted from Farmer and Sidorowich (1991) p. 381.

3 Minimizing dynamical error[25]

In the previous algorithm our objective was to find a shadowing orbit of the experimental trajectory. As we saw there, in principle, the solution is equivalent to

---

[25]Davies, (1994).

248

finding a fixed point by Newton's method, using the experimental trajectory as the initial guess. Newton-type methods have quadratic convergence, but involve a matrix inversion which is done by brute force singular value decomposition (SVD), an order-$(d_E N_r)^3$ CPU process, or by exploiting the structure of the matrix M and employing something like manifold decomposition which is unstable near homoclinic tangencies. Recall that, for those methods to work, the initial guess and the shadowing orbit must be close enough that the linear approximation made in the algorithm holds. The fact is that, near a homoclinic tangency, there may be no deterministic orbit in the linear neighborhood of the noisy trajectory.

If we relax the objective somewhat then we can use a more robust approach to the solution. Thus, finding a nearby orbit $\{u_t, t = 1, 2, \ldots\}$ which is not necessarily deterministic, but just less noisy, is equivalent to minimizing

$$\sigma_{\text{fit}}^2 + \sigma_{\text{dyn}}^2 = \sum_{t=1}^{N_r} \|\underline{y}_t - u_t\|^2 + \sum_{t=1}^{N_r-1} \|u_{t+1} - \varphi[u_t]\|^2$$

If the noise level is not too high then the experimental trajectory is probably close to some deterministic orbit, so treating the former as the initial guess for a noise minimization algorithm, involving both fitting error and dynamical error, probably works just as well without the fitting error term. In that case, the noise reduction problem is reduced to the simpler problem of minimizing dynamical error with the original noisy orbit as the initial guess. Minimizing the dynamical error σ_{dyn}^2 is the same as minimizing

$$H = \sum_{i=1}^{N_r-1} \varepsilon_i^T \cdot \varepsilon_i$$

where $\varepsilon_i = u_{i+1} - \varphi[u_i]$. One way to avoid matrix inversion, and stabilize the minimization, is to use the following steepest descent recursion

$$u_t^{(m+1)} = u_t^{(m)} - \alpha \nabla H_t; \quad u_t^{(0)} = \underline{y}_t \qquad\qquad \textbf{Eq. 6.10}$$

where α is the "step" size,

$$\nabla H_t = \frac{\partial H}{\partial u_t} = 2 \sum_{i=1}^{N_r-1} \left(\frac{\partial \varepsilon_i}{\partial u_t}\right) \varepsilon_i,$$

and both ε_t and $\partial H / \partial u_t$ are evaluated at $u_t^{(m)}$, the estimate from the m-th iteration of the procedure. Since

$$\frac{\partial \varepsilon_i}{\partial u_t} = \begin{cases} \Im & i = t - 1 \\ -J_t & i = t; \quad J_t = D_u \varphi[u_t] \\ 0 & i \neq t, t - 1 \end{cases}$$

249

we have

$$\frac{\partial H}{\partial u_1} = -2J_1\varepsilon_1$$

$$\frac{\partial H}{\partial u_t} = 2(\varepsilon_{t-1} - J_t\varepsilon_t) \quad 1 < t < N_r - 1$$

$$\frac{\partial H}{\partial u_{N_r}} = 2\varepsilon_{N_r-1}$$

Letting

$$\nabla H = \begin{bmatrix} \nabla H_1 \\ \nabla H_2 \\ \vdots \\ \nabla H_{N_r} \end{bmatrix} \quad \underline{\varepsilon} = \begin{bmatrix} \varepsilon_1 \\ \varepsilon_2 \\ \vdots \\ \varepsilon_{N_r-1} \end{bmatrix} \quad \underline{y} = \begin{bmatrix} \underline{y}_1 \\ \underline{y}_2 \\ \vdots \\ \underline{y}_{N_r} \end{bmatrix} \quad \underline{u}^{(m)} = \begin{bmatrix} u_1^{(m)} \\ u_2^{(m)} \\ \vdots \\ u_{N_r}^{(m)} \end{bmatrix}$$

and

$$D = \begin{bmatrix} -J_1 & \Im & 0 & \cdots & 0 \\ 0 & -J_2 & \Im & \cdots & 0 \\ \vdots & \vdots & \vdots & \cdots & 0 \\ 0 & 0 & 0 & -J_{N_r-1} & \Im \end{bmatrix}$$

so that

$$\nabla H = 2D^T \cdot \underline{\varepsilon},$$

the recursion defined in Eq. 6.10 can be written compactly as

$$\underline{u}^{(m+1)} = \underline{u}^{(m)} - \alpha\nabla H; \quad \underline{u}^{(0)} = \underline{y} \qquad \qquad \textbf{Eq. 6.11}$$

Gradient descent methods appear to be order-$d_E N_r$ CPU processes, but converge at best linearly. Note that the matrix D, a reduced version of the matrix M of the previous algorithm, is related to ∇H by $\nabla H = 2D^T\varepsilon$, so when D becomes singular as it will near a homoclinic tangency, the change in the 'cost function' H goes to zero. Since H is flat, that is, does not change in direction corresponding to the location of homoclinic tangencies, steepest descent will not move in the direction of the tangencies. Hence, it is possible that the procedure grinds to a halt before getting close to a deterministic trajectory.

If we assume that the deterministic orbit $\underline{u}$ is in the linear neighborhood of $\underline{y}$, then

$$\begin{aligned} \varepsilon_t &= \underline{y}_{t+1} - \varphi[\underline{y}_t] \\ &= \underline{y}_{t+1} - \varphi[\underline{y}_t] - (u_{t+1} - \varphi[u_t]) \\ &= \underline{y}_{t+1} - u_{t+1} + \varphi[u_t] - \varphi[\underline{y}_t] \\ &\approx \underline{y}_{t+1} - u_{t+1} - J_t \cdot [\underline{y}_t - u_t] \end{aligned}$$

where the second equality follows from the fact that $u_{t+1} = \varphi[u_t]$. Hence,

$$D \cdot [\underline{y} - \underline{u}] = \underline{\varepsilon}.$$

Multiplying by D^T and recalling that $\nabla H = 2D^T \varepsilon$ we have

$$D^T D \cdot [\underline{y} - \underline{u}] = D^T \underline{\varepsilon} = \frac{1}{2}\nabla H$$

so

$$\underline{y} - \underline{u} = (2D^T D)^{-1} \cdot \nabla H$$

From that we get the recursion

$$\underline{u}^{(m+1)} = \underline{u}^{(m)} - (2D^T D)^{-1} \cdot \nabla H \qquad \textbf{Eq. 6.12}$$

This recursion is a Newton-type formulation, and it is unstable because the difficulties with inverting D are not mitigated by considering $D^T D$. However, in writting the recursion in this way it is easy to see that the only difference between this recursion and Eq. 6.11, the recursion for steepest descent, is the prefactor of ∇H. That observation leads to consideration of the composite formulation

$$\underline{u}^{(m+1)} = \underline{u}^{(m)} - \frac{1}{2}(D^T D + \delta\Im)^{-1} \cdot \nabla H$$

When $\delta = 0$ the prefactor of ∇H is Newtonian. When $\delta \gg 0$, the prefactor of ∇H behaves like $(1/\delta)\Im$ which is steepest descent with $\alpha = 1/\delta$ (see Figure 6.16).

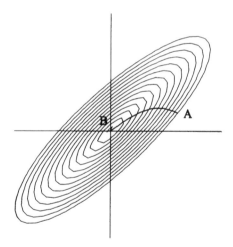

Figure 6.16 Levenberg–Marquardt method applied to a simple two-dimensional potential – well cost function with a minimum at B, and elliptical isobars as shown. Starting at the point A, the estimate for the minimium will lie somewhere on the line between A and B depending on the value of δ. If δ is small, the method is almost Newtonian and the solution is near B. If δ is large, the method is almost gradient descent and the solution is near A.
Source: Adapted from Davies (1994) p.183

The composite formulation is known as the *Levenberg–Marquardt* method[26], and it is a bit surprising that such a simple idea solves so many problems. Namely,

[26] See Fletcher, (1980).

- The formulation improves stability of the search. To see why, consider the SVD of $D = V \Sigma U^T$ and rewrite the matrix $B = (D^T D + \delta \Im)$ as

$$B = (U \Sigma^2 U^T + \delta \Im) = U(\Sigma^2 + \delta \Im) U^T$$

Clearly, the singular values of B are bounded from below by $\delta > 0$.

- Since,

$$
\begin{aligned}
B^{-1} D^T \underline{\varepsilon} &= U(\Sigma^2 + \delta \Im)^{-1} U^T U \Sigma V^T \underline{\varepsilon} \\
&= U(\Sigma^2 + \delta \Im)^{-1} \Sigma V^T \underline{\varepsilon}
\end{aligned}
$$

the correction made to $u_t^{(m)}$ in the ith singular direction has the form

$$\sigma_i / (\sigma_i^2 + \delta)$$

Thus, in singular directions where the singular values are identically zero, the method takes a zero step size. However, in singular directions where the singular values are small but finite, as will be the case toward a homoclinic tangency, the method takes a small but non-vanishing step size for a suitable choice of δ. Hence, the method does not suffer from the halting problem of the ordinary steepest descent caused by homoclinic tangencies. That appears to be enough to maintain quadratic convergence near the desired deterministic orbit.

- Because of the sparse banded structure of B, it can be efficiently inverted using Cholesky decomposition[27] to get

$$B = GG^T$$

Where G is lower triangular, and can be inverted by simple substitution methods. Hence, the algorithm is an order-$d_E N_r$ CPU process.

Parameter values
For high noise levels, initially δ should be made relatively large to increase stability of the algorithm, and then reduced on subsequent iterations. In fact, there are algorithms to set δ adaptively[28].

Example 6.3
This example compares the error reduction achieved with the Levenberg–Marquardt method to the error reduction achieved with the manifold decomposition method of the previous algorithm. Since the illustration is for comparison purposes, the true dynamical equations for the Hénon map were used in the analysis, making function estimation unnecessary. From the plots of the total dynamical and fitting errors, it is apparent that the two methods achieve similar error reduction after 10 iterations of either algorithm, but keep in mind that the Levenberg–Marquardt algorithm is an order-$d_E N_r$ CPU process, while the manifold decomposition algorithm utilizes singular value decomposition at some trajectory points, and singular value decomposition is an order-$(d_E N_r)^3$ CPU process.

---

[27] See Golub and Van Loan, Batimore, (1983).
[28] See Press, *et al.*, (1988).

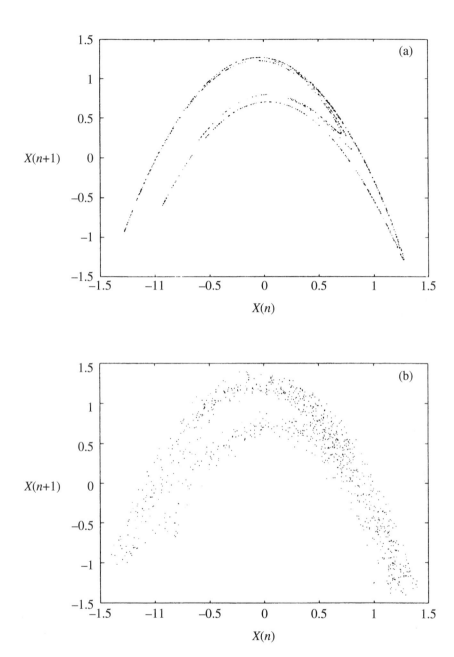

Figure 6.17 Plots of the delay coordinates of the deterministic signal (top) from the Hénon map and then the signal after 10% noise was added (bottom).

Source: Adapted from Davies (1994) p. 185.

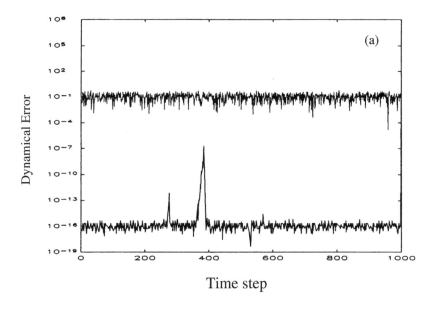

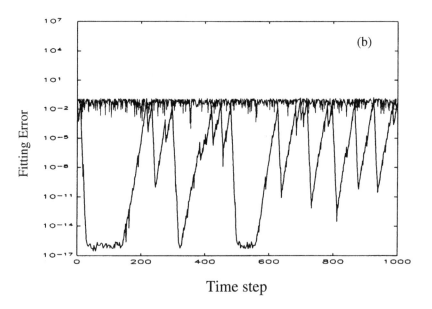

Figure 6.18 A comparison between the unfiltered noisy signal (jagged horizontal line at the top of each panel) and the same signal after 10 iterations of the manifold decomposition noise reduction algorithm (jagged line at the bottom of each panel). (a) is a plot of the dynamical error and (b) is a plot of the fitting error. The unfiltered noisy signal refers to the data plotted in Figure 6.17(b).

Source: Adapted from Davies (1994) p. 186.

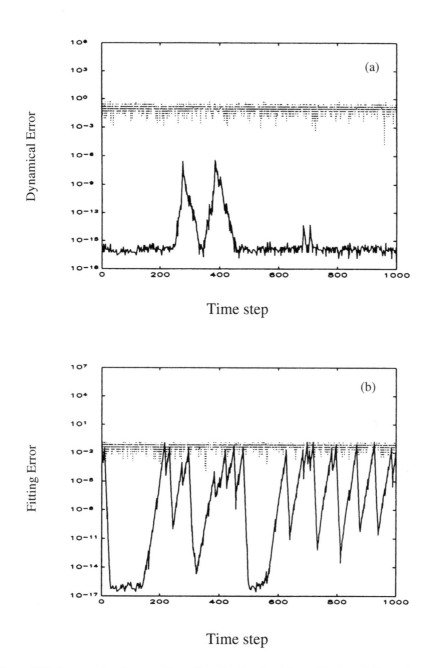

Figure 6.19 A comparison between the unfiltered noisy signal (jagged horizontal line at the top of each panel) and the same signal after 10 iterations the Levenberg–Marquardt noise reduction algorithm (jagged line at the bottom of each panel). (a) is a plot of the dynamical error and (b) is a plot of the fitting error. The unfiltered noisy signal refers to the data plotted in Figure 6.17(b).

Source: Adapted from Davies (1994) p. 187.

255

Example 6.4

This example assumes that the only information available on the system is a noisy time series. The time series was obtained by adding noise to the x-component of a trajectory from the Lorenz I system. In the example, a 10-dimensional delay reconstruction was reduced to the three most significant principal components, and then 10 iterations of function reestimation, alternating with the noise reduction algorithm produced the results in Figure 6.20 and table 6.1.

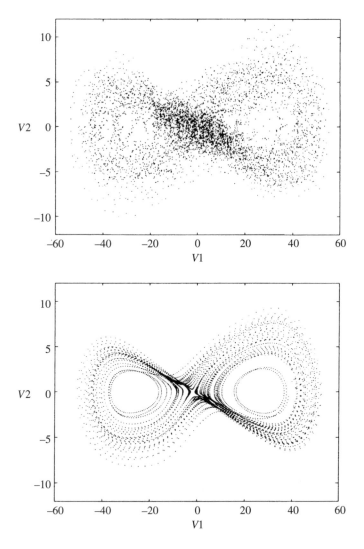

Figure 6.20 The top picture shows data from the x-component of the Lorenz equations with 10% noise added, plotted using the first two singular directions. The bottom picture shows the same data after 10 iterations the Levenberg-Marquardt noise reduction algorithm. Compare the latter with Figure 1.14.

Source: Adapted from Davies (1994) p. 190.

256

δ	0.01	0.1	0.5	1.0	2.0	5.0
σ^2_{dyn}	3.1×10^{-5}	1.6×10^{-4}	4.2×10^{-4}	7.6×10^{-4}	4.5×10^{-3}	7.2×10^{-2}
σ^2_{fit}	2.3×10^{3}	4.4×10^{2}	3.1×10^{2}	3.2×10^{2}	3.3×10^{2}	4.9×10^{2}

Table 6.1 Total dynamic error σ^2_{dyn} and total measurement (fitting) error σ^2_{fit} for a variety of step sizes (δ) for the Levenberg–Marquardt algorithm. The total dynamical error of the original noisy data was 4.88, and the total fitting error of the original noisy data was 5.48×10^3. Original noisy data was the same as that used in Figure 6.20.

Source: Adapted from Davies (1994) p. 190–191.

4 Recurrent orbits[29]

It has been shown[30] that the dynamics on typical low dimensional attractors are governed by saddle type periodic orbits[31]. The collection of low-period saddle orbits form a skeleton-like structure for the chaotic attractor and can be used to dampen the noise in experimental data as well as estimate some invariants. Although saddle periodic orbits are unstable, a nearby trajectory will often move away slowly, circulating almost periodically near the saddle orbit for a period of time. Thus, the saddle orbits can sometimes be located on the reconstructed attractor by finding recurrent experimental trajectories.

Identifying recurrent orbits

Define a reconstructed trajectory point $\underline{y}_t$ in $\Re^{d_E}$ as (m, ε)-**recurrent** if it returns to within ε of itself after m iterations of the dynamical map, that is, if $\|\underline{y}_{t+m} - \underline{y}_t\| < \varepsilon$. In the context of a delay coordinate reconstruction, the occurrance of an (m, ε)-recurrent point means that there may be m-length sections of the time series that are nearly identical.

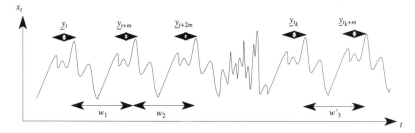

Figure 6.21 Recurrent experimental trajectory points. On the left, t, $t_i = t + m$, and $t_j = t + 2m$, are the time indices of the experimental trajectory points that define w_1 and w_2, two successive m-length recurrent sections of the original time series. On the right of the figure is the start of another recurrent section w_3, for the same saddle orbit, that begins at some later time t_k. Note that w_3 is not the image of w_2 under the dynamics g_f defined below; w_3 is a new initial point for g_f.

---

[29] Kostelich (1992).

[30] See theoretical section of this chapter.

[31] A saddle orbit has both stable and unstable manifolds. See Chapter 1, Dynamical systems.

The recurrent m-length sections of the time series can be identified as follows. For each point y_t on the experimental trajectory, we form a small neighborhood $B_\varepsilon[y_t]$ and check if the time indices t_k of any of the neighborhood points y_{t_k} occur in increments of near integer multiples k of some fixed integer m. That is, we look for $|t_j - t_i| = \kappa m$[32]. For those indices that do, we go back and check that the corresponding m-length subintervals of the original time series x_t, between y_{t_i} and y_{t_j}, also remain close relative to ε. The procedure is repeated for increasing values of the integer m until no more saddle orbits can be found.

Modeling the dynamics

Ignoring the original phase space reconstruction for the moment, treat each such m-length recurrent section of the time series as a m-dimensional vector w_i, and each *successive* recurrent section as a function of the previous one (see the caption to Figure 6.21 above and note that w_3 does not succeed w_2.) That is

$$w_{i+1} = g_f[w_i]$$

where g_f represents the dynamics near the saddle periodic orbit in question. Let w_f be an m-vector on the putative saddle periodic orbit such that $w_f = g_f[w_f]$, that is, left w_f be a fixed point of g_f. If w_i and w_{i+1} are sufficiently close to w_f, then

$$w_{i+1} - w_f = g_f[w_i] - g_f[w_f] \approx B_f \cdot [w_i - w_f]$$

where B_f is the Jacobian of g_f at w_f.

As the map g_f is unknown, B_f and w_f have to be estimated from the data. In the case of a flow, m, the orbit length of w_f can be large, and using least squares to estimate the m-by-m Jacobian is impractical. However, the redundancy in B_f, imparted by the structure of the nearly periodic w_i, can be used to reduce the dimension of the problem. Suppose $\{w_i, i = 1, \ldots, p\}$ is the collection of (m, ε)-recurrent points and we form the normalized $p \times m$ trajectory matrix Y_f whose rows are $w_i - \langle w_j \rangle$. Singular value decomposition[33] of Y_f gives

$$Y_f = V \Sigma U^T$$

where the columns of U form an orthonormal basis for the normalized w_i. Let U_q consist of the first q columns of U, that is, the most significant singular vectors of Y_f, where significant is defined as the smallest value of q such that

$$\sum_1^q \sigma_i^2 \geq C\sigma_{\text{total}}^2; \quad C \in (0, 1); \quad \sigma_{\text{total}}^2 = \sum_1^p \sigma_i^2$$

Projecting the w_i onto the subspace spanned by the columns of U_q, maps the m-vector w_i to a q-vector $v_i = (U_q)^T \cdot w_i$. The mean $\langle w_j \rangle$ gets mapped to the origin, and the fixed point w_f gets mapped to a q-vector v_f which in the new coordinates is

---

[32]κ may only be approximately integer because the sampling time τ_s may not be an exact integer multiple of the period of the saddle orbit. Note also from Figure 6.21 that k may increment by unity for a while, then jump by a large amount before incrementing by unity again as the experimental trajectory veers away from the saddle periodic orbit and then returns.

[33]See Appendix 4, Matrix decomposition.

258

near, but not at the origin. We model the dynamics in the new coordinate system as before, that is

$$v_{i+1} - v_f = G_f[v_i] - G_f[v_f] \approx D_v G_f[v_f] \cdot [v_i - v_f] = A_f \cdot [v_i - v_f]$$

Our objective now is to determine, simultaneously, values of the Jacobian A_f, the adjusted (less noisy) orbits v_i^e, and the fixed point v_f, that minimize the dynamical and fitting error sum of squares

$$\sum_i \| v_i^e - v_i \|^2 + \sum_j \| (v_{j+1}^e - v_f) - A_f \cdot [v_j^e - v_f] \|^2$$

where the first summation is over all observations near the fixed point v_f, and the second is over all observations near the fixed point which have an image. The minimization can be done using a Gauss–Newton method[34], or as was done in Example 6.5, using the Polak–Ribiere conjugate gradient method[35].

Retrieving the corrections
Since the objective was to smooth the original time series, we have to translate the corrections made in the form of the v_i^e into corrections for x_t. To do that, we must transform the coordinates of the v_i^e to coordinates in $\Re^m$, the coordinate system of the w_i, and then add back to each the mean $\langle w_j \rangle$ to get

$$w_i^{(1)} = U_q \cdot v_i^e + \langle w_j \rangle$$

The $w_i^{(1)}$ are the smoothed m-length sections of the original time series sections corresponding to the w_i.

Parameter values
- The initial guess for $v_f = 0$.
- A suitable initial guess for A_f can be obtained from an ordinary least squares fit of the noisy data to the linear model $v_{i+1} = A_f v_i$.
- The right value for ε will depend on the data set. If ε is too large, sections of the trajectory that are not close to the saddle periodic orbit will be counted as recurrent. If ε is too small, truly recurrent points will be missed. A small perturbation of the right ε will not alter the identity or classification of the recurrent sections of the time series. Using one value of ε to identify the recurrent points and another, slightly larger, value of ε to test the closeness of the recurrent sections of the time series, may improve stability. The analysis reported below used a single value of ε set at 8% of $d(A)$, the extent of the experimental attractor[36].
- The algorithm should be robust to the value of C, so one needs to verify stability for $C \in [0.5, 0.75]$.

Example 6.6
The time series for this example was obtained from a Couette–Taylor experiment[37],

---

[34]Schwetlick and Tiller (1985).
[35]Press *et al.* (1986).
[36]$d(A) = \max[x_i] - \min[x_i]$.
[37]Brandstäter and Swinney (1987).

an experimental physical system yielding weakly chaotic data. Only power spectra were used to show the efficacy of the procedure, but at least we can see (Figure 6.22) a reduction in the high frequency components of the signal which are the components that carry the high dimensional (noise) elements of the signal.

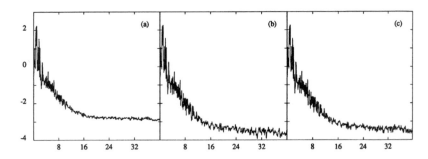

Figure 6.22 (a) Power spectrum for the raw weakly chaotic Couette–Taylor data. (b) and (c) Power spectra for the same data after noise reduction using the procedure described in the text with the first $q = 2$ and first $q = 5$ singular basis vectors, respectively. Note the reduction in the amplitude of the high frequency components.

Source: Adapted from Kostelich (1992) p. 150.

Figure 6.23 A two-dimensional projection of the reconstructed saddle orbit of the weakly chaotic Couette–Taylor data, estimated using the first $q = 5$ singular basis vectors.

Source: Adapted from Kostelich (1992) p. 151.

Problems

The method is efficient on that subset of the data that is treated, that is, that portion of the data which forms recurrent trajectories near a small number of periodic saddle orbits. Noisy data that is not part of a recurrent trajectory has to be handled with some other method.

5 Low pass embedding[38]

The approach taken in the following algorithm is quite different from that taken in any of the preceding methods, in that it uses the geometry of the attractor rather than the dynamics of the system to reduce noise.

Low pass embedding

Assume that the noise is additive with expected value zero. For $m\tau_s \gg \tau_c$[39], we form the delay vectors

$$u(t) = [x_t, \ldots, x_{t-m+1}]^T$$

and apply a low pass filter to $u(t) \in \Re^m$ to obtain a smoothed $y(t) \in \Re^{d_E}$ as follows. Let m and d_E be even. Apply the Fast Fourier Transform (FFT) of order m to $u(r)$ to obtain $m/2$ complex frequency-amplitudes. Now apply the inverse FFT of order d_E to the lowest frequency (first) $d_E/2$ of those complex amplitudes. That yields d_E evenly spaced sample values from a smoothed $u(r)$.

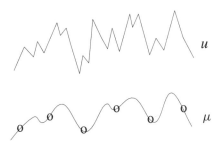

Figure 6.24 The original signal u and the smoothed signal μ after dropping the high frequency components. The six open circles represent evenly spaced sample values of μ that define the coordinates of a six-dimensional reconstructed trajectory point.

Finally, let those d_E sample values define the coordinates of the reconstructed trajectory point $y(r)$. The smoothing of $u(r)$ is a result of dropping the high frequency components of its Fourier transform. The high frequency components are assumed to carry mainly stochastic or high dimensional chaotic signals, thus, $y(r)$ contains most of the low dimensional dynamical information that is in $u(r)$.

Tangent space projection

Let $K_\varepsilon(r)$ be the number of points in $B_\varepsilon[y(r)]$, an ε-neighborhood of $y(r)$. For each neighborhood form the center of mass $c_r = \langle y(r_k) \rangle_k$, $k = 0, \ldots, K_r(\varepsilon)$, where $y(r_0) = y(r)$. Now form the pseudo tangent vectors $z(r_k) = y(r_k) - c_r$. Let Z be the 'trajectory' matrix whose rows are the $z(r_k)^T$, and use singular value decomposition[40] to obtain the representation $Z = V\Sigma U^T$.

---

[38]Sauer (1992).

[39]t_s is the sampling time, and t_c is the average cycle time of a trajectory on the attractor.

[40]See Appendix 4, Matrix decomposition.

Project the $z(r_k)$ onto U_{d_A}, the first d_A columns of U, that is, onto the singular vectors corresponding to the d_A largest singular values of $Z^{[41]}$. That yields the projections $p[z(r_k)] = (U_{d_A})^T \cdot z(r_k)$. The idea now is that the $z(r_k)$ should lie mainly in the tangent space[42] of the attractor at $\underline{y}(r)$. For small neighborhoods, the attractor and the tangent space should almost coincide. Moreover, the first d_A singular vectors span a subspace which should also almost coincide with the tangent space, while the remaining singular vectors span a subspace dominated by noise. By projecting the noisy $z(r_k)$ onto the dominant singular vectors, we nudge them toward the attractor, and in so doing, we hope to reduce the noise by squeezing out the components in directions dominated by noise (see Figure 6.25).

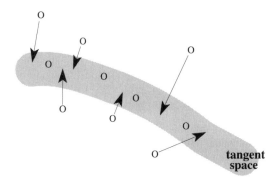

Figure 6.25 Projection of noisy tangent vectors onto the approximate tangent space defined by the first d_A right singular vectors of Z.

Back to the original coordinates

Since the objective was to smooth the original time series, we have to translate the corrections made in the form of $p[z(r_k)]$ – the projections of the $z(r_k)$ onto the tangent space – into corrections for x_t. To do that, we first get the coordinates of the $p[z(r_k)]$ in $\Re^{d_E}$, the coordinate system of the embedding, and then add back to each the center of mass c_r to get

$$y(r, r_k) = U_{d_A} \cdot p[z(r_k)] + c_r \in \Re^{d_E}$$

In order to get back to $\Re^m$ we apply an order-d_E FFT to $y(r, r_k)$ and use the resulting $d_E/2$ components as the first (low frequency) components of an order-m FFT simply by filling in the remaining high frequency components with zeroes. Applying the inverse FFT of order m to this right-zero-padded sequence results in the m-vector

$$v(r, r_k) = [v_{r,r_k}, \dots, v_{r,r_k-m+1}]^T \in \Re^m$$

which is *one* corrected version of

$$u(r_k) = [x_{r_k}, \dots, x_{r_k-m+1}]^T \in \Re^m$$

---

[41]d_A is the dimension of the manifold containing the attractor. See Chapters 3, Phase space reconstruction and 4, Fractal dimension, for ways to estimate d_A.

[42]See Chapter 1, Dynamical systems.

There may be multiple corrections contributed from the same point appearing in different neighborhoods, that is, we get multiple corrections whenever $r_i' = r_j''$.

Decorrelation and composite corrections

The correction $(v_{r,r_k-l+1} - x_{r_k-l+1})$ is an estimate of the noise in the lth coordinate of the delay vector $u(r_k)$ imputed from an analysis of a small neighborhood $B_\varepsilon[y(r)]$ of the attractor. By assumption the noise in each coordinate of a reconstructed trajectory point has an expected value of zero. Therefore, to minimize any bias imposed by the procedure, the mean of the corrections $(v_{r,r_k-l+1} - x_{r_k-l+1})$ is made to equal zero for a given neighborhood r and coordinate number l. Accordingly, the mean of these adjustments must be subtracted from the individual adjustment

$$\Delta_{r,r_k-l+1} = (v_{r,r_k-l+1} - x_{r_k-l+1}) - \langle(v_{r,r_j-l+1} - x_{r_j-l+1})\rangle_{j:y(r_j)\in B_\varepsilon[y(r)]}$$

We now drop the detailed subscripting and let $\Delta_{r,s}$ denote the corrections to x_s obtained from an analysis of the neighborhood $B_\varepsilon[y(r)]$. The final step is to average the multiple corrections that come from overlapping neighborhoods. That yields

$$\Delta_s = \langle\Delta_{r,s}\rangle_r$$

as the composite adjustment to x_s.

Iteration

The process is iterative. After the first pass of the algorithm, each point x_t of the original time series is replaced with

$$x_t^{(1)} = x_t + \alpha\Delta_t$$

The process is then repeated on the less noisy time series $x_t^{(1)}$, and so on, with step size $\alpha \in (0, 1)$, increasing slowly after each iteration.

Parameter values

- The radius of the ε-neighborhood $B_\varepsilon[y(r)]$ should be set to the level of the standard deviation of the noise.

- Choose the dimensions m and d_E even (a power of 2), and $m \gg \tau_c/\tau_s$; in Examples 6.7 and 6.8 $m = 32$ or 64, and $d_E = 16$ for $\tau_c/\tau_s \approx 20$.

- Choose the step size $\alpha \in (0, 1)$, starting low and incrementing after each iteration. In the examples below $\alpha = \min\{0.01 + 0.02(p - 1), 0.5\}$ for the pth iteration.

Remark

If there is sufficient data, computation can be reduced by analyzing the neighbohood of a point only if the point was not already included in some other neighborhood; or do the same only for every other point on the experimental trajectory.

Example 6.7

The sample data for this example was obtained by adding white noise to one coordinate of a trajectory on the Lorenz I attractor (see Figures 6.26 and 6.27, and Table 6.2).

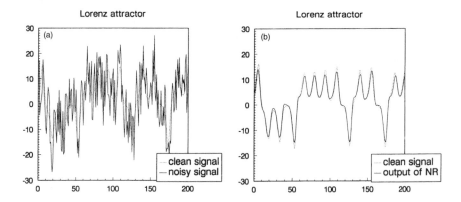

Figure 6.26 (a) The solid curve is a 200-point section of one component of the Lorenz I flow sampled at $\tau_s = 0.05$ with 100% additive white noise. The dotted curve is the Lorenz signal before noise was added. (b) The solid curve is the output of the noise reduction method applied to the same 200 points of noisy data shown in (a) using a window size of $m = 32$. The dotted curve is the clean signal as in (a).

Source: Adapted from Sauer (1992) p. 199.

Δt	Avg.pts. per oscill.	Noise level	Window m	Passes	Orig. SNR	Final SNR	Gain (dB)
0.05	20	10%	32	18	20.0	31.7	11.7
0.05	20	100%	32	58	0.0	13.8	13.8
0.05	20	100%	64	15	0.0	14.2	14.2
0.05	20	200%	64	17	−6.0	6.5	12.5
0.10	10	100%	32	19	0.0	11.7	11.7

Table 6.2 Results of noise reduction method applied to noisy Lorenz data.

Source: Adapated from Sauer (1992) p. 199.

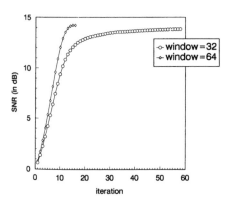

Figure 6.27 Signal-to-noise ratio after noise reduction method was applied to the Lorenz I data (with 100% additive white noise) is graphed against the number of passes through the data. The circles and diamonds refer to application of the method with window sizes of $m = 32$ and $m = 64$, respectively.

Source: Adaped from Sauer (1992) p. 199.

Example 6.8

The sample data for this example was obtained by adding white noise to one coordinate of a trajectory on the Rössler attractor (see Table 6.3).

Δt	Avg.pts per oscill.	Noise level	Window m	Passes	Orig. SNR	Final SNR	Gain (dB)
0.4	16	10%	32	20	20.0	31.8	11.8
0.4	16	100%	32	45	0.0	13.5	13.5
0.4	16	100%	64	19	0.0	14.3	14.3
0.4	16	200%	64	40	−6.0	7.2	13.2

Table 6.3 Results of noise reduction method applied to noisy Rössler data
Source: Adapted from Sauer (1991) p. 200.

6 Re-embedding[43]

The fact that the embedding theorem holds for multi-channel delay vectors[44] leads to the following idea.

SVD of the trajectory matrix Y yields an $N_r \times d_r$ trajectory matrix $P = YC$ expressed in principal components. So each row of P

$$\underline{p}(i) = [p_{i1}, p_{i2}, \dots, p_{id_r}]^T \quad i = 1, 2, \dots, N_r$$

is a point on the transformed trajectory. Truncating $\underline{p}(i)$ to its first $w < d_r$ significant principal components yields the vectors

$$\underline{q}(i) = [p_{i1}, p_{i2}, \dots, p_{iw}]^T \quad i = 1, 2, \dots, N_r$$

The components of the vectors $\underline{q}(i)$ correspond to the w largest eigenvalues of $Y^T Y$, and should represent the main features of the dynamics of the system. The truncated trajectory points $\underline{q}(t)$ can be treated as a multi-variate time series and used to form a multi-channel delay vector

$$\underline{\underline{q}}(i) = [\underline{q}(i), \underline{q}(i - \tau), \dots, \underline{q}(i - (d_r'' - 1)\tau)]^T \quad i = 1, 2, \dots, N_r$$

The technique is called **re-embedding** because it is an embedding of the original embedding if the original reconstruction was an embedding. If the original reconstruction is not an embedding (that is, it is a projection of an embedding), then the re-embedding can still produce a proper embedding for sufficiently large d_r'', where the effective embedding dimension would be $d_r'' \times w$.

Example 6.9

Compared, in Figures 6.28–6.30, are the scaling regions and convergence of the correlation integral C_2, and $\Delta(d_r)$, the estimate of the correlation exponent D_2,[45] achieved using first the ordinary delay coordinate map and then the re-embedding procedure.

[43] Fraedrich and Wang (1993).
[44] See Chapter 3, Phase space reconstruction.
[45] See Chapter 4, Fractal dimension.

In each figure the ordinary delay coordinate map uses a delay time $\tau = 2$ and both lnC_2 and $\Delta(d_r)$ are plotted against the logarithm of the normalized interpoint distance $\varepsilon/\varepsilon_0$ for a range reconstruction dimension $d_r = 1$–15. The plots for the re-embedding procedure were derived using a delay time of $\tau = 1$, a reconstruction dimension $d_r = 50$, and the first $w = 5$ principal components, from singular value decomposition of the trajectory matrix, for a range of effective reconstruction dimensions: $d_{r''} \times w$.

Figure 6.28 is a comparison of the two procedures on white noise data – a purely stochastic system. As expected, in this case neither procedure yields a scaling region with a stable correlation dimension.

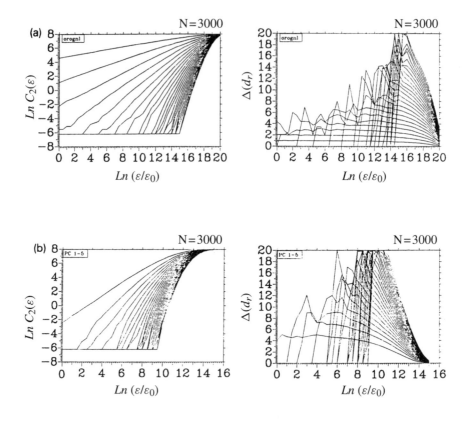

Figure 6.28 White noise data. (a) estimates derived from ordinary delay coordinate map. (b) estimates derived from re-embedding procedure. $N_r = 3,000$.

Source: Adapted from Fraedrich and Wang (1993) p. 384.

The set of plots in Figure 6.29 compare the results from the two procedures applied to noise free data from the x-component of a chaotic Lorenz attractor – a deterministic system. Here, because the data is clean, and there is lots of it, we would expect, and the plots suggest, that either procedure would yield a useful scaling region where $\Delta(d_r)$ is roughly constant.

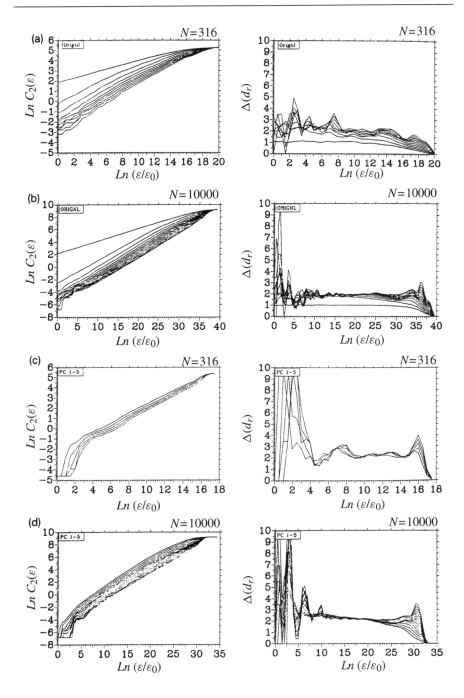

Figure 6.29 Noise free data. (a) $N_r = 316$ and (b) $N_r = 10,000$ are estimates derived from an ordinary delay coordinate map. (c) $N_r = 316$ and (d) $N_r = 10,000$ are estimates derived from re-embedding procedure.

Source: Adapted from Fraedrich and Wang (1993) p. 388–389.

The final set of plots (in Figure 6.30) compare the results from the two procedures applied to a time series from the x-component of the same chaotic Lorenz system, but in this case the noise free time series has been corrupted with additive white noise to simulate the noisy data we can expect to encounter in practice. Here the difference in the results is striking.

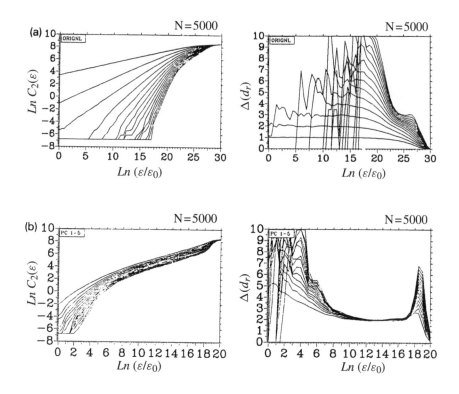

Figure 6.30 Data corrupted by additive white noise. (a) estimates derived from an ordinary delay coordinate map. (b) estimates derived from re-embedding procedure. $N_r = 5,000$.
Source: Adapted from Physica 65D, 1993, p. 385.

Using the re-embedding technique yields a reasonably large scaling region where the estimates of D_2 converge, while there is virtually no scaling region achieved when the ordinary delay coordinate map is applied to the noisy data.

7 Discrimination[46]

Here, again in a different approach, the characteristic function of the natural measure is used to discriminate among coincidental signals.

---

[46] Wright (1994).

Assume that the measurement time series x_n has been normalized so that

$$\sum_{n=1}^{N_d} x_n = 0$$

$$\sum_{n=1}^{N_d} [x_n]^2 = 1$$

and that the natural measure[47]

$$\rho_x(z) = \lim_{N \to \infty} \frac{1}{N} \sum_{n=1}^{N} \delta[z - x_n]$$

induced by the signal x_n is ergodic, where

$$\delta[u] = \frac{1}{w\sqrt{2\pi}} e^{-\frac{1}{2}(u/w)^2}; \quad 0 < w \text{ small}$$

approximates the Dirac delta function.

Let $\rho_{x,\tau}(u, v)$ denote the joint density of $x(t)$ and the time delayed variable $x(t-\tau)$. Then

$$\rho_{x,\tau}(u, v) = \lim_{N \to \infty} \frac{1}{N} \sum_{n=1}^{N} \delta[u - x_n]\delta[v - x_{n-\tau/\tau_s}]$$

The Fourier transform $\Psi_{\rho_x}(\kappa)$ of the density function ρ_x is called the characteristic function, and is defined by

$$\Psi_{\rho_x}(\kappa) = \int ds \, e^{i\kappa s} \rho_x(s) = \lim_{N \to \infty} \frac{1}{N} \sum_{n=1}^{N} e^{i\kappa x_n}$$

Note that the characteristic function only confers amplitude information. We get no information on the coupling of $x(t)$ and $x(t - \tau)$. To recover that information, we form something like the characteristic function version of the autocorrelation function. Namely

$$\Psi_{\rho_{x,\tau}}(\kappa_1, \kappa_2) = \int du dv \, e^{i\kappa_1 u} e^{i\kappa_2 v} \rho_{x,\tau}(u, v) = \lim_{N \to \infty} \frac{1}{N} \sum_{n=1}^{N} e^{i\kappa_1 x_n} e^{i\kappa_2 x_{n-\tau/\tau_s}}$$

The two-dimensional time lagged characteristic function $\Psi_{\rho_{x,\tau}}(\kappa_1, \kappa_2)$ is a bit unwieldy. However, we still retain a significant amount of information in a one-dimensional profile of $\Psi_{\rho_{x,\tau}}(\kappa_1, \kappa_2)$ obtained by letting $\kappa = \kappa_1 = \kappa_2$. The result is the one-dimensional time lagged characteristic function

$$\Psi_{\rho_{x,\tau}}(\kappa) = \lim_{N \to \infty} \frac{1}{N} \sum_{n=1}^{N} e^{i\kappa x_n} e^{i\kappa x_{n-\tau/\tau_s}} = \lim_{N \to \infty} \frac{1}{N} \sum_{n=1}^{N} e^{i\kappa(x_n + x_{n-\tau/\tau_s})}$$

The special one-dimensional time lagged characteristic functions $\Psi_{\rho_{x,\tau}}(\kappa)$ seem to

---

[47] Alternative way to define the natural measure. See also Chapter 1, Dynamical systems.

carry enough information on a signal, or combined signals, to discriminate between sources, provided we look at it over a range of values for κ. As can be seen from Figures 6.31 and 6.32, the graph of $\Psi_{\rho_{x,\tau}}(\kappa)$ is quite distinctive for signals emanating from different sources (Figure 6.31), or for different values of the lag time for the same system (Figure 6.32).

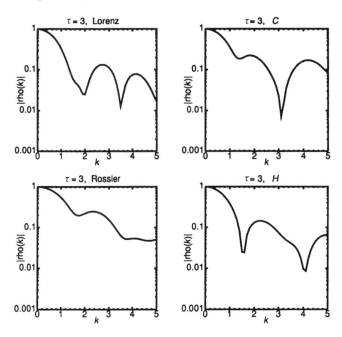

Figure 6.31 The absolute value of the time lagged characteristic functions of signals from the Lorenz system, the Rössler system, and two experimental systems of electronic circuits labeled 'C' and 'H.' In each case the time delay $\tau = 3.0$, and the signal source is indicated at the top of the graph.

Source: Adapted from Wright (1994) p. 265.

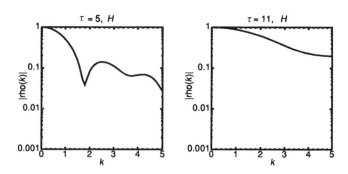

Figure 6.32 The absolute value of the time lagged characteristic function of the signal from experimental circuit 'H' for four different values of the time delay τ.

Source: Adapted from Wright (1994) p. 266.

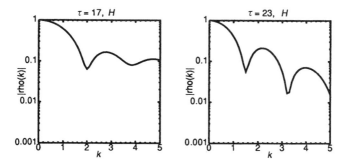

Figure 6.32 Continued.

Now, suppose that $x(t)$ is actually the combined signal from two independent processes $u(t)$ and $v(t)$, that is

$$x(t) = u(t) + v(t) \qquad \text{Eq. 6.13}$$

It is well known that the characteristic function of the sum of independent variables is the product of their characteristic functions. Hence, in the present case

$$\Psi_{\rho_{x,\tau}}(\kappa) = \Psi_{\rho_{u,\tau}}(\kappa) \cdot \Psi_{\rho_{v,\tau}}(\kappa)$$

and in fact, if

$$x(t) = \alpha u(t) + \beta v(t)$$

then

$$\Psi_{\rho_{x,\tau}}(\kappa) = \Psi_{\rho_{u,\tau}}(\alpha\kappa) \cdot \Psi_{\rho_{v,\tau}}(\beta\kappa) \qquad \text{Eq. 6.14}$$

and this works for any linear combination of any number of independent variables. The usefulness of the factorization property, Eq. 6.14, of the characteristic function for independent signals can be appreciated in the following context. If one has a 'library' of characteristic functions for known signals, various different combinations of products of these characteristic functions can be compared to the characteristic function of an unknown signal of interest. A match means we know what the probable constituents of the unknown signal are.

Problems

The overwhelming problem with this procedure is the *absence* of a 'library' of relevant characteristic functions when we start from the twin premises that we know very little about the systems of interest and clean data is simply not available. Having said that, the method can be very useful for identifying similarities or dissimilarities between known and unknown systems, or between various unknown systems.

Example 6.10

For this example, Eq. 6.13 was used to combine the signals from the systems depicted in Figure 6.31. In Figure 6.33, panel (b), the footprint of the correct product of time lagged characteristic functions (Eq. 6.14) clearly identifies the original mixed signal in panel (a).

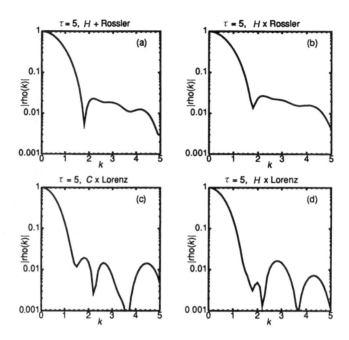

Figure 6.33 The absolute value of the left hand side of Eq. 6.14 is shown in (a) for the signal which is the sum of the 'H' system and the Rössler system. The absolute value of the right hand side of the equation is shown in (b − d) for the signals indicated at the top of each panel. The time delay used was $\tau = 5$.

Source: Adapted from Wright (1994) p. 269.

Figure 6.34 is the same as Figure 6.33 except that the value of the lag time has been changed. The shapes of all the characteristic functions have changed, but in such a way that the footprint of the correct product of characteristic functions continues to accurately identify the original mixed signal.

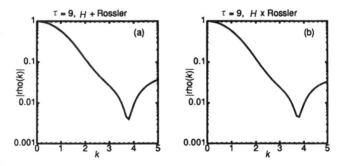

Figure 6.34 The absolute value of the left hand side of Eq. 6.14 is shown in (a) for the signal which is the sum of the 'H' system and the Rössler system. The absolute value of the right hand side of the equation is shown in (b − d) for the signals indicated at the top of each panel. The time delay used was $\tau = 9$.

Source: Adapted from Wright (1994) p. 270.

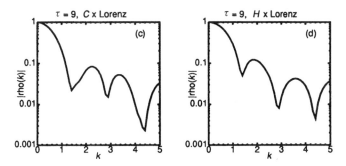

Figure 6.34 Continued.

Example 6.11

Suppose that $x(t)$ is generated by the mixed signal

$$x(t; \mu) = u(t) + v(t; \mu)$$

where $u(t)$ and $v(t; \mu)$ are independent and μ is a system parameter of v. The same technique can be used to detect a change in the system parameter, that is, it can be used to check for non-stationarity. The plots in each panel of Figure 6.35 show an easily identifiable change in the footprint of the time lagged characteristic function for only a slight change in one of the parameters of the Lorenz system. The figure also shows that the method's ability to identify the change in the footprint is not affected by a change in the lag time τ; although the shapes are different for different values of τ, the change is still easy to see.

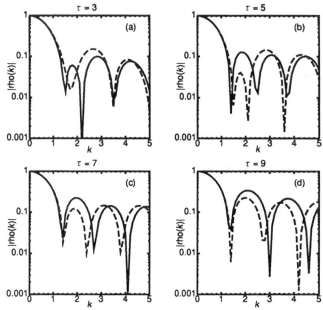

Figure 6.35 The absolute value of the time lagged characteristic function for a signal that is pure Lorenz. The solid and dashed curves are for two different values of a parameter in the Lorenz equations.

Source: Wright (1994) p. 271.

273

Example 6.12

This example shows that not only is the method robust with respect to the time delay τ for a pure signal, but also that its ability to detect small changes in system parameters is not impeded by the presence of other strong signals whether they be chaotic or random (Figure 6.36).

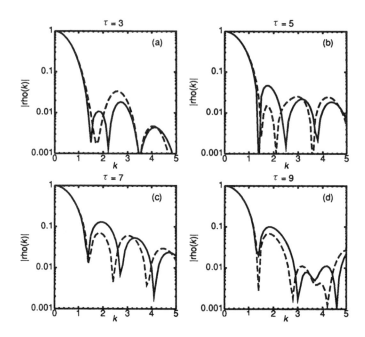

Figure 6.36 The absolute value of the time lagged characteristic function for a signal that is the sum of the Lorenz and Rössler equations. The solid and dashed curves are for two different values of a parameter in the Lorenz equations.

Source: Adapted from Wright (1994) p. 272.

REFERENCES AND FURTHER READING

Anosov, D. V. (1967) 'Geodesic flows and closed Riemannian manifolds with negative curvature', *Proc. Steklov Inst. Math.* 90.

Auerbach, D., Cvitanovic, P., Eckmann, J., Gunaratne. G. and Procaccia, I. (1987) 'Exploring chaotic motion through periodic orbits', *Physical Review Letters* 58.23, pp. 2387–2389.

Bowen, R. (1975) '*w*-limit sets for axiom *A* diffeomorphisms', *Journal of Differential Equations* 18, pp. 333–339.

Brandstäter, A. and Swinney, H. L. (1987) 'Strange attractors in weakly turbulent Couette-Taylor flow', *Physical Review A* 35, pp. 2207.

Casdagli, M., Eubank, S., Farmer, J. D. and Gibson, J. (1991) 'State space reconstruction in the presence of noise', *Physica D* 51, pp. 52–98.

Cvitanovic, P. (1988) 'Invariant measurement of strange sets in terms of cycles', *Physical Review Letters* 61.24, pp. 2729–2732.

Davies, M. (1994) 'Noise reduction schemes for chaotic time series', *Physica D* 79, pp. 174–192.

Farmer, D. and Sidorowich, J. (1991) 'Optimal shadowing and noise reduction', *Physica D* 47, pp. 373–392.

Fletcher, R. (1980) *Practical Methods of Optimization*, John Wiley & Sons, New York.

Fraedrich, K. and Wang, R. (1993) 'Estimating the correlation dimension of an attractor from noisy and small datasets based on re-embedding', *Physica D* 65, pp. 373–398.

Golub, G. H. and Van Loan, C. (1983) *Matrix Computations*, John Hopkins University Press, Baltimore.

Grebogi, C., Ott, E. and Yorke, J. A. (1988) 'Unstable periodic orbits and the dimensions of multifractal chaotic attractors', *Physical Review A* 37.05, pp. 1711–1724.

Hammel, S. (1990) 'A noise reduction method for chaotic systems', *Physics Letters A* 148.08, pp. 421–428.

Hammel, S., Yorke, J. and Grebogi, C. (1987) 'Do numerical orbits of chaotic processes represent true orbits?', *J Complexity* 3, pp. 136–145.

Kostelich, E. (1992) 'Problems in Estimating Dynamics from Data', *Physica D* 58, pp. 138–152.

Kostelich, E. and York, J. (1988) 'Noise reduction in chaotic systems', *Physical Review A* 38.03, pp. 1649–1652.

Kostelich, E. and York, J. (1990) 'Noise reduction: the simplest dynamical system consistent with the data', *Physica D* 41, pp. 183–196.

Linsay, P. (1991) 'An efficient method of forecasting chaotic time series using linear interpolation', *Physical Letters A* 153.06, pp. 353–356.

Pawelzik, K. and Schuster, H. (1991) 'Unstable periodic orbits and prediction', *Physical Review A* 43.04, pp. 1808–1812.

Press, W. H., Flannery, B. P., Teukolsky, S. A. and Vetterling, W. J. (1986) *Numerical Recipes*, Cambridge University Press, Cambridge.

Press, W. H., Flannery, B. P., Teukolsky, S. A. and Vetterling, W. J. (1988) *Numerical Recipies in C*, Cambridge University Press, Cambridge.

Schwetlick, H. and Tiller, V. (1985) 'Numerical methods for estimating parameters in non-linear models with errors in the variables', *Technometrics* 27, 17.

Provenzale, A., Smith, L., Vio, R. and Murante, G. (1992) 'Distinguishing between Low Dimensional Dynamics and Randomness in Measured Time Series', *Physica D* 58, pp. 31–49.

Sauer, T. (1992) 'A Noise Reduction Method for Signals from Nonlinear Systems', *Physica D* 58, pp. 193–201.

Schreiber, T. and Grassberger, P. (1991) 'A simple noise reduction method for real data', *Physics Letters A* 160.05, pp. 411–418.

Tufillaro, N., Solari, H. and Gilmor, R. (1990) 'Relative rotation rates: Fingerprint for strange attractors', *Physical Review A* 41.1, pp. 5717–5720.

Wright, J. (1994) 'Signals associated with nonlinear dynamical systems: identification and monitoring', AIP. Proc 2nd ONR Tech Conf on NLD & full spectral, pp. 260–274.

7

Concluding remarks
(a plan of attack)

The question we set out to answer is whether or not a system under study is chaotic, and if so, can a model be developed that gives reasonable predictions of its evolution over time. We have covered a lot of material on concepts and methods in the previous six chapters, and the reader may justifiably feel bewildered by the absence of any discussion on how to proceed from a collection of algorithms to a sensible approach to applying those algorithms so as to yield an answer to the above question. The methods developed in the previous chapters address the modelling and prediction parts of the question. We will go some way toward addressing the existence (of chaos) part of the question in these concluding remarks, but it would be misleading to suggest that there exists a general step-by-step recipe for applying some combination of the methods that will definitively prove the existence of chaos in an experimental situation.

Our task would be simpler if we were sure that the system under study was either deterministic or stochastic, but in the experimental situation even a deterministic system yields observations with a noise (stochastic) component. Moreover, in practice we never get to see the actual system; just a noisy time series from which we estimate system invariants, models of system dynamics, and prediction error. System invariants and prediction error are the statistics we use to characterize the nature of the system under study. These statistics constitute a footprint left in the wake of the evolving system. Since the measured footprint is subject to estimation error and noise, the appropriate question to ask is whether that footprint looks more like the footprint of a deterministic system than the footprint of a stochastic system. While we cannot prove the existence of chaos in a real system, we can accumulate enough statistical evidence to confidently accept or reject the hypothesis that there is a significant chaotic component to the observed data.

The *ansatz* for estimating system invariants and system dynamics is an accurate reconstruction of the system's trajectory from an observed noisy time series. Hence, the quality of the reconstructed trajectory may depend heavily on the success of the noise reduction techniques used as well as the reconstruction parameters themselves. Notwithstanding its importance, we leave aside the topic of noise reduction for the moment and begin with the development of a two-phase approach to the investigation which ignores the noise issue. Later, we will add a third phase to incorporate noise reduction into the analysis.

In phase one of the approach, we reconstruct a trajectory for the system using rough estimates of the reconstruction parameters as discussed in chapter three. From the reconstructed trajectory we estimate system invariants, a model for the dynamics, and the resulting prediction error. (Recall that the algorithm sections of the previous chapters contain numerous methods for state space reconstruction, noise reduction and estimation of system invariants. In many cases the different algorithms of a given chapter explore different characteristics of a chaotic system to measure the same quantity. Some of these will work better than others on specific systems and data sets, but since the nature of the system and the extent to which the data is contaminated are unknown, it is advisable to use as many methods as possible in the procedure described here[1].)

The objective of phase two is to test the hypothesis that there is a significant chaotic component to the observed data as reflected in the statistics (footprints) computed in phase one. To do that we use the methods of Appendix G. Thus, we begin with surrogate data sets generated from phase randomized Fourier transforms of the original time series data. Next, using the same reconstruction parameters as were used in phase one, each surrogate time series is subjected to state space reconstruction. Finally, using the same algorithms as were used in phase one, we estimate the invariants and prediction error from the reconstructed trajectory of each surrogate data set. The result is a set of statistics for each surrogate time series, each member of which is an observation on an invariant or the prediction error of a *stochastic process*. The ensemble of observations for each statistic is then used to construct a confidence level for testing the null-hypothesis: *the invariant or prediction error statistic of the original data could have been produced by the stochastic process represented by the surrogate data.* Convincing evidence for determinism would be a rejection of that null-hypothesis for a majority of the statistics.

The results of the analysis undertaken in phases one and two can be quite sensitive to the reconstruction parameters. A bad choice of reconstruction parameters can easily make a chaotic system indistinguishable from a stochastic system. Now, a chaotic dynamical system should produce reasonably stable estimates of system invariants, and a material drop in prediction error, when the reconstruction parameters are near their correct values. We use that fact to approach a good representation of the system's true trajectory by repeating the calculations of phases one and two as we make moderate adjustments to the reconstruction parameters, searching for an area in parameter space where the estimates of system invariants are stable and the prediction error has dropped to a low level. Moreover, that same area in parameter space should also coincide with the maximum number of statistics from the original data failing the stochastic (null) hypothesis.

---

[1]In most cases, the code implementing the algorithm is probably available from the authors of the referenced research papers.

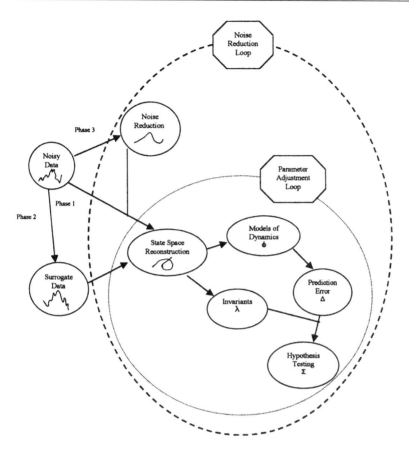

Figure 7.1 Schematic of the three phase process discussed in the text.

There are two obvious reasons why the above procedure might fail to indicate determinism in the data. The first reason is that the system of interest is in fact a random system indistinguishable from the one represented by the surrogate data. The second reason is that the noise component of the observed signal is too high to be able to distinguish it statistically from the stochastic signals of the surrogate time series. For example, if the noise component of the data is too high relative to the deterministic signal, then the prediction error will be bounded from below by the average noise and the adjustments to the reconstruction parameters will not produce the anticipated drop in prediction error. It is for the latter reason that we add a third phase to the procedure, bringing into play the noise reduction algorithms of chapter six. In phase three we replace the original data used in phase one with a noise reduced data set, and then proceed with phases one and two as before. If there is a deterministic system underlying the original time series, and the deterministic signal is only contaminated by a moderate level of noise, then a noise reduction algorithm may be able to retrieve the true orbit, or a shadowing orbit that captures the true dynamics of the system, although it may take several iterations of phases one, two and three to achieve that result.

In summary, the procedure embodied in phases one through three above, when applied to a noisy time series from a chaotic dynamical system, should yield a significant body of statistical evidence supporting the deterministic character of the system. However, there are important caveats. The first concerns noise reduction. Namely, any noise reduction procedure necessarily destroys or eliminates information through transformation of the original data. Thus, the use of such a procedure carries an inherent risk of creating nonsense. For example, a shadowing orbit identified by such a procedure is a deterministic trajectory that closely follows the experimental trajectory of the system. If the underlying system is in fact random, then clearly the dynamics of the shadowing orbit have nothing to do with the true dynamics of the underlying system.

The second caveat concerns the null hypothesis of the statistical tests. Rejection of the null hypothesis means simply that it is highly unlikely that the original data set emanated from the stochastic process represented by the surrogate data sets. It is of course possible that the original data were generated by a different stochastic mechanism, so that the null hypothesis is rejected because the underlying process, although random, is distinguishable from the stochastic process represented by the phase randomization of the Fourier transform of the original data. The point is that the conclusion of the hypothesis test is relative to a specific class of stochastic process, and not to the universe of all stochastic processes. It therefore makes sense to consider other plausible stochastic mechanisms for the null hypothesis. For example, there exist numerous well known econometric models of stochastic processes that purport to explain market behavior. Those models, ARIMA and its elaborations, can also be used to generate surrogate data sets for phase two by estimating model parameters from the original data, and then, employing a random number generator to simulate the innovations (stochastic component) of the model.

The strength of the conclusions reached about the nature of the system underlying the observed data is proportional to the number of alternative classes of stochastic processes considered, and is only limited by our imagination and resources. However, resources are finite and there must be a trade-off between confidence and cost. Generating surrogate data is simple and cheap, but the analysis encompassed by phases one through three is complicated and computationally expensive.

A

Data requirements

Estimates of system invariants will depend on the *amount of quality* data that we have to make those estimates. By quality we mean the amount of relevant non-redundant information carried in each observation of the measurement time series. As a system is sampled (observed) more and more frequently, the time between observations approaches the irrelevance time[1] of the system, and each new observation contains less and less new information. Another way to look at that effect is to observe, as we did in chapter 4: Fractal Dimension, that oversampling a system just reinforces the trivial one-dimensional temporal-distribution of points on a reconstructed trajectory, and eventually we learn nothing new about the divergence of nearby trajectories or the recurrence properties of those trajectories; effects which determine the system invariants. We can increase the amount of information we have on the system in two ways. The first is to obtain a concurrent independent set of observations on the system and use the resulting multi-variate time series in place of the scalar time series as the basis for the phase space reconstruction. The second way to obtain more information is to increase the time spanned by the scalar measurement time series. In other words, the time series either has to go back further in time, or further forward in time, to add any new information.

However, in the presence of noise, just increasing the size of the set of observational data may not be enough to improve the accuracy of the estimates of system invariants. To see why, suppose that the experimental trajectory points are uniformly distributed in the region of the state space occupied by the attractor, then each trajectory point would lie in a small d_A-dimensional[2] hypercube with edge size L_0, and volume equal to $[L_0]^{d_A}$. Since there are N_r trajectory points, the total volume of the attractor would be $N_r[L_0]^{d_A}$, and the average distance between the trajectory points would be L_0.

---

[1] See Chapter 3, Phase space reconstruction.

[2] d_A is the dimension of the manifold containing the attractor.

Since $d(A)$ is the diameter[3] of the attractor, the volume of state space occupied by the attractor is also approximately $[d(A)]^{d_A}$. Hence.

$$N_r[L_0]^{d_A} \approx [d(A)]^{d_A}$$

or

$$L_0 = \frac{d(A)}{[N_r]^{\frac{1}{d_A}}}$$

In the general situation, the appropriate dimension to use is D_1[4] and not d_A, and L_0 is the approximate minimum accessible length scale for a data set with N_r points. If L_0 is below the noise level, then increasing N_r adds little to our knowledge of the attractor's fine structure.

The next two sections develop rough bounds on minimum data set sizes needed to estimate some invariants. Keep in mind that if the system of interest is chaotic and low dimensional, we can often get good estimates with small data sets. However, since the system of interest is unknown, we need larger data sets to obtain a degree of confidence in the estimates.

Fractal Dimension: minimum data set size[5]

The computation of the correlation exponent D_2 assumes that $\text{Log}\, C_2(N_r, \varepsilon)$[6] is linear in $\text{Log}(\varepsilon)$ for small ε, that is

$$\text{Log}\, C_2(N_r, \varepsilon) \approx a + D_2 \,\text{Log}\, \varepsilon \qquad \textbf{Eq. A.1}$$

and since $C_2(N_r, \varepsilon)$ is proportional to $N(\varepsilon)$, the number of pairs of trajectory points in a sphere of radius ε, Eq. A.1 gives

$$N(\varepsilon) \approx b\varepsilon^{D_2} \qquad \textbf{Eq. A.2}$$

If $d(A)$ is the extent of the reconstructed attractor, then the total number of pairs of reconstructed trajectory points is

$$N(d(A)) \approx \frac{1}{2}N_r^2 \qquad \textbf{Eq. A.3}$$

From Eq. A.2 and Eq. A.3 we get

$$N(\varepsilon) = \frac{N_r^2}{2}\left[\frac{\varepsilon}{d(A)}\right]^{D_2}$$

---

[3] $d(A) = \max[x_i] - \min[x_i]$

[4] D_1 is the information dimension discussed in Chapter 4, Fractal dimension.

[5] Eckmann and Ruelle (1992).

[6] We assume an embedding, suppress the dependence of C_2 on d_r, and make explicit the dependence of C_2 on N_r.

We assume ε is small so that

$$\rho = \frac{\varepsilon}{d(A)} \ll 1$$

For statistical reasons we need $N(\varepsilon) \gg 1$. Hence, we must have

$$2 \operatorname{Log} N_r > D_2 \operatorname{Log} \frac{1}{\rho}$$

or

$$D_2 < \frac{2 \operatorname{Log} N_r}{\operatorname{Log} \frac{1}{\rho}}$$

If we assume that $\rho \approx 0.1$, then a correlation dimension estimate of 6 or more is meaningless when $N_r \leq 1000$. Similarly, if the fractal dimension of the system's attractor is around 10, then we will need more than 100,000 trajectory points to estimate it with any confidence.

Lyapunov Exponents: minimum data set size[7]

Estimating Lyapunov exponents requires the construction of tangent vectors between a reference point and a number of nearby points inside a small region (sphere) of radius ε. For ε small, the number of points $M(\varepsilon)$ contained in an ε-neighborhood of a reference point is

$$M(\varepsilon) \approx c\varepsilon^{D_2}$$

and since

$$M(d(A)) = N_r$$

we get

$$M(\varepsilon) \approx N_r \left(\frac{\varepsilon}{d(A)}\right)^{D_2}$$

Assuming $M(\varepsilon) \gg 1$ and $\rho = \frac{\varepsilon}{d(A)} \ll 1$ we must have

$$\operatorname{Log} N_r > D_2 \operatorname{Log}\left(\frac{1}{\rho}\right)$$

Hence, if $\rho \cong 0.1$, then a confident estimate of the Lyapunov exponents for a six-dimensional system will require more than 10^6 experimental trajectory points. Note that the data requirements for the Lyapunov exponents are the square of the requirements for the correlation dimension.

---

[7]Eckmann and Ruelle (1992).

Normalization

In general, to avoid numerical problems due to origin and scale, it is best to normalize the data before analyzing it. The jth term of the normalized time series then has the form

$$\frac{x_j - \langle x_i \rangle}{\Delta x}$$

where

$$\langle x_i \rangle = \frac{1}{N_d} \sum x_j$$

and

$$\Delta x = \left[\frac{1}{N_d} \sum \left| x_j - \langle x_i \rangle \right|^p \right]^{1/p} \qquad p = 1, 2$$

REFERENCE

Eckmann, J.-P. and Ruelle, D. (1992) 'Fundamental limitations for estimating dimensions and Lyapunov exponents in dynamical system stationarity' *Physica D* 56. pp. 185–187.

B

Nearest neighbor searches

In the previous chapters of the book we have often spoken of near neighbors of an experimental trajectory point as if they were arbitrary points on the attractor which could lie on different nearby trajectories. In fact, all the trajectory points reconstructed from a single time series will constitute a single experimental trajectory, so all near neighbors of a reference point[1] will necessarily be on the same trajectory as the reference point. In practice this will not cause severe difficulties provided there is ample data, since most near neighbors will come from parts of the experimental trajectory that are separated by relatively long periods of time. Nearby points from the same trajectory, separated by relatively long periods of time, will be practically independent and behave much like nearby points on different trajectories.

m-Nearest neighbors

Objective: given N_r experimental trajectory points in $\mathfrak{R}^{d_r}$, we want the identities of the m experimental trajectory points closest to an arbitrary reference point in $\mathfrak{R}^{d_r}$. More often than not, the reference point will be one of the experimental trajectory points. To differentiate between vectors and their components, let $\underline{y}_k(i)$ denote the value of the kth component of the vector $\underline{y}(i)$.

k-d Trees[2]
The k-d tree is a recursive binary partitioning of $\mathfrak{R}^{d_r}$. The root node, at the top of the tree, first partitions $\mathfrak{R}^{d_r}$ into two subspaces by running a hyperplane in $\mathfrak{R}^{d_r-1}$ through $\mathfrak{R}^{d_r}$. The partitioning hyperplane is constructed by fixing a single coordinate, say β, of the general point in $\mathfrak{R}^{d_r}$ at some value, say α. The resulting two subspaces are represented in the tree by two offspring nodes of the root node, the 'left' of which represents all trajectory points whose βth coordinate is less than α, and the 'right' of which is the complement. Similarly, the subspaces represented by the offspring nodes

---

[1]A reference point (orbit), also known as a fiducial point (orbit), is an experimental trajectory point (orbit).

[2]Friedman *et al.* (1977); Bently and Friedman (1979).

are then further partitioned, each by a hyperplane in $\mathfrak{R}^{d_r-1}$ constructed by fixing one coordinate of the general point in that subspace at some value. The process continues until a node represents a subspace containing no more than a fixed number Ω of trajectory points. At that stage the partitioning stops and the terminal node represents a subspace called a 'bucket.'

It turns out that the optimal partitioning at each node is achieved by cutting $\mathfrak{R}^{d_r}$ along the coordinate with maximum 'spread'[3], in such a way that an equal number of trajectory points falls into each of the two subspaces created by the partition. Hence, the value α, of the β-coordinate, at which the partition cuts the subspace is equal to the median of the β-coordinate values of all the trajectory points of the subspace. Thus, as we descend the tree from the root node, each successive node effectively represents an increasingly fine partitioning of $\mathfrak{R}^{d_r}$ into hypercubes. Finally, at the bottom of the tree, the terminal nodes represent hypercubes (buckets) containing subcollections of approximately Ω trajectory points (see Figure B.1).

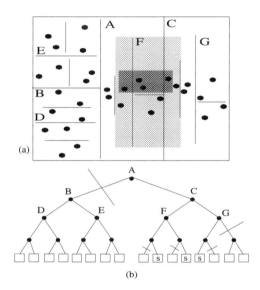

Figure B.1 Panel (a) is a two-dimensional state space showing the trajectory points and the partitioning (vertical and horizontal lines), and panel (b) is the abstract tree showing the nodes, branches, and buckets (boxes at the bottom of the tree) corresponding to the state space partitioning. The 'query rectangle' is shown as a dark-shaded region near the centre of panel (a). The search starts at the root node A, and since the query rectangle is entirely to the right of the vertical line A in panel (a), the left subtree of node A (with node B as root) can be pruned from the search. This is illustrated in panel (b) by the perpendicular line through the branch from node A to node B. The search continues through both branches of node C, both branches of node F, but only the left branch of node G. A total of three buckets are searched; these buckets are contained in the light-shaded region in the planar representation in panel (a) and marked by an 's' in the tree representation panel (b).

Source: Adapted from Bently and Friedman (1979) p. 397.

[3]As determined by, for example, the standard deviation of that coordinate for the subcollection of trajectory points represented by that node.

A search for the m nearest neighbors of some reference point then proceeds by descending the tree, from the root node, to the nodes whose descendent terminal nodes represent hypercubes containing the m closest trajectory points to the reference point. In practice, the search is carried out by exploring only those branches in the tree that represent a hypercube, containing the volume element (neighborhood) defined by the m nearest neighbors. If the neighborhood overlaps both branches of a node, then both branches have to be searched.

Performance
The time taken to build the tree is proportional to $N_r Ln N_r$, and searches take on the order of $Ln N_r$ time. Search times are minimized with the max norm, and terminal node bucket sizes Ω of between 4 and 32 points.

ε–Neighborhoods

Objective: given N_r experimental trajectory points in $\mathfrak{R}^{d_r}$, we want the identities of all experimental trajectory points in the ε-neighborhood[4] of an arbitrary reference point in $\mathfrak{R}^{d_r}$. More often than not, the reference point will be one of the experimental trajectory points. To differentiate between vectors and their components, let $\underline{y}_k(i)$ denote the value of the kth component of the vector $\underline{y}(i)$.

k-d Trees[5]
The occupied nodes of the decision tree are in a one-to-one correspondence with the reconstructed trajectory points $\underline{y}(i)$. The top of the tree is level 0. At level 0 there is a single node called the root node. At level L there are 2^L nodes. We start populating the tree by inserting, that is, assigning, $\underline{y}(1)$ to the root node. We then take the remaining $N_r - 1$ trajectory points in succession, inserting $\underline{y}(i)$ into the first empty node of the tree as follows. At the root node, we compare the first coordinate of $\underline{y}(i)$ with that of $\underline{y}(1)$. If $\underline{y}_1(i) < \underline{y}_1(1)$ we branch to the left descendent of the root node, otherwise we branch to the right descendent. At the next level, level 1, we compare $\underline{y}_2(i)$ with the second coordinate of the trajectory point residing at that node and branch to level 2 using the same kind of branching rule that we used at level 0, unless the node has no resident, in which case we place $\underline{y}(i)$ at that node. In general, at level L, we compare $\underline{y}_{1+(L \mod d_r)}(i)$ with the same coordinate of the vector residing at that node and branch to the next lower level as before, unless the node is vacant, in which case we insert $\underline{y}(i)$ there.

The first step in identifying the members of a reference point's ε-neighborhood is to prune the tree. For specificity, let $\underline{y}(n)$ be the reference point. Starting at the root node, we initiate a recursive search down the tree. At each step of the search we will be at some node at level L in the tree. If the node is not vacant, we examine coordinate $k = 1 + (L \mod d_E)$ of the resident vector $\underline{y}(j)$. If $\underline{y}_k(j) < \underline{y}_k(n) - \varepsilon$, then $\|\underline{y}(j) - \underline{y}(n)\| > \varepsilon$, so we know that $\underline{y}(j)$ is not in the ε-neighborhood of $\underline{y}(n)$. In fact, from the way the tree was constructed, we also know that $\|\underline{y} - \underline{y}(n)\| > \varepsilon$ for

---

[4]The ε-neighborhood of a point y is the set of all points u such that $\|y - u\| < \varepsilon$. Assume that $\| \cdot \|$ is either the Euclidean or max norm.

[5]Bingham and Kot (1989).

287

all points y assigned to the left subtree of $y(j)$. Hence, the left subtree of $y(j)$ can be eliminated from further consideration since its residents cannot be members of the ε-neighborhood of $y(n)$. Similarly, if $y_{\underline{k}}(j) > y_{\underline{k}}(n) + \varepsilon$, then we can eliminate the right subtree of $y(j)$ from further consideration. Of course, if $|y_{\underline{k}}(j) - y_{\underline{k}}(n)| < \varepsilon$, then we must look down both subtrees from $y(j)$. When we arrive at a vacant (terminal) node, we return to the nearest ancestor node that has an unexplored subtree which has not been eliminated. When no such ancestor exists, the pruning is complete, and the resident points of the nodes that were actually visited form a shortlist for the members of $y(n)$'s ε-neighborhood. The true members of that ε-neighborhood are those points $y(\bar{j})$ on the shortlist that satisfy $\|y(n) - y(j)\| < \varepsilon$.

Parameter values

There is no guarantee that a given ε-neighborhood will contain any points. If neighborhoods with a given population are needed, then the m-Nearest Neighbor algorithm above should be used. Neighborhoods that have m members, *on average*, can be obtained with this algorithm by setting $\varepsilon = d(A)[m/N_r]^{1/d_A}$, where d_A is the dimension of the manifold containing the attractor and $d(A) = \max[x_i] - \min[x_i]$ is the extent of the attractor[6].

Performance

The time taken to build the tree is proportional to $N_r Ln N_r$. Apparently, the search times are proportional to the number of points found in a hypercube with edge size ε. If that is the case then, all else being constant, the search time is proportional to $N_r \varepsilon^{d_A}$. Search times are minimized with the max norm.

Linked-lists[7]

The idea here is to cover the attractor with hypercubes and associate all the trajectory points in a given hypercube with a linked-list. Then the members of a neighborhood of an arbitrary point on the attractor are among the linked-lists of the hypercubes that intersect the neighborhood of the reference point.

Without much loss of generality, we simplify the presentation by assuming a two-dimensional state space reconstruction. Thus, $y_1(i) = x_i$ and $y_2(i) = x_{i-1}$. Define the attractor extent $d(A) = c_{max} - c_{min}$, where $c_{max} = \max[x_i]$ and $c_{min} = \max[x_i]$. Let L be a one-dimensional array with N_r elements in a one-to-one correspondence with the experimental trajectory points. Cover the attractor with a D by D lattice of boxes, each of whose sides is of length $\delta = d(A)/D$. Let B be a D by D array in a one-to-one correspondence with the boxes in the lattice so that the natural numbering of the elements B_{ij} of the array B confer a numbering (i, j) on the boxes of the lattice (Figure B.2).

---

[6] See Appendix A, Data requirements, for the reasoning.
[7] Grassberger (1990).

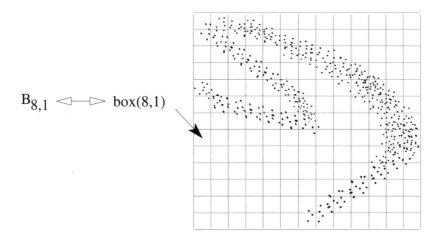

$B_{8,1} \Longleftrightarrow box(8,1)$

Figure B.2 On the right of the figure, a 13×13 lattice of boxes covering the experimental attractor. These boxes are in a one-to-one correspondence with a 13×13 array B that confers a natural numbering on the boxes as shown on the left of the figure.

Define the coordinate transformation

$$g(z) = \text{least integer upper bound of } (z - c_{min})/\delta \qquad \text{**Eq. B.1**}$$

and note that if

$$i = g[\underline{y}_1(n)]$$

and

$$j = g[\underline{y}_2(n)]$$

then $\underline{y}(n)$ lies in box (i, j) of the covering lattice.

After initializing the arrays L and B to zero, the trajectory points are organized into linked-lists as follows. Read in the trajectory points $\underline{y}(n)$ one at a time. Use the coordinate transformation g to identify the box (i, j) that contains $\underline{y}(n)$. If $B_{ij} = 0$, then set $B_{ij} = n$, and repeat the process for the next point $\underline{y}(n + 1)$. If $B_{ij} = k > 0$, then examine L_k. If $L_k = k' > 0$, then examine $L_{k'}$. Continue in this way until $L_{k''} = 0$, for some k''. Then set $L_{k''} = n$, and go back and repeat the process from the start with the next trajectory point. Thus, the first point on the experimental trajectory that falls into box (i, j) starts a linked-list to all subsequent trajectory points falling into that box, and that linked-list starts from the array element B_{ij} and continues into the array L (see Figure B.3).

289

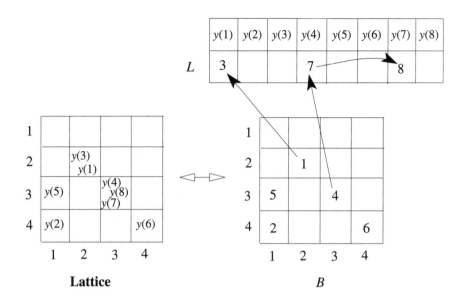

Figure B.3 On the left, the way the trajectory points fall into the lattice boxes. On the right, the way they get assigned to the arrays B and L.

To find the members of an ε-neighborhood of a reference point $\underline{y}(n)$, we first use the coordinate transformation g to identify the box (i, j) that contains $\underline{y}(n)$. Then, we go through the linked-list corresponding to each box (α, β) with

$$i - q \leq \alpha \leq i + q$$
$$j - q \leq \beta \leq j + q$$

where q is the least integer upper bound of ε/δ. Those are the lattice boxes that intersect the ε-neighborhood of $y(n)$, and the corresponding linked-lists start at the array element $B_{\alpha\beta}$ and continue into the array L. That gives us a shortlist for the neighborhood points. Finally, we obtain the true members of the ε-neighborhood of $\underline{y}(n)$ by selecting those $\underline{y}(k)$ on the shortlist that actually satisfy $\|\underline{y}(n) - \underline{y}(k)\| \leq \varepsilon$.

Remarks

- If we choose a value for δ rather than a lattice size D, then the storage requirements could be excessive. We can choose δ and still hold D at a desired value by 'wrapping' the two-dimensional lattice several times around a 2-torus. Numerically, that is achieved by redefining the coordinate transformation, Eq. B.1, to be

$$g(z) = \text{least integer upper bound of } \{(z - c_{\min})/\delta \mod D\}$$

- Even though the algorithm was developed around a two-dimensional state space, we might still choose a two-dimensional lattice for higher dimensional reconstructions. The reason for using a D by D (two-dimensional) lattice is to reduce the complexity and resource requirements of the algorithm by sorting the points based only on two coordinates of the state space vectors, that is, the attractor is projected onto a two-dimensional subspace and a point falls into box(i, j) if one of its coordinates, say its first, falls in the ith interval of the grid, and another coordinate, say its last, falls into the jth interval of the grid. Usually, a more uniform population of the lattice is achieved by uniformly spacing the projection axes. For example, for a five-dimensional reconstruction and a three-dimensional lattice, project the experimental trajectory onto the 1st, 3rd, and 5th coordinates. Higher dimensional grids might be advantageous in higher dimensional phase space reconstructions, and the necessary modifications are straightforward.

Parameter values

The following recommendations are made with the computation of the correlation integral in mind.

- Choose $\delta = \varepsilon = 2^{-i}$. Looking at distances that are powers of 2 improves performance, as the computer's masking and shifting functions can be used to efficiently extract the exponent i[8].
- Choose $D \approx \sqrt{N_r}$, and if necessary 'wrap' the lattice around a torus as described above.

Performance

The algorithm appears to compute the correlation integral $C(N_r, \varepsilon)$ in $N_r + Q$ time where Q is the number of point pairs with interpoint distance less that ε. Computation times are minimized with the max norm.

Neighborhood geometry

The differences in the results obtained when using a spherical neighborhood (Euclidean norm) or a cubical neighborhood (max norm) are not material for a given ε or m in the above algorithms. The issue there is really performance, as the max norm is generally faster than the Euclidean norm. There are, however, applications where changes in the neighborhood shape and size can make a material difference in the quality of the results.

Shells[9]

In some situations it may not be desirable to include, in a neighborhood, points which are too close to a reference point. For example, to estimate tangent maps (Jacobians) of a system near a reference point, we need a lot of tangent vectors. The tangent vectors are obtained by constructing small vectors between the reference point and its near neighbors. Using neighbors that are too close can cause two problems. The first is that in a sample of very near neighbors there may be a predominance of points that lie

---

[8]The same technique is used in the naive algorithm presented in Appendix C, Correlation integral.

[9]Zeng *et al.* (1991).

on the same trajectory segment as the reference point. In that case many of the tangent vectors will lie nearly tangent to the trajectory segment containing the reference point. As a result the tangent map may estimate well in the direction of the trajectory segment containing the reference point, but estimate badly in other directions. The effect can be reduced by requiring a minimum evolution time between neighbors as was done in calculating the correlation integral.

The second problem arises from the presence of noise in the data. If the neighborhood extent is smaller than the noise level, then tangent vectors constructed from points in the neighborhood will generally be randomly oriented. Increasing the radius ε of the neighborhood should help, but in the presence of measurement error or additive noise, it is best to use a shell rather than a sphere for the tangent space neighborhood. A shell is just a neighborhood $B_\varepsilon[\underline{y}(i)]$ with an internal neighborhood $B_{\varepsilon_{min}}[\underline{y}(i)]$ removed. Thus, the neighborhood of a trajectory point $\underline{y}(i)$ becomes the shell

$$\{\underline{y}(k)|\varepsilon_{min} \leq \|\underline{y}(k) - \underline{y}(i)\| \leq \varepsilon\}$$

Parameter values

- The inner radius for the shell is chosen to be $\sigma_{noise}\sqrt{d_r}$ where σ_{noise}^2 is the noise variance as estimated from, say, the noise floor defined by the small singular values of a global SVD[10].

- In the example below, the upper bound ε on the shell is initially set to 5% of $d(A)$ so as to get the neighborhood populations $K_\varepsilon(i) \geq 10$. If for some $\underline{y}(i)$, $K_\varepsilon(i) < 10$, then ε is increased to 10% of $d(A)$. If $K_\varepsilon(i)$ is still less than 10, then the point $\underline{y}(i)$ is dropped, and the next point $\underline{y}(i + 1)$ is considered.

Example

Clearly, using a shell neighborhood, the Jacobian-based estimates[11] of the Lyapunov spectrum reported in the table below are not bad for relatively small ($N_r = 5000$) noisy (precision $10^{-1} - 10^{-2}$) data sets. The method even picks up negative exponents which are of the same order as λ_1. However, when the magnitudes of the negative exponents are not comparable to λ_1, the estimates are not reliable (see Figure B.4).

---

[10]See Appendix D, Matrix decomposition.
[11]See Chapter 5, Lyapunov exponents.

System	Reported λ_i (in the absence of noise)	Computed λ_i (in the presence of noise)
Lorenz ($\tau = 20\tau_s$)	1.50	1.63 ± 0.15
($\sigma = 16, b = 4.0, R = 45.92$)	0.00	0.05 ± 0.25
($N_d = 5000, 10^{-1}$ precision)	-22.46	-3.59 ± 0.41
Rossler ($r = 12\tau_s$)	0.090	0.096 ± 0.008
($a = 0.15, b = 0.2, c = 10$)	0.00	-0.006 ± 0.004
($N_d = 5000, 10^{-1}$ precision)	-9.8	-0.735 ± 0.057
Mackay–Glass ($\tau = 9\tau_s$)	0.0071	0.0075 ± 0.0007
($a = 0.2, b = 0.1, c = 10, T = 30$)	0.0027	0.0030 ± 0.0010
($N_d = 10{,}000, 10^{-2}$ precision)	0.000	-0.0027 ± 0.0010
	-0.0167	-0.0156 ± 0.0006
	-0.0245	-0.0394 ± 0.0064
Mackay–Glass ($\tau = 9\tau_s$)	0.00956 ± 0.00005	0.00938 ± 0.00040
($a = 0.2, b = 0.1, c = 10, T = 23$)	0.00000	0.00008 ± 0.00020
($N_d = 5000, 10^{-2}$ precision)	-0.0119 ± 0.0001	-0.0160 ± 0.0010
	-0.0344 ± 0.0001	-0.0734 ± 0.0227
Mackay–Glass ($\tau = 9\tau_s$)	0.00956 ± 0.00005	0.00946 ± 0.00008
($a = 0.2, b = 0.1, c = 10, T = 23$)	0.00000	0.00064 ± 0.00049
($N_d = 30{,}000, 10^{-4}$ precision)	-0.0119 ± 0.0001	-0.0134 ± 0.0011
	-0.0344 ± 0.0001	-0.0572 ± 0.0135

Figure B.4 Jacobian-based estimates of the Lyapunov exponent spectrum for various known model systems-see Chapter 1, Dynamical systems, for the equations of motion of these systems. The parameters used in the different systems, the total number of data points N_d, the precision of the data (from 10^{-1} to 10^{-4}), and the delay time τ of the reconstruction are given in the table. Error bars are produced by varying ε, ε_{min} and τ, the delay time of the state space reconstruction.

REFERENCES

Bently, J. and Friedman, J. (1979) 'Data structures for range searching', *Computing Surveys*, Vol. 11, No. 4, pp. 397–409.

Bingham, S. and Kot, M. (1989) 'Multidimensional trees, range searching, and a correlation dimension algorithm of reduced complexity', *Physics Letters A* 140.06, pp. 327–330.

Friedman, J., Bentley, J., and Funkel, R. (1977) 'An algorithm for finding best matches in logarithmic expected time', *ACM Transactions on Mathematical Software* 3.03, pp. 209–226.

Grassberger, P. (1990) 'An optimized box-assisted algorithm for fractal dimensions', *Physics Letters A* 148.01, pp. 63–68.

Zeng, X., Eykholt, R., and Pielke, R. A. (1991) 'Estimating the lyapunov exponent spectrum from short time series of low precision', *Physical Review Letters* 66.25, pp. 3229–3232.

C

Correlation integral

The correlation integral is the basis of numerical estimates of entropy and fractal dimension for a trajectory of a general system. The theoretical order-2 correlation integral $C_2(\varepsilon)$[1] is the probability that two randomly chosen points on the trajectory of a system will fall within a distance ε of each other. Computationally, the order-2 correlation integral for a d_r-dimensional state space reconstruction is defined in two nearly equivalent ways. Namely,

$$C^{(1)}(\varepsilon; d_r) = \frac{1}{[N_r]^2} \sum_{i,j} \Theta(\varepsilon - \|\underline{y}(i) - \underline{y}(j)\|)$$

and

$$C^{(2)}(\varepsilon; k, d_r) = \frac{1}{N_r(N_r - 1)} \sum_{i,j:|i-j|>k\geq 0} \Theta(\varepsilon - \|\underline{y}(i) - \underline{y}(j)\|)$$

where N_r is the number of reconstructed trajectory points[2], Θ is the Heaviside function

$$\Theta(z) = \begin{cases} 0 & \text{if } z \leq 0 \\ 1 & \text{if } z > 0 \end{cases}$$

and $\|z\|$ is a norm for the vector $z = [z_1, z_2, \ldots, z_{d_r}]$, which we will assume is either the max norm

$$\|z\| = \max_i |z_i|$$

or the Euclidean norm

$$\|z\| = \left[\sum_i [z_i]^2 \right]^{\frac{1}{2}}$$

---

[1]$C_2(\varepsilon; d_E, N_r) \to C_2(\varepsilon)$ as $N_r \to \infty$.
[2]The dependence of $C^{(1)}$ and $C^{(2)}$ on N_r is suppressed in the notation.

In the case of a dynamical system, $C_2(\varepsilon)$ is determined by the recurrence properties of the dynamics on the attractor of the system. As the size of the data set gets smaller, there is less and less opportunity for the recurrence property to influence the distribution of points on the attractor. Because it counts all pairs of points within a distance ε of one another, including pairs that consist of a point and itself, $C^{(1)}(\varepsilon; d_r)$ is artificially high for small values of ε, in fact, $C^{(1)}(\varepsilon; d_r) \geq 1/N_r$, which reduces the useful range over which it can be used to estimate the correlation dimension. $C^{(2)}(\varepsilon; k, d_r)$ attempts to correct for the anomaly in $C^{(1)}(\varepsilon; d_r)$ by eliminating from consideration pairs of points that are close in space simply because they are close in time. Perversely, the price paid for that correction is a decrease in the amount of data available for the calculation. Since $C^{(1)}(\varepsilon; d_r)$ and $C^{(2)}(\varepsilon; k, d_r)$ tend to $C_2(\varepsilon)$ as $N_r \to \infty$, for a reasonably large sized data set and moderate value of k, there should be no practical difference in the results obtained from using any of the above definitions for computing the correlation integral. By a moderate value of k we mean $k \approx \tau_c/\tau_s$ where τ_s is the sampling time of the time series and τ_c is the average cycle time

$$\tau_c = \frac{[\text{ length of the time series (units of time)}]}{[(\#\text{of times the times series crosses its average})/2]}$$

Evidently, from the definition of $C^{(1)}(\varepsilon; d_r)$ and $C^{(2)}(\varepsilon; k, d_r)$, estimating the correlation integral can be organized so that it amounts to counting the number of nearby experimental trajectory points in an ε-neighborhood[3] of each experimental trajectory point. In Appendix B we discuss several algorithms for finding near neighbors. Those algorithms should be used to compute the correlation integral as they minimize nearest neighbor search times. The method outlined below is a highly inefficient brute force approach that computes all the interpoint distances on the attractor, and then ignores the large distances among them. We present the method for comparison, and because it does illustrate an efficient way to sort and count interpoint distances once they have been calculated.

Consider

$$C^{(2)}(\varepsilon; 0, d_r) = \frac{1}{N_r(N_r - 1)} \sum_{i,j:|i-j|>0} \Theta(\varepsilon - d_{ij}) \qquad \textbf{Eq. C.1}$$

where $d_{ij} = \|\underline{y}(i) - \underline{y}(j)\|$ is the distance between the experimental trajectory points $\underline{y}(i)$ and $\underline{y}(j)$. Since, $d_{ij} = d_{ji}$, we only need to compute the distance once for $i < j$. Hence, Eq. C.1 becomes

$$C^{(2)}(\varepsilon; 0, d_r) = \frac{2}{N_r(N_r - 1)} \sum_{i,j:i<j} \Theta(\varepsilon - d_{ij}) \qquad \textbf{Eq. C.2}$$

Now, note that a computer's floating point representation of d_{ij} is of the form

$$d_{ij} = \pm mb^e$$

where the exponent e is an integer, the base b is also an integer (typically a power of 2), and the mantissa m is a fraction such that $1 \geq m > 1/b$. Thus, if $e = k$, then

$$2^{k-1} < d_{ij} < 2^k$$

[3] An ε-neighborhood of a point y is the set of points u which satisfy $\|y - u\| < \varepsilon$.

By using a computer's shifting and masking functions we can retrieve the exponent e without having to employ any floating-point operations. Hence, we can efficiently count the interpoint distances by incrementing an array B_k where

B_k = number of pairs (i, j) such that $i < j$ and $2^{k-1} < d_{ij} < 2^k$

Then by letting

$$S_n = \sum_{k=-\infty}^{n} B_k$$

we can rewrite Eq. C.2 as

$$C^{(2)}(2^n; 0, d_r) = \frac{2S_n}{N_r(N_r - 1)} \qquad \textbf{Eq. C.3}$$

Performance
Very slow. Takes 15–30 minutes of CPU time to sort out 20K points on an IBM 370/165[4].

REFERENCE

Grassberger, P. and Procaccia, I. (1983) 'Measuring the strangeness of strange attraction', *Physica D* 9, pp. 189–208.

---

[4] Grassberger and Procaccia (1983).

D

Matrix decomposition

The reader is directed to the general references for a full treatment of the subject of matrix decomposition[1]. Here, for convenience, we review some of the relevant material on linear algebra and singular value decomposition.

Let N_d denote the number of data points x_i in the measurement time series, that is $i = 1, 2, \ldots, N_d$. Assume that we have used the measurement time series to obtain a (d_E, τ)-reconstruction[2] of the phase space of the system. Then the number N_r of reconstructed trajectory points $\underline{y}_i \in \mathfrak{R}^{d_E}$ is given by

$$N_r = N_d - (d_E - 1)\tau/\tau_s$$

Define the $N_r \times d_E$ **trajectory matrix** by

$$Y = \frac{1}{\sqrt{N_r}} \begin{bmatrix} \underline{y}_1^T \\ \underline{y}_2^T \\ \vdots \\ \underline{y}_{N_r}^T \end{bmatrix}$$

where $N_r^{-1/2}$ is a convenient normalization. We will assume that the columns of Y have zero mean, that is

$$\sum_{i=1}^{N_r} x_{i+j} = 0; \quad j = 0, \ldots, d_E - 1$$

The last assumption implies that the center of mass of the experimental trajectory (attractor) is located at the origin and does not alter the results in principle.

Let the column vectors $\underline{s}_i \in \mathfrak{R}^{N_r \times 1}$ be such that $\{\underline{s}_i^T Y | i = 1, \ldots, k\}$ is a set of linearly independent vectors in $\mathfrak{R}^{d_E}$. Without loss of generality we can assume that the vectors $\{\underline{s}_i^T Y | i = 1, \ldots, k\}$ are orthogonal, and it follows that

$$\underline{s}_i^T Y = \sigma_i \underline{c}_i^T \qquad \text{Eq. D.1}$$

[1] Also see Broomhead and King (1986); and Albano et al. (1988).
[2] See Chapter 3, Phase space reconstruction.

where $\{\underline{c}_j | j = 1, \ldots, d_E\}$ is a complete orthonormal basis for $\mathfrak{R}^{d_E}$. Since the $\underline{c}_j$ are orthonormal

$$\underline{s}_i^T Y \left(\underline{s}_j^T Y \right)^T = \sigma_i \sigma_j \delta_{ij} \qquad \text{Eq. D.2}$$

$$\delta_{ij} = \begin{cases} 0 & i \neq j \\ 1 & i = j \end{cases}$$

From Eq. D.2, $\underline{s}_i$ and σ_i^2 must be, respectively, eigenvectors and eigenvalues of the $N_r \times N_r$ real symmetric **structure matrix** $\theta = YY^T$. It follows that

$$\theta \underline{s}_i = \sigma_i^2 \underline{s}_i \qquad \text{Eq. D.3}$$

and that $\{\underline{s}_i | i = 1, \ldots, k\}$ form part of a complete orthonormal basis $\{\underline{s}_i | i = 1, \ldots, N_r\}$ for $\mathfrak{R}^{N_r}$.

Since

$$\theta = \frac{1}{N_r} \begin{bmatrix} \underline{y}_1^T \underline{y}_1 & \underline{y}_1^T \underline{y}_2 & \cdots & \underline{y}_1^T \underline{y}_{N_r} \\ \underline{y}_2^T \underline{y}_1 & \underline{y}_2^T \underline{y}_2 & \cdots & \underline{y}_2^T \underline{y}_{N_r} \\ \vdots & \vdots & \cdots & \vdots \end{bmatrix}$$

the structure matrix θ is an array of scalar products of all pairs of points on the experimental trajectory. Equally, it gives the covariances between all patterns that have appeared in the reconstruction window defined by a delay coordinate map Φ.

Note that by Eq. D.1

$$\left(\underline{s}_i^T Y \right)^T = \sigma_i \underline{c}_i$$

so premultiplying by Y gives

$$YY^T \underline{s}_i = \theta \underline{s}_i = \sigma_i Y \underline{c}_i$$

But by Eq. D.3

$$\theta \underline{s}_i = \sigma_i^2 \underline{s}_i$$

so

$$\sigma_i^2 \underline{s}_i = \sigma_i Y \underline{c}_i$$

or

$$Y \underline{c}_i = \sigma_i \underline{s}_i \qquad \text{Eq. D.4}$$

Premultiplying Eq. D.4 by Y^T gives

$$Y^T Y \underline{c}_i = \sigma_i Y^T \underline{s}_i = \sigma_i^2 \underline{c}_i$$

where the last equality follows from Eq. D.1 again. Hence, $\underline{c}_i$ and σ_i^2 are, respectively,

300

eigenvectors and eigenvalues of the $d_E \times d_E$ real symmetric ***covariance matrix*** $\Xi = Y^T Y$. Ξ is called the covariance matrix because on expansion (for $\tau = \tau_s$) we get

$$
\Xi = \frac{1}{N_r} \sum_{i=1}^{N_r} \underline{y}_i \underline{y}_i^T
$$

$$
= \frac{1}{N_r}
\begin{bmatrix}
\sum_1^{N_r} x_i x_i & \sum_1^{N_r} x_i x_{i+1} & \cdots & \sum_1^{N_r} x_i x_{i+d_E-1} \\
\vdots & \vdots & \cdots & \vdots \\
\sum_1^{N_r} x_{i+d_E-1} x_i & \sum_1^{N_r} x_{i+d_E-1} x_{i+1} & \cdots & \sum_1^{N_r} x_{i+d_E-1} x_{i+d_E-1}
\end{bmatrix}
$$

which is the time averaged pairwise covariance between the vector components of the experimental trajectory points $\underline{y}_i$. Now, the embedding vectors $\underline{y}_i$ define a set of points in d_E-space with a multivariate distribution whose variables are the d_E components of the reconstructed vectors. Hence, Ξ is the covariance matrix of that multivariate distribution. The off diagonal elements of Ξ measure the ***linear dependence*** among the components of the $\underline{y}_i$.

Since we have assumed that the columns of Y have zero mean, the off diagonal elements of Ξ are also proportional to the ***correlation coefficients*** of the distribution of Y. A vanishing correlation coefficient implies ***linear independence***. For large data sets

$$
\lim_{N_r \to \infty} \left(Y^T Y \right)_{ij} = \sigma^2 g[(j - i)\tau]
$$

where g is the ***autocorrelation function*** defined by

$$
g(t) = \frac{\sum_i x_i x_{i+t}}{\sum_i x_i^2}
$$

and $\sigma^2 = \sum_i x_i^2$

Note that $g(0) = 1$ and $g(t) \to 0$ for large t, so that $Y^T Y \approx \Xi$ if the delay time of the embedding is large. Hence, if the embedding window size $\tau_w = (d_E - 1)\tau$ is too large, $Y^T Y$ always has rank d_E, and it appears that we are looking at noise. The reason is that the first q components contain the functional relationship, but the remaining $d_E - q$ uncorrelated elements start to look random and fill out the additional phase space volume.

If the embedding window is too small, then all the elements of the covariance matrix approach 1. In that case, one eigenvalue has the value d_E and the remaining $d_E - 1$ eigenvalues approach zero corresponding to the case where the trajectory is projected onto the main diagonal of the phase space.

By symmetry the non-zero eigenvalues of the structure matrix are equal to the non-zero eigenvalues of the covariance matrix so that rank $\theta = $ rank $\Xi = k \le d_E$. Thus $\mathfrak{R}^{N_r}$ can be decomposed into a subspace of dimension k and its orthogonal compliment. The first subspace is spanned by $\{\underline{s}_i | i = 1, \ldots, k\}$ such that $\underline{s}_i^T Y$, a linear combination over the $\underline{y}_i$, gives rise to a unique basis vector $\underline{c}_i \in \mathfrak{R}^{d_E}$. The remaining $\{\underline{s}_i | i = k+1, \ldots, N_r\}$ span the kernel or null-space of Y, that is, they map to the origin of the embedding space. The implication is that, if the x_i are noise free, then as $N_r \to \infty$ there are k linearly independent vectors in the embedding space, and that, therefore, the space containing the embedded manifold has dimension k.

Let

$$S = [\underline{s}_1, \underline{s}_2, \dots, \underline{s}_{d_E}] \in \mathfrak{R}^{N_r \times d_E}$$
$$C = [\underline{c}_1, \underline{c}_2, \dots, \underline{c}_{d_E}] \in \mathfrak{R}^{d_E \times d_E}$$

and

$$\Sigma = \text{diag}(\sigma_1, \sigma_2, \dots, \sigma_{d_E}); \quad \sigma_1 \geq \sigma_2 \geq \cdots \geq \sigma_{d_E} \geq 0$$

From Eq. D.4

$$YC = S\Sigma$$

The orthonormality of C then implies that

$$Y = S\Sigma C^T \qquad\qquad\qquad \textbf{Eq. D.5}$$

which is the usual form of the well known *singular value decomposition (SVD)* of the (trajectory) matrix Y. The columns of S and C, respectively the eigenvectors of the structure matrix θ and the covariance matrix Ξ, are collectively referred to as the *singular vectors* of Y. The columns of S, called the *left singular vectors* of Y, are orthonormal in the column space of Y. The columns of C, called the *right singular vectors* of Y, form an orthonormal basis for the row space of Y, that is, the embedding space. The diagonal elements of Σ are called the associated *singular values* of Y.

The projection YC of the trajectory matrix onto the orthonormal basis of the embedding space is called the *matrix of principal components*, and in that context the eigenvectors $\{\underline{c}_i\}$ are called the *principal axes*. Note that since C is orthonormal, it is a pure rotation. Thus, the projection YC does not change distances between the trajectory points. In particular, it does not change the dimension of the attractor. Since

$$C^T Y^T Y C = \Sigma^2$$

the principal components of the trajectories are uncorrelated, and the variance of the ith principal component is σ_i^2. Moreover, σ_i^2 is also the mean square projection of the experimental trajectory onto the corresponding principal axis $\underline{c}_i$. Therefore, the $\underline{c}_i$ and σ_i give the direction and extent that the experimental trajectory explores the embedding space on average.

REFERENCES

Albano, A., *et al.* (1988) 'Singular value decomposition and the Grassberger–Procaccia algorithm', *Physical Review A* 38.06, pp. 3017–3026.

Broomhead, D. and King, G. (1986) 'Extracting qualitative dynamics from experimental data', *Physica D* 20, pp. 217–236.

E

Linearized dynamics

Having reconstructed a sample trajectory of the system, we can obtain a model for the system's equation of motion, that is, an approximation for φ, the function governing the dynamics in the embedding space. To do that we use the fact that (the assumed) differentiable dynamics are linear in small regions to get a *local* approximation to φ, and then patch these together to obtain the global model for φ.

As usual let $\underline{y}_t$ be an experimental trajectory point reconstructed from an observed time series x_t using a d_E-dimensional delay coordinate map Φ. Let y_{t_k} be the kth neighbor of $\underline{y}_t$ in $B_\varepsilon[\underline{y}_t]$, a small neighborhood of $\underline{y}_t$. Define z_{t_k} by

$$y_{t_k} = \underline{y}_t + z_{t_k}; \qquad y_{t_k}, \underline{y}_t, z_{t_k} \in \mathfrak{R}^{d_E}$$

Since z_{t_k} is small

$$
\begin{aligned}
\varphi[y_{t_k}] &= \varphi[\underline{y}_t + z_{t_k}] \\
&= \varphi[\underline{y}_t] + D_{\underline{y}_t}\varphi \cdot z_{t_k} \\
&= (\varphi[\underline{y}_t] - D_{\underline{y}_t}\varphi \cdot \underline{y}_t) + D_{\underline{y}_t}\varphi \cdot y_{t_k} \\
&= b_t + J_t \cdot y_{t_k}
\end{aligned}
$$

where b_t is a d_E-dimensional vector and J_t, the Jacobian of φ at $\underline{y}_t$, is a $d_E \times d_E$ matrix. Now, let L_t denote the linear map describing the dynamics of the system in a small region around $\underline{y}_t$, that is

$$y_{t_k+1} = \varphi[y_{t_k}] \approx b_t + J_t \cdot y_{t_k} \equiv L_t[y_{t_k}]$$

Provided the above equation is overdetermined, we can use least squares methods to solve for the L_t. Note that, while the maps L_t are linear, the implied global map

$$\varphi[\underline{y}_t] = L_t[\underline{y}_t]$$

is in general nonlinear.

Remarks

- Because it is simple and keeps one honest, it is standard operating procedure when fitting models to data to split the data into two sets, a training set and a testing set. Build the model using the training set, then measure its accuracy using the testing set. For a good model, the lack of agreement between model and test data (the out-of-sample error) should not be significantly worse than the lack of fit to the training data (the in-sample error).

- We recommend examining the singular values of the 'trajectory' matrix of the tangent vectors (z_{t_k} above) before trying to estimate J_t. If some of the singular values are very small, then it is best to first project onto the significant principal axes, then estimate J_t, and then 'lift' J_t back to $\mathfrak{R}^{d_E}$. Choose the significant singular axes to be the first k, where k is such that $\sigma_1/\sigma_k \leq 10$ and $\sigma_1/\sigma_{k+1} > 10$.

Parameter values

See the discussion on ε-neighborhoods and neighborhood geometry in Appendix B, Nearest neighbor searches.

F

Locating cycles and equilibrium points

As noted elsewhere, equilibrium points and periodic orbits play important roles in defining the long term behavior of dynamical systems whether they be chaotic or not. For that reason it makes sense to know where a dynamical system's periodic orbits and equilibrium points are, and in this appendix we give several algorithms for locating them.

Newton's method

Consider $H : \mathfrak{R}^n \to \mathfrak{R}^n$ with $H(\underline{x}^*) = 0$. Now, approximate $\underline{y}^{(i)} = H[\underline{x}^{(i)}]$ with the linearization

$$\Delta \underline{y}^{(i)} = DH[\underline{x}^{(i)}] \cdot \Delta \underline{x}^{(i)} \qquad \text{Eq. F.1}$$

where

$$\Delta \underline{y}^{(i)} = \underline{y}^{(i+1)} - \underline{y}^{(i)}$$
$$\Delta \underline{x}^{(i)} = \underline{x}^{(i+1)} - \underline{x}^{(i)}$$

and $DH[\underline{x}^{(i)}]$ is the Jacobian of H at $\underline{x}^{(i)}$. If we choose $\underline{x}^{(i+1)}$ such that $\underline{y}^{(i+1)} = 0$ in Eq. F.1, we obtain

$$-\underline{y}^{(i)} = DH[\underline{x}^{(i)}] \cdot [\underline{x}^{(i+1)} - \underline{x}^{(i)}] \qquad \text{Eq. F.2}$$

or, assuming $(DH[\underline{x}^{(i)}])^{-1}$ exists,

$$\begin{aligned} \underline{x}^{(i+1)} &= (DH[\underline{x}^{(i)}])^{-1} \cdot [DH[\underline{x}^{(i)}] \cdot \underline{x}^{(i)} - \underline{y}^{(i)}] \\ &= \underline{x}^{(i)} - (DH[\underline{x}^{(i)}])^{-1} \cdot \underline{y}^{(i)} \qquad \text{Eq. F.3} \\ &= \underline{x}^{(i)} - (DH[\underline{x}^{(i)}])^{-1} \cdot H[\underline{x}^{(i)}] \end{aligned}$$

It is unlikely that we are lucky enough to obtain $\underline{y}^{(i+1)} = 0$ exactly, but if $\underline{x}^{(0)}$ is chosen sufficiently close to $\underline{x}^*$, then we hope $\underline{x}^{(i)} \to \underline{x}^*$ quickly.

Note that if H is not known it has to be estimated, and this procedure will inherit the errors of any approximation to H.

Equilibrium points

At an equilibrium point $\underline{x}^*$ of a continuous time dynamical system, the vector field $F[\underline{x}^*] = 0$, so Newton's method can be applied to $H = F$.

At a fixed point $\underline{x}^*$ of a discrete time dynamical system, $f[\underline{x}^*] = \underline{x}^*$, so Newton's method can be applied to $H[\underline{x}^*] = f[\underline{x}^*] - \underline{x}^*$.

Periodic solutions

Any point $\underline{x}$ on a periodic orbit (cycle) of period Γ corresponds to a fixed point of $f^\Gamma[\underline{x}]$, or equivalently, to a root of $H[\underline{x}, \Gamma] = f^\Gamma[\underline{x}] - \underline{x}$. As before, we can use Newton's method to get the roots of H. To wit, linearizing H gives

$$
\begin{aligned}
\Delta \underline{y}^{(i)} &= D_x H[\underline{x}^{(i)}, \Gamma^{(i)}] \cdot \Delta \underline{x}^{(i)} + D_\Gamma H[\underline{x}^{(i)}, \Gamma^{(i)}] \cdot \Delta \Gamma^{(i)} \\
&= [D_x f^{\Gamma^{(i)}}[\underline{x}^{(i)}] - \Im] \cdot \Delta \underline{x}^{(i)} + F[f^{\Gamma^{(i)}}[\underline{x}^{(i)}]] \cdot \Delta \Gamma^{(i)}
\end{aligned}
\qquad \textbf{Eq. F.4}
$$

where

$$
\Delta \underline{x}^{(i)} = \underline{x}^{(i+1)} - \underline{x}^{(i)}
$$

and

$$
\Delta \Gamma^{(i)} = \Gamma^{(i+1)} - \Gamma^{(i)}
$$

Now, given $\underline{x}^{(i)}$, $\Gamma^{(i)}$, and $\underline{y}^{(i)}$, we want to choose $\underline{x}^{(i+1)}$ and $\Gamma^{(i+1)}$ such that $\underline{y}^{(i+1)} = 0$. That gives the iteration formula

$$
-\underline{y}^{(i)} = [D_x f^{\Gamma^{(i)}}[\underline{x}^{(i)}] - \Im] \cdot \Delta \underline{x}^{(i)} + F[f^{\Gamma^{(i)}}[\underline{x}^{(i)}]] \cdot \Delta \Gamma^{(i)}
$$

which is a system of n equations in the $n + 1$ unknowns, namely, $\Gamma^{(i+1)}$ and the n components of the vector $\underline{x}^{(i+1)}$. If we restrict $\Delta \underline{x}^{(i)}$ to lie orthogonal to the trajectory, then $(F[\underline{x}^{(i)}])^T \cdot \Delta \underline{x}^{(i)} = 0$, and we obtain a system of $n + 1$ equations in $n + 1$ unknowns

$$
\begin{bmatrix} -\underline{y}^{(i)} \\ 0 \end{bmatrix} = \begin{bmatrix} D_x f^{\Gamma^{(i)}}[\underline{x}^{(i)}] - \Im & F[f^{\Gamma^{(i)}}[\underline{x}^{(i)}]] \\ (F[\underline{x}^{(i)}])^T & 0 \end{bmatrix} \cdot \begin{bmatrix} \Delta \underline{x}^{(i)} \\ \Delta \Gamma^{(i)} \end{bmatrix}
\qquad \textbf{Eq. F.5}
$$

which can be iterated in the usual way.

Recurrent points[1]

As noted, Newton's method requires both a good approximation to the function H in Eq. F.1 and Eq. F.4, and a good approximation to the root of H to initialize the

---

[1]Mindlin and Gilmore (1992).

algorithm. We can get a pretty good idea of the location of attracting (or saddle) points and periodic orbits with the following simple procedure, even without an estimate of the system's equation of motion or a phase space reconstruction.

The idea is that if the system passes close to a saddle orbit of period p, then it is likely to circulate near the saddle orbit for a while, and measurements $x(i)$ will reflect that periodicity by nearly repeating after p units of time – for a while. Hence, for some small ε and some contiguous sequence of values of i the following inequality will hold

$$|x(i) - x(i + p)| < \varepsilon \qquad\qquad \textbf{Eq. F.6}$$

So one way to see the location and periodicity of the attractors saddle orbits is by plotting the 'truth' value of Eq. F.6 as a function of i and p. Such a plot is called a *close return map or recurrence plot* (Figure F.1).

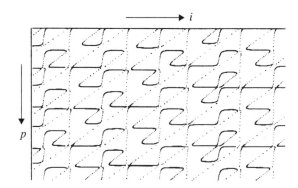

Figure F.1 Close returns map or recurrence plot of the Belousov–Zhabotinsky chemical reaction – see Chapter 4, Fractal dimension, for the equations of motion of this system. In the figure a pixel (i, p) is colored black if $|x(i) - x(i + p)| < \varepsilon$, and white if $|x(i) - x(i + p)| > \varepsilon$. ε is about 2% of the diameter $d(A)$ of the attractor.
Source: Mindlin and Gilmore (1992) p. 231.

Another visual aid is the *close returns histogram* (Figure F.2) defined by

$$H(p) = \sum_i \Theta(\varepsilon - |x(i) - x(i + p)|)$$

where Θ is the Heaviside function

$$\Theta(z) = \begin{cases} 1 & \text{if } z > 0 \\ 0 & \text{otherwise} \end{cases}$$

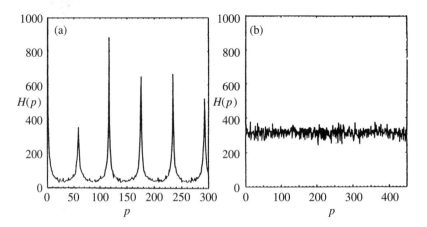

Figure F.2 Close returns histograms. (a) chaotic time series; (b) random time series.
Source: Mindlin and Gilmore (1992) p. 233.

REFERENCE

Mindlin, G. and Gilmore, R. (1992) 'Topological analysis and synthesis of chaotic time series', *Physica D* 58, pp. 229–242.

G

Surrogate data and non-parametric statistics

Specificallly, we are concerned with measuring the statistical significance and accuracy of models, parameters, forecasts, and invariants of a noisy dynamical system. More generally, we are interested in the nature of the process (system) that gives rise to a set of measurements.

The well known approach to structuring statistical questions that we will follow here will be to postulate a class of models for the system, and then ask if it is reasonable for us to reject the hypothesis that that class of systems gave rise to the observed measurements. If the hypothesized model was sensible to begin with, then whatever answer we get to that question increases our knowledge of the system. How much our knowledge increases of course depends on the hypothesis made and whether or not it is rejected. For example, we might postulate white noise for the process, and hope that that notion is convincingly rejected on the grounds that it would be extremely unlikely that a white noise process could have produced the observed measurements. If the hypothesis is rejected, then we still do not know what process produced the observed data, but we do know that it is unlikely to have been white noise. On the other hand, if the proposed model is not an unlikely source of the observed data, then we cannot reject it, and while we do not know if it actually is the source of the data, we do know that it is not inconsistent with the data. In that case we have some justification in using the proposed model to develop statistical statements about the system.

Let $\{p_i^{(0)}, i = 1, \ldots, N_r\}$, denote a set of measurements on the system, and let $\wp$ symbolize the true process that generated those observations. If the system that we are looking at is really stochastic, then $\wp$ follows a pure probability law. If the system that we are looking at is a dynamical system contaminated by additive noise, then $\wp$ is the sum of a deterministic process and a stochastic process. The previous chapters of the book are devoted to estimating invariants and developing a model for the *deterministic* component of $\wp$. In practice, $\wp$ will always contain a noisy element and in this appendix we probe the nature and extent of that noisy element. The two key questions we will want to answer are:

1. Are the observed data more consistent with a stochastic process or a noisy dynamical system?
2. If the data are consistent with a noisy dynamical system, then what are the statistical properties of estimates of system characteristics computed from the observed data? Formally, we are interested in the properties of some function $h[\{p_i^{(0)}\}] = g_0^*(\wp)^{[1]}$, which is meant to estimate some feature $g(\wp)$ of the process $\wp$ that generated the data set.

The first question refers to the standard hypothesis test whose constituents are a **null hypothesis** and a **discriminating statistic**. The null hypothesis is $\wp$, the model of the process postulated to be the source of the observed data. The discriminating statistic is simply h, a function of the observed data, and is normally chosen to have some physical meaning. For example, if the observed data are a time series of market prices, then the null hypothesis might be that the data is the product of a 'random walk' (stochastic) process, and the discriminating statistic might be the profit of a trading strategy. Other choices for the discriminating statistic would include fractal dimension, Lyapunov exponents, entropy, and prediction error[2].

The second question also involves a model for the process $\wp$ and a function h of the observed data, except in this case the intent is often to get a confidence interval for the value of the function.

The bootstrap[3]

Recall that $g_0^*(\wp)$ is our estimate of the characteristic $g(\wp)$ of $\wp$. Suppose we wish to estimate the accuracy of $g^*(\wp)$ as measured by a functional $\sigma_{g^*}(\wp)^{[4]}$. For example, we could measure the accuracy of $g^*(\wp)$ by its standard deviation

$$\sigma_{g^*}(\wp) = \{E_\wp[g^*(\wp) - E_\wp g^*(\wp)]^2\}^{1/2}$$ **Eq. G.1**

Since the expectation uses the law of $\wp$ to average, the obvious problem with evaluating Eq. G.1 is of course that we do not know $\wp$. A second problem is that $\sigma_{g^*}(\wp)$ may be very difficult to evaluate analytically. Now, note that even though we do not know $\wp$, we do have an estimate of the law of $\wp$ in $\wp^*$, the *empirical* distribution implied by the actual observed measurements $\{p_i^{(0)}\}$. Further, note that we could evaluate $\sigma_{g^*}(\wp)$ numerically by replacing expectations with averages, provided $\wp$ is stable and we have a reasonable number of independent observations like $g_0^*(\wp)$. We know that we are only going to get the one observation $g_0^*(\wp)$, but (in what is the closest thing to the elusive free lunch that we know) we could use $\wp^*$ to manufacture independent trials from $\wp$.

To get the independent data sets, we begin by using a random number generator to sample *with replacement* N_r times from the *stochastic* component of $\wp^*$. That

---

[1]In the sequel we will use the star (*) to denote empirical quantities.

[2]For an example using the prediction error as the discriminating statistic, see the algorithms section of Chapter 3, Phase space reconstruction. For an example using the fractal dimension, see the BDS algorithm in Chapter 4, Fractal dimension. See also the cross-correlation algorithm in Chapter 5, Lyapunov exponents, and the discrimination algorithm in Chapter 6, Noise reduction.

[3]Efron and Tibshirani (1986); Efron (1979); Hinkley (1988).

[4]The dependence on the number of observations N_r is suppressed in the notation.

is, we sample with replacement from the stochastic part of $\{p_i^{(0)}, i = 1, 2, \ldots, N_r\}$. That yields a **surrogate** data set for the stochastic component of $\wp^*$, which, when combined with any deterministic component, produces a set $\{p_i^{(1)}, i = 1, 2, \ldots, N_r\}$ called a **bootstrap** sample. Using the bootstrap sample we obtain a bootstrap estimate g_1^*. Note that $g_1^* = h[\{p_i^{(1)}\}]$ is a sample, or realization, of $g(\wp^*)$. If we repeat the operation N_B times, then we obtain N_B observations $g_j^*, j = 1, \ldots, N_B$. The final step is to replace the expectations with averages to get

$$S_{g*} = \left[\frac{\sum_{j=1}^{N_B} (g_j^* - \langle g_i^* \rangle_i)^2}{N_B - 1} \right]^{1/2}$$

where

$$\langle g_i^* \rangle_i = \frac{1}{N_B} \sum_{i=1}^{N_B} g_i^*$$

and

$$S_{g*} \xrightarrow[N_B \to \infty]{} \sigma_{g*}(\wp^*)$$

where $\sigma_{g*}(\wp^*)$ is the standard deviation of $g^*(\wp^*)$.

In the general case, we want an estimate of

$$E_\wp R[g^*(\wp), g(\wp)]$$

where R is any sensible functional that compares the random variable $g^*(\wp)$ with some attribute $g(\wp)$ of a process $\wp$. The actual form of R is not important. The essence of the bootstrap method is in replacing the unknown process $\wp$ with $\wp^*$ constructed from the original data, then sampling from $\wp^*$ to simulate a sample from $\wp$, and then averaging over $\wp^*$, that is, over the bootstrap samples, to arrive finally at the **non-parametric bootstrap** estimate

$$E_\wp R[g^*(\wp), g(\wp)] \approx E_{\wp*} R[g^*(\wp^*), g(\wp^*)] \approx \langle R[g_i^*, g(\wp^*)] \rangle_i$$

For example, from Eq. G.1

$$R^2[g^*(\wp), g(\wp)] = E_\wp [g^*(\wp) - E_\wp g^*(\wp)]^2$$

Other examples include

$$E_\wp g^*(\wp) \approx E_{\wp*} g^*(\wp^*) \approx \frac{1}{N_B} \sum_{i=1}^{N_B} g_i^*$$

and

$$P[g^*(\wp) < a] = E_\wp I_{\{g^*(\wp) < a\}} \approx \frac{1}{N_B} \sum_{i=1}^{N_B} I_{\{g_i^* < a\}}$$

where $P[S]$ is the probability that S is 'true', and $I_S = 1$ if S is 'true' and 0 otherwise.

311

If we had reason to believe that the probability law governing the stochastic component of $\wp$ was a member of a parameterized family F_θ, we can adapt the above methodology by estimating θ with θ^* using the original data, and then sampling from F_{θ^*}. In that case, the method is called **parametric bootstrapping**. Moreover, $p^{(0)} = \{p_1, p_2, \ldots, p_{N_d}\}$, the original data, need not be the product of a single process $\wp$. We could have, say, two processes $\wp$ and ϑ which generate an original data set $\{p_1, p_2, \ldots, p_m, v_1, v_2, \ldots, v_n\}$. The non-parametric bootstrap is then obtained by drawing randomly with replacement m times from the stochastic component of the set $\{p_1, p_2, \ldots, p_m\}$ and n times from the stochastic component of the set $\{v_1, v_2, \ldots, v_n\}$ to obtain the first bootstrap sample $[p^{(1)}, v^{(1)}] = \{p'_1, p'_2, \ldots, p'_m, v'_1, v'_2, \ldots, v'_n\}$, and so on.

We emphasize that the sampling done to create the surrogate data sets is done on the stochastic component of the observation set $\{p^{(0)}\}$. What constitutes the stochastic component depends on the null hypothesis. If the null hypothesis is that $\wp$ is a noisy dynamical system, then the deterministic part of $\{p^{(0)}\}$ should be removed before sampling. If the null hypothesis is that $\wp$ is stochastic, then the ordering in the sample $\{p^{(0)}\}$ is coincidental and the sampling is done directly on $\{p^{(0)}\}$.

Hypothesis testing

In hypothesis testing the situation is as follows. Our hypothesis is that $\wp \in \wp_H$[5], and we will reject that notion if g_0^* is a rare value for $g^*(\wp)$. Specifically, let's say if for some small α of our choosing, A satisfies

$$\alpha = P\{g^*(\wp) \geq A\}$$

then we reject the hypothesis whenever $g^*(\wp) \geq A$.

In practice we need to find $\wp^*(H) \in \wp_H$ that is in some sense close to $\wp^*$, and that is usually done by using the original data to estimate the parameters[6] of the family $\wp_H$. Then, using $\wp^*(H)$, we generate N_B bootstrap samples g_i^*, and compute the value of A^* for which

$$\alpha \approx \frac{1}{N_B} \sum_{i=1}^{N_B} \Theta(g_i^* - A^*)$$

and reject the hypothesis whenever the original statistic $g_0^* \geq A^*$.

Confidence intervals

For some small α of our choosing, choose A^* such that

$$\alpha \approx \frac{1}{N_B} \sum_{i=1}^{N_B} \Theta(A^* - g_i^*)$$

and B^* such that

$$1 - \alpha \approx \frac{1}{N_B} \sum_{i=1}^{N_B} \Theta(B^* - g_i^*)$$

---

[5]$\wp_H$ denotes a class of parameterized processes.
[6]See also Phase randomization.

Then a central $1 - 2\alpha$ percent confidence interval for $g^*(\wp^*)$ is $[A^*, B^*]$.

The number of bootstrap samples N_B needed to get reasonably stable results seems to be $50 \leq N_B \leq 250$ depending on the objective; the lower end of the range for estimating the standard error of $g^*[\wp^*]$, and the upper end of the range for estimating confidence intervals.

Phase randomization[7]

We have already seen one method of generating surrogate data in the bootstrap where these were obtained by sampling with replacement from the empirical distribution. As an alternative we can obtain the surrogate data by randomizing the phases of the Fourier amplitudes obtained from the original time series data. Specifically, at each frequency $\omega \in \{-N_d/2, \ldots, 0, \ldots, N_d/2\}$, the complex amplitudes of the Fourier transform are multiplied by $e^{i\phi(\omega)}$, where $\phi(\omega)$ is chosen randomly from the interval $[0, 2\pi]$ and $\phi(-\omega) = -\phi(\omega)$. The surrogate data is then obtained by applying the inverse Fourier transform to the phase randomized amplitudes[8]. The resulting signal will have the same Fourier amplitude spectrum as the original signal, and therefore, they will both have the same autocorrelation function.

Remarks
In some cases where it is desirable to maintain some of the macro structure, one can randomize the phases of the amplitudes only at the high frequencies, or only at frequencies where the amplitude has relatively low power.

Problems
- Note that spurious high frequency effects are introduced if $x(1)$ is not approximately equal to $x(N_d)$. If the series is sufficiently large and stationary, it should not be difficult to truncate it such that $x(1) \approx x(N'_d)$ and N'_d is of the same order as N_d.

- The influence of the Central Limit theorem can cause the surrogate data to become Gaussian. As a result, a hypothesis test may be able to discriminate between the surrogate data and the original data, even when the original data is random but non-Gaussian. The problem can sometimes be mitigated by using a *histogram transform* on the original data. That is, we generate a Gaussian sequence $v(i)$, $i = 1, \ldots, N_d$, and sort in numerical order to get $v(j_k)$, $k = 1, \ldots, N_d$. Then we sort the original time series in numerical order to get $x(i_k)$, $k = 1, \ldots, N_d$. The correspondence $x'(i_k) = v(j_k)$ defines an invertible nonlinear transform of the original series $x(n)$ to a new one $x'(n)$ which follows the trends of the original time series, but has an approximately Gaussian distribution of amplitudes. The transformation apparently preserves local phase space neighborhoods and system invariants when the system is deterministic. When the system is random, the transformed data will also be random, but Gaussian. Finally, the transformed time series $x'(t)$ is subjected to the phase randomization procedure discussed above to obtain surrogates for statistical comparison with $x'(t)$.

---

[7]Theiler *et al.* (1992); Kennel and Isabelle (1992).

[8]Setting $\phi(-\omega) = -\phi(\omega)$ ensures that the inverse transform is real valued.

REFERENCES

Efron, B. (1979) 'Computers and the theory of statistics: thinking the unthikable', *SIAM Review* 21.04, pp. 460–480.

Efron, B. and Tibshirani, R. (1986) 'Bootstrap methods for standard errors, condifence intervals, and other measures of statistical accuracy', *Statistical Science* 1.01, pp. 54–57.

Kennel, M. and Isabelle, S. (1992) 'Method to distinguish possible class from colored noise and to determine embedding parameters', *Physical Review A* 46.06, pp. 3111–3118.

Theiler, J., Eubank, S., Longtin, A., Galdrileian, B., and Farmer, J. D. (1992) 'Testing for nonlinearity in time series: the method of surrogate data', *Physica D* 58, pp. 77–94.

General references and further reading

Abraham, N. B. *et al.* (1989) *Measures of Complexity and Chaos*, Plenum, New York.

Abraham, R. H. and Shaw, C. D. (1992) *Dynamics: The Geometry of Behavior*, Addison-Wesley, Redwood City.

Arrowsmith, D. K. and Place, C. M. (1990) *An Introduction to Dynamical Systems*, Cambridge University Press.

Berge, P., Pompeau, Y. and Vidal, C. (1984) *L'Ordre dans le Chaos*, Herman, Paris.

Billingsley, P. (1965) *Ergodic Theory and Information*, Wiley, New York.

Brocker, Th. and Janich, K. (1982) *Introduction to Differential Topology*, Cambridge University Press, Cambridge.

Brockwell, P. and Davies, R. (1987) *Time Series: Theory and Methods*, Springer-Verlag, New York.

Casdagli, M. and Eubank, S. eds. (1992) *Non Linear Modeling and Forecasting (SFI Studies in the Sciences of Complexity Vol XII)*, Addison-Wesley, Reading, MA.

Chillingworth, D. R. J. (1976) *Differential Topology with a View to Applications*, Pitman, London, Research Notes in Mathematics.

Devijver, P. and Kittler, J. (1982) *Pattern Recognition: A Statistical Approach*, Prentice Hall, New York.

Dongarra, J., Moler, C., Burch, J. and Stewart, G. (1979) *LINPACK Users Guide*, Society for Industrial and Applied Mathematics, Philadelphia.

Efron, B. and Tibshirani, R. (1993) *An Introduction to the Bootstrap*, Chapman and Hall, London.

Fletcher, R. (1987) *Practical Methods of Optimization, 2nd edition*, Wiley, Chichester.

Fletcher, T. J. (1972) *Linear Algebra Through Its Applications*, Van Nostrand Reinhold, London.

Fukunaga, K. (1972) *Introduction to Statistical Pattern Recognition*, Academic, New York.

Fuller, W. (1987) Measurement Error Models, Wiley, New York.

Gill, P., Murray, W., Saunders, M. and Wright, M. (1986) *Users Guide for NEPSOL (4): A Fortran Package for Non Linear Programming*, Systems Optimization Laboratory, Stanford Univ Tech Report SOL 86-2.

Gill, P., Murray, W. and Wright, M. (1981) *Practical Optimization*, Academic, New York.

Goldberg, D. (1989) *Search Optimization and Machine Learning*, Addison-Wesley, Reading, MA.

Golub, G. and Van Loan, C. (1983) *Matrix Computations*, North Oxford Academic, Oxford.

Goof, G. and Hartmanin, J. (1977) *Matrix Egiensystem Routines, EISPACK Guide Extension, Lecture Notes in Computer Science, Vol 51*, Springer-Verlag, Berlin.

Guckenheimer, J. and Holmes, P. (1983) *Non Linear Oscillations, Dynamical Systems, and Bifurcations of Vector Fields*, Springer, New York.

Hall, P. and Heyde, C. (1980) *Martingale Limit Theory and its Applications*, Academic Press, New York.

Halmos, P. (1958) *Finite Dimensional Vector Spaces*, D. Van Nostrand, New York.

Hirsch, M. W. and Smale, S. (1979) *Differential Equations, Dynamical Systems, and Linear Algebra*, Academic Press, Orlando.

Holland, J. (1992) *Adaption in Natural and Artificial Systems*, MIT Press, Cambridge, MA.

Katok, A. and Hasselblatt, B. (1995) *Introduction to the Modern Theory of Dynamical Systems*, Cambridge University Press, Cambridge.

Kelso, J. A. S., Mandell, A. J. and Schleshinger, M. F., eds. (1988) *Dynamic Patterns in Complex Systems*, World Scientific, Singapore.

Khinchin, A. (1957) *Mathematical Foundations of Information Theory*, Dover Publications, New York.

Kifer, Y. (1986) *Ergodic Theory of Random Transformations*, Birkhauser, Basel.

Kim, J. H. and Stringer, S., eds. (1992) *Applied Chaos*, Wiley, New York.

Knuth, D. E. (1973) *The Art of Computer Programming*, Addison-Wesley, New York.

Kullback, S. (1959) *Information Theory and Statistics*, Wiley, New York.

Lang, S. (1964) *Second Course in Calculus, 2nd edition*, Addison-Wesley, Reading, MA.

Lawson, C. L. and Hanson, R. J. (1974) *Solving Least Squares Problems*, Prentice Hall, Englewood Cliffs, N.J.

Lee, Y. C., ed. (1988) *Evolution, Learning and Cognition*, World Scientific, Singapore.

Martin, N. F. G. and England, J. W. (1981) *Mathematical Theory of Entropy*, Addison-Wesley, Reading, MA.

Mason, J. C. and Cox, M. G., eds. (1987) *Algorithms for Approximation*, Carendon Press, Oxford.

Mayer-Kress, G. (1986) *Dimension and Entropies of Chaotic Systems*, Springer, Berlin.

McCauley, J. L. (1993) *Chaos, Dynamics and Fractals*, Cambridge University Press, Cambridge.

Morrison, F. (1991) *The Art of Modelling Dynamical Systems*, John Wiley.

Palis, J. and de Melow, W. (1982) *Geometric Theory of Dynamical Systems: An Introduction*, Springer, New York.

Panchev, S. (1971) *Random Functions and Turbulance*, Pergamon, Oxford.

Pandit, S. M. and Yu, S.-M. (1983) *Time Series and System Analysis with Applications*, Wiley, New York.

Pao, Y.-H. (1989) *Adaptive Pattern Recognition and Neural Networks*, Addison-Wesley, Reading, MA.

Press, W., Flannery, B., Teukolsky, S. and Vetterling, W. (1986) *Numerical Recipes: The Art of Scientific Computing*, Cambridge University Press, Cambridge.

Priestley, M. (1988) *Non Linear and Non Stationary Time Series Analysis*, Academic Press, San Diego, CA.

Priestley, M. (1981) *Spectral Analysis and Time Series*, Academic Press, New York.

Rice, J. R. (1969) *Approximation of Functions, Vol 1 & 2*, Addison-Wesley, Reading, MA.

Rosenfeld, A. and Kak, A. (1981) *Digital Signal Processing, 2nd edition*, Academic, New York.

Ruelle, D. (1989) *Chaotic Evolution and Strange Attractors*, Cambridge University Press, Cambridge.

Shannon, C. and Weaver, W. (1949) *The Mathematical Theory of Communication*, Illinois University Press, Urbana.

Silverman, B. W. (1986) *Density Estimation for Statistics and Data Analysis*, Chapman and Hall, London.

Sinai, Ya. G. (1976) *Introduction to Ergodic Theory*, Princeton University Press, Princeton.

Stewart, G. W. (1973) *Introduction to Matrix Computions*, Academic Press, New York.

Stoer, J. and Bulirsch (1980) *An Introduction to Numerical Analysis*, Springer-Verlag, New York.

Thomson, J. M. T. and Bishop, S. R. (1994) *Nonlinearity and Chaos in Engineering Dynamics*, John Wiley, Chichester.

Thomson, J. M. T. and Stewart, H. B. (1986) *Nonlinear Dynamics and Chaos*, Wiley, Chichester.

Tong, H. (1990) *Non Linear Time Series Analysis: A Dynamical Systems Approach*, Oxford University Press, Oxford.

Traub, J. F. *et al.* (1983) *Information, Uncertainty and Complexity*, Addison-Wesley, Reading, MA.

Walter, P. (1982) *An Introduction to Ergodic Theory*, Springer, Berlin.

Weigland, A. S. and Gershenfeld, N. A., eds. (1993) *Time Series Prediction: Forecasting the Future and Understanding the Past [SFI studies in the science of complexity Vol XV]*, Addison-Wesley.

Whitman, G. B. (1974) *Linear and Nonlinear Waves*, Wiley – Interscience, New York.

Wilkinson, J. and Reinsch, C. (1971) *Linear Algebra*, Springer, Berlin.

Yaglom, A. M. (1962) *An Introduction to the Theory of Stationary Random Functions*, Prentice Hall, Englewood Cliffs, NJ.

Young, P. (1984) *Recursive Estimation and Time Series Analysis: An Introduction*, Springer-Verlag, New York.

Index

Introduction

The index covers Chapters 1 to 7 and Appendices A to G. Index entries are to page numbers, those suffixed 'n' referring to footnotes. Alphabetical arrangement is word-by-word, where a group of letters followed by a space is filed before the same group of letters followed by a letter, eg 'random systems' will appear before 'randomness'. Initial articles, conjunctions and prepositions are ignored in determining filing order.